Kinetics and Thermodynamics
of
Tumor Markers
(CA 19-9- CA 15-3- AFP-Sialic Acid) 11

Prof.Dr.Sami AlMudhaffar and

Dr .Salwa Hameed Nasir Al-Rubae

Dr.Bilal Jasir Mohammed Al-Rawi

Contents

Subject   Page No.

PartA

This part   10

Abbreviation   11

Summary   14

Chapter One   16

1.1. Introduction   16

1.2. Benign Breast Tumors   17

1.2.1. Fibroadenoma   17

1.2.2. Fibrocytic disease   17

1.3. Malignant Breast Tumors   18

1.3.1. Incidence   18

1.3.2. Etiology and Risk Factors in Breast Cancer   18

1.3.2.1. Genetic Factors   19

1.3.2.2. Increasing Ages   19

1.3.2.3. Family History   19

1.3.2.4. Proliferative Breast Disease   19

1.3.2.5. Personal Cancer History   20

1.3.2.6. Menstrual and Reproductive Factors   20

1.3.2.7. Environmental Factors   20

1.3.2.8. Hormonal Influences   20

1.3.2.9. Dietary Factors   20

1.3.2.10. Lactation   21

1.3.3. Histopathology of Breast Cancer   21

1.3.3.1. In Situe Carcinoma (Non-Spreading Type)   21

1.3.3.2. Infiltrating Carcinoma (Invasive Carcinoma)   22

1.3.4. Staging System of Breast Cancer   22

1.3.5. Treatment of the Breast Tumors 25

1.3.5.1. Surgical Therapy 25

1.3.5.2. Conservation Breast Cancer Surgery 25

1.3.5.3. Mastectomy 25

1.3.5.4. Radiotherapy 25

1.3.5.5. Chemotherapy 25

1.3.5.6. Endocrine Therapy 26

1.3.6. Detection and Diagnosis 26

1.3.6.1. Self Examination 26

1.3.6.2. Mammography 26

1.3.6.3. Ultrasound 26

1.3.6.4. Magnetic Resonance Imaging (MRI) 27

1.3.6.5. Needle Biopsy/Cytology 27

1.3.6.6. Triple Assessment 27

1.4. Tumor Markers 27

1.4.1. Definition of Tumor Marker 27

1.4.2. Routes of Tumor Marker Production 28

1.4.3. Classification of Tumor Markers 28

1.4.3.1. Chemical Tumor Markers 29

1.4.3.2. Genetic Tumor Markers 30

1.4.4. Clinical Applications of Tumor Markers 32

1.4.5. Tumor Markers in Breast cancer 35

1.4.5.1. CEA 38

1.4.5.2. TPA 38

1.4.5.3. TPS 38

1.4.5.4. CA549 38

1.4.5.5. CA 27.29 39

1.4.5.6. CA 125 39

1.4.5.7. Mammary Antigen 39

1.4.5.8. Galectin-4 40

1.4.5.9. Cathepsin-D 40

1.5. Carbohydrate Antigen 15-3 (CA 15-3) 40

1.5.1. Structure of CA 15-3 41

1.5.2. CA 15-3 Expression 41

1.5.3. Biosynthesis of CA 15-3 42

1.5.4. Methodology 42

1.5.5. Biology of CA15-3   42

1.5.6. Clinical Application   43

1.6. Carbohydrate Antigen 19-9 (CA 19-9)   44

1.6.1. Marker Definition   44

1.6.2. Methodology   44

1.6.3. Screening   45

1.6.4. Monitoring Response to Treatment   45

1.6.5. Clinical Application   46

ChapterTwo:
CA 15-3 in Human Breast Tumor   47

Materials and Methods   48

2.1. Materials   48

2.1.1. Chemicals   48

2.1.2. Instruments   48

2.1.3. Patients   49

2.1.4. Preparation of Blood Samples   50

2.1.5. Collections of Specimens   50

2.1.6. Preparation of Phosphate–Buffered Saline   50

2.1.7. Preparation of Breast Tumors Tissues Homogenates   50

2.1.8. Statistical Analyses   51

2.2. Methods   51

2.2.1. Protein Determinations   51

2.2.2. Determination of CA 15-3 Levels in Sera of Breast Tumors Patients   51

2.2.3. Preliminary Test of CA 15-3 Binding to 125I -Anti CA15-3 Antibody in Breast Tumor Homogenate   53

2.2.4. Factors Effecting of 125I-Anti CA-3 Antibody Binding to CA 15-3 in Breast Tumors Homogenates   54

2.2.4.1. The Effect of Different Amounts of Protein Concentration of the Tumor Homogenate on the Binding with 125I-Anti CA15-3 Antibody   54

2.2.4.2. The Effect of 125I -Anti CA15-3 Antibody Concentration on the Binding   55

2.2.4.3. The Effect of pH on the Binding

2.2.4.4. Time Course of the Binding of 125I -Anti 15-3
Antibody to CA 15-3 in Breast Tumor Homogenate   56

2.2.4.5. The Effect of Different Halides on the Binding   57

2.2.4.6. The Effect of Monovalent and Divalent Cations on
the Binding   58

2.2.4.7. Recovery of CA 15-3   59

2.3. Results and Discussion   60

ChapterThree: Purification of CA15-3   76

Materials and Methods   77

3.1. Materials   77

3.1.1. Chemicals   77

3.1.2. Instruments   77

3.1.3. Patients   77

3.2. Methods   77

3.2.1. Isolation of CA15-3 by Sepharose CL-4B Column   77

3.2.1.1. Preparation of the Column   77

3.2.1.2. Preparation of Phosphate Buffered Saline   78

3.2.1.3. Preparation of the Gel   78

3.2.1.4. Void Volume Determination   78

3.2.1.5. Column-Calibration   79

3.2.1.6. Separation Procedure   79

3.2.1.7. Dialysis for Concentration   81

3.3. The Choice of the Optimum Conditions for the
Binding of the Partially Purified CA15-3 to 125I-Anti
CA15-3 Antibody   81

3.3.1. Optimum Protein Concentration   82

3.3.2. Influence of 125I-Anti CA15-3 Antibody on the
Binding   82

3.3.3. Optimum pH   83

3.3.4. Optimum Temperature   83

3.3.5. The Effect of Incubation Time   84

3.3.6. Stability of CA15-3 at –20°C   85

3.4. Results and Discussion   85

Chapter Four: Kinetic and Thermodynamic Studies of Binding CA15-3 to its antibody

Introduction   96

Materials and Methods   96

4.1. Materials   96

4.1.1. Chemicals   96

4.1.2. Instruments   96

4.2. Methods   96

4.2.1. Kinetic Studies   97

4.2.1.1. The Time-Course of the Binding of 125I-anti CA15-3
     Antibody with CA15-3 in Breast Tumor   97
   Homogenate

4.2.1.2. Determination of Kinetic Parameters of 125I-Anti
     CA 15-3 Antibody Binding with Partially Purified   97
     CA 15-3 in Benign and Malignant Breast Tumors

4.3. The Thermodynamic Studies of
   125I-Anti CA15-3 Antibody Binding to Partially Purified   100
   CA15-3 in Benign and Malignant Breast Tumors

Results and Discussion   102

Chapter Five: Characterization of complexes of CA15-3   120

Introduction   121

Materials and Methods   122

5.1. Materials   122

5.1.1. Chemicals   122

5.1.2. Instruments   122

5.1.3 Buffers and Reagents 122

5.2. Methods 122

5.2.1. Gel Filtration Technique for Separation of Free and Bound 125I -Anti CA 15-3 Antibody 122

5.2.1.1. Preparation of the Column 122

5.2.1.2. Preparation of the Gel and Determination of Void Volume 122

5.2.1.3 Separation Procedure of (125I-Anti CA 15-3 Antibody/CA15-3) Complex 123

5.2.2. The UV Spectrum of (125I-Anti CA 15-3 Antibody/CA15-3) Complex from Benign and Malignant Breast Tumors 124

5.2.3. The UV. Spectrum of 125I-Anti CA 15-3 Antibody 124

5.2.4. The UV Spectrum of Partially Purified CA 15-3 125

5.2.5.Factors Affecting the Absorption Properties of (125I-Anti CA 15-3 Antibody/CA 15-3) Complex from Benign and Malignant Breast Tumors 125

5.2.5.1. The pH Effect on the Complex 125

5.2.5.2. Effect of Solvent Polarity on UV Spectra of the Complex 126

5.2.5.3. Spectrophotometric pH Titration on the Complex 127

5.2.5.4. The Effect of NaCl Concentration on the Thermal Stability of the Complex by UV Spectral Studies 127

5.2.5.5. Effect of Urea, KCl and (Urea, KCl) Mixture on the Spectrum of the Complex 128

Results and Discussion 129

Chapter Six: Immuno radiometric assay 143

Abstract 144

Introduction 145

Materials and Methods   147

6.1. Materials   147

6.1.1. Chemicals   147

6.1.2 Instruments   147

6.1.3 Patients and Blood Samples   147

6.2. Methods   147

6.2.1. Determination of CA19-9 Levels in Sera of Patients
       with Benign and Malignant Breast Tumors   147

6.2.2. Preliminary Test of the Binding of CA19-9 with 125I-
       Anti CA19-9 Antibody in Breast Tumors Homogenates   149

6.2.3. Factors Effecting of 125I-Anti CA19-9 Antibody
       Binding to CA19-9 in Breast Tumors Homogenates   150

6.2.3.1. Effect of Protein Concentration on the Binding   150

6.2.3.2. Influence of 125I-Anti CA19-9 Antibody
         Concentration on the binding   150

6.2.3.3. Effect of pH on the Binding   151

6.2.3.4. Effect of Temperature on the Binding   151

6.2.3.5. Effect of Incubation Time on the Binding   153

6.2.3.6. Effects of Different Halides on the Binding   154

6.2.3.7. Effects of Monovalent and Divalent Cations on the
         Binding   155

6.2.3.8. Recovery of CA19-9   156

Results and Discussion   157

Conclusion   168

References   170

Part A

CA15-3 and CA19-9 in Breast Tumors

Prof.Dr.Sami AlMudhaffar and
Dr .Salwa Hameed Nasir1. Al-Rubae

This Part

Deals with Determination of carbohydrate antigen 15-3 (CA15-3) and carbohydrate antigen 19-9 (CA 19-9) levels in sera of patients with benign and malignant breast tumors , Development of IRMA assay for the determination of CA15-3 and CA19-9 from cytosolic tissues of benign and malignant breast tumors ,Characterization of the binding of 125I-anti CA15-3 antibody with isolated human – CA15-3 in benign and malignant breast tumors, such as those of binding capacity and the effect of various factors (pH, temperature, time, halides, salts, CA15-3 and its antibody concentration) ,Determination of kinetic and thermodynamic parameters of the binding of partially purified CA15-3 with its specific antibody and Spectroscopic studies on (125I-anti CA15-3 antibody / CA15-3) complex in breast tissue .

Further more the level of sera CA 15-3 in patients with benign and malignant breast tumors (preoperative) was measured by Immunoradiometric Assay (IRMA). Data analysis showed, concentrations of CA 15-3 were significantly higher in pre-and post-menopausal malignant breast tumors ($P < 0.0001$) and significantly lower in benign breast tumors, compared with healthy subjects.

The results obtained revealed higher incidence of CA 15-3 in two groups of malignant breast tumors than those in benign breast tissues and in the supernatant fraction more than the pellet fraction .The binding of 125I –anti CA 15-3 antibodies with CA 15-3 was studied in three groups: benign breast tumor (Fibroadenoma), pre-and post- menopausal malignant breast tumors (IDC). The optimum conditions observed for the binding were as follows:

CA 15-3 in concentration tissue homogenate 100µg. mL-1 for groups II and I while it was 200 µg.mL-1 for group III. 125I -anti CA 15-3 antibody concentrations: 0.175µg.mL-1 for group I and II, whereas it was 0.140 µg.mL-1 for group III. Temperatures of incubation were: 45oC for groups I and III, 15 oC for group II, while time of incubation was 90 min for both group I and III, 30 min. for group II. The optimum pH was 7.0 for group I, 7.6 for group II and 7.8 for group III. The use of different halides was shown to increase the binding between CA 15-3 and 125I –anti CA 15-3 antibody in both group II and III, while inhibition occurred on the binding in group I.

Abbreviations

Ab	Antibody
Ab*	Labeled antibody
Ag	Antigen
(Ab-Ag)	125I-anti CA15-3 antibody / CA 15-3) Complex
B	Bound
Bmax.	Maximal binding capacity
BSA	Bovine Serum Albumin
(B/T) %	Percentage of bound over total
CA15-3	Carbohydrate antigen 15-3
CA19-9	Carbohydrate antigen 19-9
CA27.29	Carbohydrate antigen 27.29
CA125	Carbohydrate antigen 125
CA549	Carbohydrate antigen 549
CAF	Cyclophosphamide Doxorubicin (Adriamycin) 5-Fluorouracil
CEA	Carcino-Embryonic Antigen
Ci	Curi
CMF	Cyclophosphsamide Methotrexate 5-Fluorouracil
cpm	Counts per minute
DCIS	Ductal Carcinoma in Situ
DMSO	Dimethyl sulphoxide
ε	Absorption coefficient
EDTA	Ethylenediammine tetraaceticacid
EG	Ethylene Glycol
ELISA	Enzyme-Linked immunosorbent assay
F	Free
FNA	Fine needle aspiration
h-CA15-3	Human carbohydrate antigen 15-3
His	Histidine
ICR	Iraqi Cancer Registry
IDC	Infiltrating ductal carcinoma

ILC	Infiltrating lobular carcinoma
IRMA	Immunoradiometric Asaay
J	Joule
K	Kelvin
Ka	Affinity constant
Kav	Partition coefficient
Kd	Equilibrium dissociation constant
KD	Kilodalton
KJ	Kilo joule
LCIS	Lobular carcinoma in situ
Mab	Monoclonal antibody
μ	Micro (10-6 x)
Max	Maximum
μg	Microgram
μL	Microliter
MCA	Mucin-like carcinoma associated antigen
MRI	Magnetic Resonance Image
MSTI	Molar surface tension increment
λmax	Maximum wavelength
M.Wt	Molecular weight
nm	Nanometer
N.M.R.	Nuclear Magnetic Resonance
P	Probability
PEG	Polyethylene glycol
P53	Tumor suppressor gen
PR	Progesterone receptor
PAGE	Polyacrylamide gel electrophoresis
RIA	Radioimmuno assay
r.p.m	Round per minute
RRA	Radio active iodine ablation
SD	Standard deviation
Ser	Serine
SDS - PAGE	Sodium dodecyle sulfate – poly acrylamide gel electrophoresis
T	Total

Thr	Threonine
TNM	Tumor, Node, Metastasis, Staging System
TPA	Tissue polypeptide antigen
TPS	Tissue polypeptide specific antigen
Trp	Tryptophane
Tyr	Tyrosine
UV	Ultraviolet
WHO	World Health Organization

1. The level of CA15-3 was determined in sera of (16) premenopausal malignant breast tumors patients, (12) postmenopausal malignant breast tumors patients, and (20) benign breast tumors patients matched with one group of (10) healthy women as control by Immunoradiometric Assay (IRMA). The data obtained demonstrated highly significant increase ($P<0.0001$) in patients with malignant breast tumors, whereas slightly increase ($P<0.05$) in patients with benign breast tumors when matched with normal women.

2. A modified Imunoradiometric Assay (IRMA) was used for determination of cytosolic carbohydrate antigen 15-3, using 125I-anti CA15-3 antibody and found to be suitable for the assessment of those antigens in benign and malignant breast tumors. The data revealed an increment of CA15-3 in the cytosolic fraction in comparison to the nuclear fraction.

3. CA 15-3 was isolated from cytosolic of human benign and malignant breast tumors homogenate by gel filtration techniques. The binding characteristics of the partially purified CA15-3 from benign and malignant breast tumors homogenate with 125I-anti CA15-3 antibody were investigated.

4. Kinetic parameters of the binding 125I-anti CA15-3 antibody with partially purified CA15-3 from benign and malignant breast tumors homogenate were determined at five different temperatures, the results indicated that the binding reaction was time and temperature –dependent process. However the time – course data for the binding followed the pseudo –first order kinetic.

5. The thermodynamic studies of the 125I-anti CA15-3 antibody binding to the partially purified CA15-3 from benign and malignant breast tumors were studied. The thermodynamic parameters of the standard state (ΔG_o, ΔH_o, ΔS_o) and the transition state (ΔG^*, ΔH^*, ΔS^*) and activation energy (Ea) were determined.

6. The complex formed (125I-anti CA15-3 antibody / CA15-3) of partially purified CA15-3 from benign and malignant breast tumors and 125I-anti CA15-3 antibody, were investigated by UV methods. Different factors affecting the absorption band were extensively studied, such as pH, solvent perturbation, and denaturation agents. The heat stability and spectroscopic titration were also studied.

7. The level of CA19-9 was determined in sera of (10) premenopausal malignant breast tumors patients, (10) postmenopausal malignant breast tumors patients, and (10) benign breast tumors patients matched with one group of (10) healthy women as control by Immunoradiometric Assay (IRMA). The data obtained demonstrated significant increase ($P<0.05$) in patients with benign and premenopausal malignant breast tumors, whereas highly significant increase ($P<0.005$) in patients with postmenopausal malignant breast tumors when matched with normal women.

8. An Immunoradiometric Assay (IRMA) for the determination of cytosolic CA19-9 was developed, using 125I-anti CA19-9 antibody and found to be suitable for assessment of those antigens in benign and malignant breast tumors. The data revealed an increment of CA19-9 in the cytosolic fraction in comparison to the nuclear fraction.

Chapter one
1.1 Introduction

The breast is constantly responding to changes in hormonal, nutritional, genetic, psychological, and environmental stimuli such as radiation that cause continual cellular changes (1). As a result of these changes, breast tumors (abnormal breast tissue) may develop either benign (noncancerous) or malignant (cancerous) (2). The major significance of the benign processes less in the need to separate them from malignancies. The World Health Organization (WHO) classifies tumors of the breast (1981) according to histological aspects (3) (Table 1.1).

Table (1.1): Histological classification of breast tumors (3)

I. Epithelial Tumors	
A. Benign	
1.	Intraductal papilloma
2.	Adenoma of the nipple
3.	Adenoma
a.	Tublar
b.	Lactating
B. Malignant	
1.	Non invasive
a.	Intraductal carcinoma
b.	Lobular carcinoma
2.	Invasive
a.	Invasive ductal carcinoma
b.	Invasive lobular carcinoma
3.	Paget's disease of the nipples
II. Mixed Connective Tissue and Epithelial Tumors	
a.	Fibroadenoma

Table (1.1): Continued.

b.	Phyllodes tumor (cystosarcoma phyllodes)
c.	Carcinosarcoma
III. Miscellaneous Tumors	
a.	Soft tissue tumors
b.	Skin tumors
c.	Tumors of haemopcietia and lymph tissues
IV. Unclassified Tumors	
V. Mammary Dysplasia / Fibrocystic Change	
VI. Tumor like Lesion	
a.	Duct ectasia
b.	Inflammatory pseudotumors

1.2. Benign Breast Tumors

1.2.1. Fibroadenoma

This is the most common benign tumor of the female breast. It is a new growth composed of both fibrous and glandular tissue (4). These tumors are commonly found in younger women between the ages 20-35 years (5). It increases in size, during pregnancy (6). It is less likely to develop after menopause. An epidemiological study suggests that fibroadenoma represents a long-term risk for breast carcinomas and that risk is increased in women with ductal hyperplasias, or a family history of breast carcinoma (7).

1.2.2. Fibrocystic disease

This is an ill-defined condition of the breast where palpable lumps can be felt and is usually associated with pain and tenderness that fluctuates with the menstrual cycle (8). Fibrocystic changes are the most common, occurring in approximately 60% of premenopausal women. Women with this disease usually have a freely movable, palpable mass, at time it may cause pain, particularly when women are in the premenopausal phase of the menstrual cycle, however, breast pain can be caused by lesions other than fibrocystic changes. The palpable lesion may appear to increase and decrease is size cyclically; usually achieving its maximum size in the premenstrual phase of the menstrual cycle. Cystic disease is frequently accompanied by varying degrees of epithelial hyperplasia in adjacent ducts and lobules (9). In patients who have one particular form of fibrocystic disease (the proliferate form), the incidence of cancer increased very slightly (10).

1.3. Malignant Breast Tumors

1.3.1. Incidence

Breast cancer is the most common malignant tumors in women (11), and it is the leading malignancy affecting women in North America and Europe. In 2000, approximately 184200 new cases of invasive breast cancer were diagnosed in the United States. The number of noninvasive breast cancer is hard to verify, but it probably account for and an additional 20000 to 30000 new cases; thus, the number of invasive and noninvasive breast cancer treated in 2000 approximately 200000 (12).

In Iraq, according to the results of Iraqi Cancer Registry (ICR), breast cancer accounting for (31.11%) and remained the commonest tumors in the year 2000 (13-15), it was also shown that breast cancer was the first among the commonest ten cancers in Iraq. Figure (1.1) represents the population of breast cancer in Iraq for the last nine years. As shown in the same figure, the population of breast cancer increased more than double in the last nine years (16).

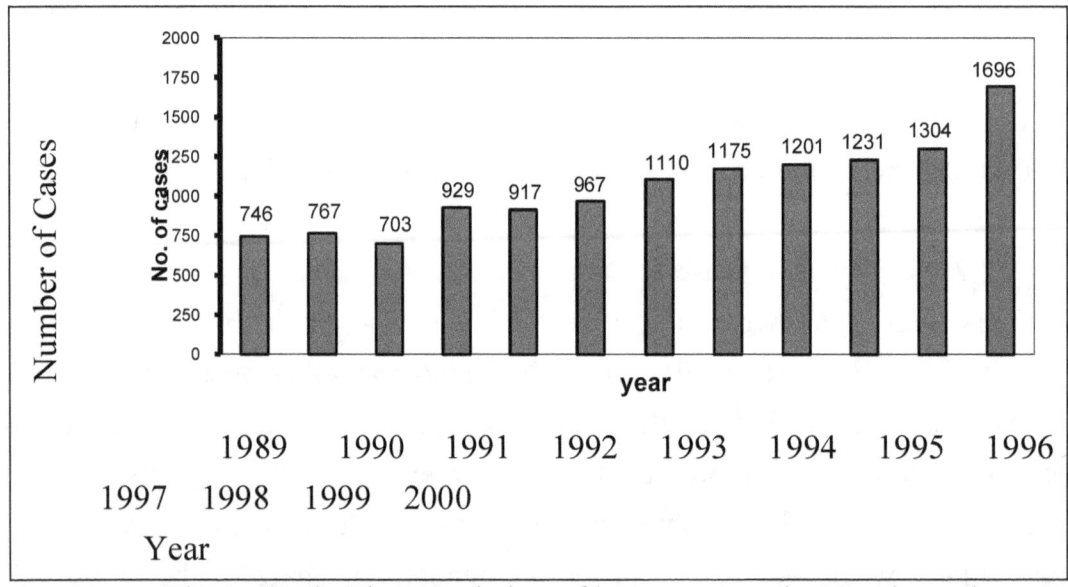

1989   1990   1991   1992   1993   1994   1995   1996
1997   1998   1999   2000
Year

Figure (1.1): The population of breast cancer in Iraq through (1989-2000)(13-16).

1.3.2. Etiology and Risk Factors in Breast Cancer

Numerous risk factors have been associated with the development of breast cancer, such as genetic, environmental, hormonal, and nutritional.

Despite all available data on breast cancer risk factors, 75% of women with this cancer have not exposed to any risk factors (17).

1.3.2.1. Genetic Factors

Breast cancer is the result of mutations in one or more critical genes. Two genes in women on chromosome 17 have been implicated. The most important gene is called BRCA-1; the other is the P53 gene. A third gene is BRCA-2 on chromosome 13 (11).

1.3.2.2. Increasing Ages

Breast cancer is uncommon before age 25 years, but then there is a steady rise to the time of menopause, followed by a slower rise throughout life. The average age at the diagnosis is 64 years (4).

1.3.2.3. Family History

The overall relative risk of breast cancer in women with a positive family history in a first –degree relative (mother, daughter, or sister) is 1.7. Premenopausal onset of the disease in a first–degree relative is associated with three-fold increase in breast cancer risk, whereas postmenopausal diagnosis increases relative risk by only 1.5. When the first degree has relative bilateral disease, there is fivefold increase in risk. The relative risk for a woman whose first-degree relative developed both bilateral and premenopausal breast cancer is nearly nine. No increased risk has been demonstrated when only a second-degree relative (aunt, cousin, or grandmother) has had breast cancer (17).

1.3.2.4.Proliferative Breast Disease

The diagnosis of certain condition after breast biopsy is also associated with an increased risk for the subsequent development of invasive breast cancer(17) . Women with proliferative disease of the breast with a typical hyperplasia (atypia) are at increased risk for developing breast cancer (five-fold increase), however. The risk for atypia is greater in patient with a strong family history of breast cancer (11-fold increase) (11) .

1.3.2.5. Personal Cancer History

A personal history of breast cancer is significant risk factor for the subsequent development of a second, new breast cancer. This risk has been estimated to be as high as 1% per year from the time of diagnosis of the initial cancer. Women with a history of endometerial, ovarian, or colon cancer also have a higher likelihood of developing breast cancer than those with no history of these malignancies (17).

1.3.2.6. Menstrual and Reproductive Factors

Early onset of menarche (<12 years old) has been associated with a modest increase in breast cancer risk (two fold or less). Women who undergo menopause before age 30 have a twofold reduction in breast cancer risk when compared to women who undergo menopause after age 55. A first full-term pregnancy before age 30 appears to have a protective effect against breast cancer, whereas a late first full-term pregnancy or nulliparity may be associated with higher risk, There is also suggestion that lactation protects against breast cancer development (17).

1.3.2.7. Environmental Factors

Women exposed to therapeutic radiation or after atom bomb exposure have a higher rate of breast cancer. Risk increases with younger age and higher radiation doses (4). Radiation is believed to cause 1-2% of all cancer deaths(18,19). In respect to pollution studies, which have shown that there is a well-established correlation between many pollutants and cancer, it has been estimated, that 1% of cancer deaths is due to air, water and land pollution (20).

1.3.2.8. Hormonal Influences

Endogenous estrogen excess, or more accurately, hormonal imbalance, clearly plays a significant role. Many risk factors mentioned-long duration of reproductive life, nulliparity, and late age at first child-imply increased exposure to estrogen peaks during the menstrual cycle (4).

1.3.2.9. Dietary Factors

Diets that are high in fat have been associated with an increased risk factor for breast cancer (21). It has been suggested that differences in dietary fat content may account for the variations in breast cancer incidence observed among different countries. Sala et.al, illustrated that certain macronuterients and food such as protein, carbohydrate and meat intake influence of risk of breast cancer through their effects on breast tissue morphology (22). Data from prospective studies have confirmed that the relationship exists between alcohol intake and risk of developing breast cancer (23). Alterations in endrogenous estrogen levels secondary to obesity may enhance breast cancer risk (17).

1.3.2.10. Lactation

In the search of practical methods to prevent breast cancer, lactation has strong evidence as a potentially modifying factor especially at early age and for long period. There is a significant reduction in the risk of breast cancer associated with lactation for more than two years. This effect appeared to be limited to premenopausal women (24). Lactation may reduce the risk of breast cancer by interrupting ovulation or by modifying pituitary and ovarian hormone secretions. Direct physical changes in the breast that accompany milk production may also contribute to prevent the effect (25).

1.3.3. Histopathology of Breast Cancer

Most cancer of the breast is a carcinoma of the epithelial cells that line breast ducts and lobules. Rarer forms of cancer occurring in the breast arise from the stromal cells that surround the epithelial glands (12). Breast cancer is a complex, devastating diseases and the most frequently diagnosed cancer in women. It is the single leading cause death for women of age 20-59 years (26).

There are a various type of breast carcinomas according to (WHO) classification (3) :

1.3.3.1. In Situ Carcinoma (Non-Spreading Type)

A. Ductal Carcinoma In Situ (DCIS)

The malignant cells in this disease are confined to the ductal basement membrane (27). DCIS usually occurs without forming a mass because there is no scirrhous component (11). DCIS is also known as intraductal carcinoma.

B. Lobular Carcinoma In Situ (LCIS)

LCIS is composed of smaller lobular or acinar cells and fills the terminal breast lobule with a homogenous proliferation; most clinicians currently regard LCIS as a risk factor for the development of invasive breast cancer. LCIS is usually not treated, but affected women are placed under frequent   surveillance (28,29).

1.3.3.2. Infiltrating Carcinoma (Invasive Carcinoma)

A. Infiltrating Ductal Carcinoma (IDC)

Most invasive carcinoma of the breast is ductal in origin (11). Infiltrating (invasive) breast carcinoma differs from intraductal carcinoma (ductal carcinoma in situ) by the presence of stromal invasion, through which tumor cells spread not only locally but also regionally and distantly via vascular lymphatic space (30).

B. Infiltrating Lobular Carcinoma (ILC)

This type is rare, from about (5%-10%) of breast cancer. It is begins in the milk-secreting glands of the breast. It is often multicentirc, several areas of thickening may occur in one or both breasts. It is characterized by the presence of small and relatively uniform tumors cell growing singly around lobules involved by in situ lobular neoplasia (31).

1.3.4. Staging System of Breast Cancer (11)

The standard staging system for breast cancer is the TNM system table (1.2). The TNM classification devised by the International Union Against Cancer (UICC) and accepted by the American Joint Commission on Cancer Staging a world standard (32). Another system is the Colombia Clinical Classification (CCC), formulated by Haagensen and Stout (33). Although this system was a valuable precursor and is easier to remember than the TNM, it is a less precise classification where stage A represents a tumor confined to the breast; stage B include tumors with clinical axillary lymph node enlargement; stage C represents the presence of grave prognostic sings in the breast; and stage D indicates metastatic disease. The TNM based on the clinical features of tumor (T), the regional lymph nodes (N), and the presence or absences of distant metastases (M) (34). The purposes of staging are the following (23):

- Plane a therapeutic strategy that most appropriate for the patient.
- Allow for more intelligent prognostication of the disease statues of the patient.
- Permit comparison of therapeutic results obtained from different sources by different means.

Table (1.2): TNM System and Stage Grouping (35).

	T	Primary tumor
	T O	No evidence of primary tumor
	T is	Carcinoma in situ
	T 1	Tumor 2 cm or less in greatest diameter
	T 2	Tumor more than 2 cm but less than 5 cm in greatest diameter
	T 3	Tumor more than 5 cm in greatest diameter
	T 4	Tumor of any size with direct extension to chest wall or skin
	N	Regional lymph nodes

Table (1.2): Continued.

N0	No regional lymph node metastases		
N1	Metastasis to movable ipsilateral axillary lymph node(s)		
N2	Metastasis to ipsilateral lymph node(s) fixed to one another or to other structures		
N3	Metastasis to ipsilateral internal mammary lymph node(s)		
M	Distant metastasis		
Mo	No distant metastasis		
M1	Distant metastasis (including metastasis to ipsilateral supraclavicular lymph nodes)		
Stage 0	Tis	No	Mo
Stage I	T1	No	Mo
Stage IIA	T0	N1	Mo
	T1	N1	Mo
	T2	N0	Mo
Stage IIB	T2	N1	Mo
	T3	N0	Mo
Stage IIIA	T0	N2	Mo
	T1	N2	Mo
	T2	N2	Mo
	T3	N1/N2	Mo
Stage IIIB	T4	Any N	Mo
	Any T	N3	Mo
Stage IV	Any T	Any N	M1

1.3.5. Treatment of the Breast Tumors

The goal of any oncologic treatment is to maximize the cure and at the same time optimize the quality of life (36).

1.3.5.1. Surgical Therapy

Surgical treatment represents most frequently used and the most successful sign method of cancer therapy currently available. More patients are cured of cancer by surgery than by any other therapeutic modality (37).

1.3.5.2. Conservation Breast Cancer Surgery

It is aimed at removing the tumor plus a rim of at least (1 cm) of normal breast tissue. This is commonly referred to as a wide local excision or lymphectomy (38).

1.3.5.3. Mastectomy

Removal of all breast tissue, choice may not be offered if the lesion is too large, multi-focal, and lobular or, in the surgeon's opinion, so close the nipple that it is likely to cause distortion (39).

1.3.5.4. Radiotherapy

Palliative radiotherapy may be advised for locally advanced cancers with distant metastasis in order to control ulceration, pain, and other manifestation in the breast and regional nodes (40). Radiotherapy is especially useful in the treatment of the isolated bony metastasis, chest wall recurrence and brain metastasis (41,42).

1.3.5.5. Chemotherapy

Breast cancer is responsive to all major classes of cytoxic drugs: alkylating agents, antimetabolites, mitotic inhibitor, and the antitumor antibiotic(43). Among the most active are alkylating agent including cyclophosphamide and thaitepa, and anthracyclines such as doxorubicin. The antimetabolites methotrexate (MTX) and 5–flourouracil (5-FU) are also active. Numerous combination of chemotherabutic agents have been evaluated in the treatment of metastatic breast cancers such as: CMF, CAF (44-47).

1.3.5.6. Endocrine Therapy

Hormonal therapy is the initial treatment of metastatic disease in patients with ER or PR positive tumors. Tamoxifen, a nonsteroidel antiestragon was approved first for the treatment of metastic breast cancer over 20 years ago, is usually the first agents of choice because of its favorable toxicity profile.(48,49)

1.3.6. Detection and Diagnosis (50)

Although an accurate history and clinical examination are still the most important method of detecting breast disease, there are a number of investigations that can assist in the diagnosis as follows:

1.3.6.1. Self Examination

All women over age 20 should be advised to examine their breast monthly–premenopausal women should perform the examination 7-8 days after the menstrual period. The breast should be inspected initially while standing before a mirror with the hands at the side, overhead, and pressed firmly on the hips to contract the pectoralis muscles. Masses asymmetry of breasts and slight dimpling of the skin may become apparent as a result of these maneuvers. Next, in a supine position, each breast should be carefully palpated with the fingers of the opposite hand. Some women discover small breast lumps more readily when their skin is moist while bathing or showering. Most women do not practice self-examination, and its value is controversial. Clearly, however, it is not harmful, it is inexpensive, and it may be beneficial (35).

1.3.6.2. Mammography

Soft tissue x-rays are taken by placing the breast in direct contact with ultrasensitive film and exposing it to low-voltage, high-amperage x-rays. The dose of radiation in approximately 0.1 Gy and therefore mammography is a very safe investigation (50).

1.3.6.3. Ultrasound

Ultrasound is particularly useful in young women with dense breasts in whom mammograms are difficult to interpret, and in distinguishing cysts for solid lesions. It can also be used to localize impalpable breast lumps (50).

1.3.6.4. Magnetic Resonance Imaging (MRI)

MRI is of increasing interest to breast surgeons in a number of settings, it can be useful to distinguish scar from recurrence in women who have had previous breast conservation therapy for cancer (although it is not accurate within 9 months of radiotherapy because of abnormal enhancement); it is the gold standard for imaging the breast of women with implants; it may prove useful as a screening tool in high-risk women; and it is being evaluated in the management of the axilla in both primary breast cancer and recurrent disease (50).

1.3.6.5. Needle Biopsy/Cytology

Histology can be obtained by using a fine needle such as a trucut or corecut biopsy device under local anesthesia. Cytology is obtained by using a 21 or 23 gauge needle and 10 mL syringe with multiple passes throughout the lump without releasing the negative pressure in the syringe. The aspirate is then smeared on to a slid, which is air-dried. Fine needle aspiration cytology (FNAC) is the least invasive technique to obtain a cell diagnosis and is very accurate if both operator and cytologist are experienced. However, false negatives do occur mainly through sampling error, and invasive cancer cannot be distinguished from in situ disease (50).

1.3.6.6. Triple Assessment

In any patients who presents with a breast lump or other symptoms suspicious of carcinoma, the diagnosis should be made by a combination of clinical assessment, radiological imaging and tissue sample taken for either cytological or histological analysis (50).

1.4. Tumor Markers
1.4.1. Definition of Tumor Marker (51)

A tumor marker is a substance present in or produced by a tumor or by the tumor's host in response to the tumor's presence that can be used to differentiate a tumor from normal tissue or to determine the presence of a tumor based on measurement in the blood or secretions. Such a substance can be found in cells, tissue or body fluids. It can be measured qualitatively or quantitatively by chemical, immunological, or molecular biological methods to identify the presence of a cancer.

Tumor markers are the biochemical for immunological counterparts of the differentiation state of the tumor. In general, tumor markers represents re-expression of substances produced normally in embryoginically closely related tissues. Few markers are specific for a single individual tumor (tumor-specific markers); most are found with different tumor of the same tissue type (tumor-associated markers). They are present in higher quantities in cancer tissue or in blood from cancer patients than in benign tumors or in the blood of normal subjects (51).

1.4.2. Routes of Tumor Markers Production

Benign tumors are generally well differentiated. The cells in a benign tumor are similar to the cells of the normal tissue, and the tumor markers produced are the products found in the normal tissue. They may be found in increased amounts in the circulation depending on the size of the tumor.

Malignant tumors may produce substance may associated with normal cell, or they may be different. As a zygote is transformed into an embryo, which then evolves into a fetus, the rapidly dividing cells became differentiated into specialized tissues by selective gene expression. The genes expressed are responsible for the production of hormones, enzymes, receptors, structural proteins, and cell metabolism. When a normal cell is transformed into tumor cell, gene expression changes. The affected cell may lose its ability to synthesize some specific cell products, or it may manufacture greatly increased amounts. The cell may be less specialized than the tissue it evolved from and assume the characteristics of the less well-differentiated cells of the embryo, synthesizing proteins found in the embryo but not in a normal adult. Cell proliferation rates change as the metabolic rate of the cells increases. After the cell is transformed, it loses growth control and begin to divide rapidly. The cells lose contact inhibition and invade the primary site. They then invade the adjacent organs and blood and lymph system, which may carry the cells to distance organs. The cell may then lodge in a capillary bed and begin to invade the new site. As this invasion process takes place, new proteins are produced that actively aid in the invasion. These proteins can also be used as markers. (52)

1.4.3. Classification of Tumor Markers

Tumor markers may be classified into chemical and genetic tumor markers. (52)

1.4.3.1. Chemical Tumor Markers

Table (1.3) summarizes the chemical tumor markers classified according to biochemical characteristics, and their associated malignancy. The table shows the low specificity of tumor marker for cancer (52).

Table (1.3): Chemical tumor markers (51).

Marker	Example	Associated Malignancy
Enzyme	Alcohol	Liver
	Alkaline phosphatase	Bone, Liver, Leukemia, Sarcoma
	Alkaline phosphatase Placental	Ovarian, Lung, trophoplastic gastrointestinal, seminoma, Hodgkin's
	Amylase	Pancreas, Various
	Aryl Sulfatase B	Colon, breast
	Galactosyi transferase	Colon, bladder, gastrointestinal, Various
	Neuron-Specific enolase	Lung (small-cell), neuroblastoma, carcinoid, melanoma, Pheochromocytoma,
	Prostate-specific antigen (PSA)	Prostate Various (Large bowel. Lung, ovarian)
	Telomerase	Colorectal, Breast, etc.
	Sialyl transferenase	Colon, Breast, Lung
Hormone	ACTH	Gushing's syndrome, lung (small – cell)
	Antidiuretic hormone	Lung (small - cell) adrenal cortex,
	Calcitonin	Medullary thyroid
	Growth hormone hCG	Pituitary adenoma, renal, lung, Embryonal choriocarcinoma, testicular
	Human placental	Trophoblastic, gonads, lung, breast
	Parathyroid hormone	Liver, renal, breast, lung, various
	Prolactin	Pituitary adenoma, renal, lung. Breast
	Vasoactive intestinal	Pancreas, bronchogenic,
	Peptide 5	Pheochroinocytom neuroblastoma

Table (1.3): Continued.

Oncofetal Antigen	□-Feto protein	Hepato cellular, germ line (non-
	□-oncofeta antigen	Colon
	Carcino fetal ferritin	Liver
	CEA	Colorectal, gastrointestinal,

	Pancreatic oncofetal	Pancreatic
	Sequamous cell antigen	Cervical, lung, skin, head and neck
	Tennessee antigen	Colon, gastrointestinal, bladder
Mucin	CA 125	Ovarian, endometrial
	CA 15-3 (Episialin)	Breast, Ovarian
	CA 27-29	Breast
	MCA	Breast, ovarian
	Du-PAN-2	Pancreatic, ovarian, gastrointestinal,
Blood group related antigen	CA 19-9	Pancreatic, hepatic; gastrointestinal
	CA 19-5	GastrointestinaT, pancreatic, ovarian
	CA50	Pancreatic, colon, gastrointestinal
	CA 27.4	Ovarian, breast, colon,
	CA 242 1	Ovarian, breast, colon,
Protein	$\square$2-Microglobulin	Multiple myeloma, $\square$-cell
	C-peptide	Insulinoma
	Ferritin	Liver, lung, breast, leukemia
	Immunoglobuin	Multiple myeloma, lymphomas
	Melanoma associated antigen	Melanoma
	Pancreas associated antigen	Pancreatic, stomach
	Pregnancy specific antigen	Trophopiastic, germ cell
	Prothrombin precursor	Meato cellular
	Tumor associated trypsin inhibitor	Ovarian
Others	Estrogen and progesteron	Breast
	Catecholamine	Neuroblastoma, pheochromocytoma
	Hydroxy proline	Bone metastasis (breast) multiple
	Lipid-associated sialic	Gastrointestinal, lung. Rheumatoid
	Polyamine	Brain, various

1.4.3.2. Genetic Tumor Markers (51)

Two classes of genes are implicated in the development of cancer: Oncogenes (Cell activation genes–table 1.4) and suppressor genes (genes involved in the recognition and repair of damaged DNA-table 1.4). Oncogenes are derived from proto-oncogenes, which may be activated by dominate mutations. The type of mutation could be point mutation, insertion, deletion, translocation, and inversion. Most oncogenes code for proteins that function at the same stage of activation of cells for proliferation, and there activation leads to cell division. Most oncogenes are associated with hematological malignancies, such as Leukemia and to a lesser extent, solid tumors.

The other class of tumor genes the suppressor genes has been isolated from mostly solid tumors. The oncogenicity of suppressor genes is derived from the loss of the gene rather than their activation, as with oncogenes. The major tumor suppressor gene, P53, functions to repair damaged DNA by apoptosis (programmed cell death).

Repair is mediated by activation of the production of P21, which blocks the cell cycle in late G1 to allow repair to take place. The loss of function of this gene may result in the inability of the DNA repair process and lead to the development of tumorgensis.(53)

The exciting promise of using detection of oncogenes and suppressor genes, for the diagnosis, determining the prognosis, and predicting the response to chemotherapy remains to be realized. However, oncogenes detection remains an experimental approach to human cancer, with great expectations not yet fulfilled. The ability to develop cancer by detection of mutations in tumor suppressor genes raises ethical questions that remain to be resolved.

Table (1.4): Classification of genetic tumor markers (52) .

Marker	Example	Associated Malignancy
Oncogene	N-ras mutation	Acute myeloid leukemia neuroblastoma
	K-ras mutation	Leukemia, lymphoma
	C-myc	b-and T-cell lymphoma., small cell
	C-erb B-2	Breast, ovarian, gastrointestinal
	C-abllber	Chronic myelocytic leukemia
	N-myc	Neuroendocrine
	bcL-2	Leukemia, iymphoma
Suppressor Gene	VHL mutation	Kidney
	APC mutation	Colorectal

PI 6 (cd Kn2)	Bladder, glioblastoma, melanoma
WT1 mutation	Wilms', tumor
Loss of	Wilms', breast, hepatoblastoma
BRCA2, PB1	Breast
RB1 mutation	Retinoblastoma, osteosarcoma small-
PI 6 E-cadheim	Breast
BRCA1 mutation	Neurofibromatosis 1 Melanoma, breast
P53 mutation	Breast, colorectal, lung, liver, renal cell,
DCC mutation	Colorectal
NF2 mutation	Neurofibro matosis2, meningioma

1.4.4. Clinical Applications of Tumor Mrkers (54)

The potential uses of tumor markers are summarized in table (1.5). In general, tumor markers may be used for diagnosis and prognosis of carcinomas and for monitoring the effects of therapy as well as targets for localization and therapy. Ideally, a tumor marker should be produced by tumor cells and be detectable in body which fluids should not be present in healthy people or in benign conditions. Therefore, it could be used for screening for the presence of cancer in symptomatic individuals in general population.

Most tumor markers are present in normal, benign, and cancer tissues. They are not specific enough to be used for screening cancer. In situations where the incidence of cancer is high among certain populations, screening might be possible.

Table (1.5): Clinical Usefulnees of tumor markers (55).

	Biochemical properties	Molecular weight	Primary clinical applications
Alpha-fetoprotein (AFP)	Glycoprotein, 4% carbohydrate; considerable	~70 KD	Diagnosis and monitoring of primary hepatocellular carcinoma and germ

	homology with albumin		cell tumors. Prognosis of germ cell tumors.
Cancer antigen 125 (CA 125)	Mucin identified by monoclonal antibodies	~200 KD	Monitoring ovarian carcinoma. Prognosis after chemotherapy
Cancer antigen 15-3 (CA 15.3, BR 27.29)	Mucin identified by monoclonal antibodies	>250 KD	Monitoring breast cancer
Cancer antigen 72.4 (CA 72.4)	Glycoprotein identified by monoclonal antibodies	~48 KD	Monitoring gastric carcinoma
Cancer antigen 19-9 (CA 19-9)	Glycolipid carring the Lewisa blood group determinate	~1,000 KD	Monitoring pancreatic carcinoma
Carcinoem bri-yonic antigen (CEA)	Family of glycoproteins , 45%-60% carbohydrate	~180 KD	Monitoring gastrointestinal and other adenocarcinomas
CYFRA 21-1	Fragments of cytokeratin	~30 KD	Monitoring bladder and lung carcinoma
Estrogen receptor	Nuclear transcription	65 KD	Predicting response to endocrine therapy in breast cancer

Table (1.5): Continued.

Human chorionic gonadotrophin (hCG)	Glycoprotein hormone consisting of tow non-	~36 KD	Diagnosis and monitoring non-seminomatous germ cell tumors ,

	covalently bound subunits (α and β)		choriocarcinomas , hydtidiform moles , seminomas. Prognosis of germ cell tumors.
Neuron specific enolase (NSE)	Dimer of the enzyme enolase	~87 KD	Monitoring small cell lung carcinoma , neuroblastoma , apudoma.
Placental alkaline phosphatase (PLAP)	Heat-stable isoenzyme of alkaline phosphatase	~86 KD	Monitoring of germ cell tumors (seminomas)
Progesterone receptor	Nuclear transcription factor	A from: 94 KD B from: 120 KD	Predicting response to endocrine therapy in breast cancer .
Prostate specific antigen (PSA)	Glycoprotein serine protease	~36 KD	Diagnosis , screening and monitoring prostatic carcinoma
Squamous cell carcinoma antigen (SCC)	Glycoprotein sub-fradion of tumor antigen T4	48 KD	Monitoring squamous cell carcinomas
Tissue polypeptide antigen (TPA)	Fragments of cytokeratin 8,18 and 19	~22 KD	Monitoring bladder and lung carcinoma
Tissue polypeptide specific antigen (TPS)	Fragment of cytokeratins 18	~22 KD	Monitoring metastatic breast carcinoma

1.4.5. Tumor Markers in Breast Cancer (56)

Several tumor markers have been investigated for one or more clinical use in breast cancer (Table 1.6).

Tumor-associated antigens (TAAs) that have been associated with breast cancer include carcinoembryonic antigen (CEA); tissue polypeptide antigen (TPA), tissue polypeptide-specific antigen (TPS), gross cystic disease protein (GCDP); prostate specific antigen (PSA); and the products of the MUC-1 gene. The MUC-1 gene encodes a cell-associated mucin-like protein. Secretary epithelial cells such as breast epithelial cells express this antigen. Several assays detect the MUC-1 gene products, but they are not identical. These proteins have been identified by monoclonal antibodies to breast cancer cell lines, breast cancer tissue, or human milk fat globule membranes. Assays that detect circulating MUC-1 products include CA 15-3, CA 27-29, CA 549, breast cancer mucin (BCM), mammary serum antigen (MSA), and mucin-like carcinoma-associated antigen (MCA) (57).

The results obtained with these assays may not be identical, presumably due to reactivity of different antibodies to different epitopes, and/or different sensitivities and specificities that result from different assay configurations (46).

More recently, markers of tumor biology have been investigated in breast cancer (Table 1.6), and molecules related to angiogenesis, adhesion, invasion, and metastases. Several, but not all of these are indeed, detected with immunologic assays, and could arguably be designated as TAAs.

Table (1.6): Tumor markers that have been investigated in breast cancer(56)

| Tumor-associated antigens |
| Carcinoembryonic antigen (CEA) |
| Products of or related to products of the MUC-1 gene |
| CA 15-3 |

CA 27-29
CA 549
Breast cancer mucin (BCM)
Mammary serum antigen (MSA)

Table (1.6): Continued.

Mucinous carcinoma antigen (MCA)
Tissue polypeptide antigen (TPA)
Tissue polypeptide-specific antigen (TPS)
Gross cystic disease protein (GCDP)
Prostate-specific antigen (PSA)
Markers of tumor biology
Extra-ceullular domain (ECD) of c-erbB-2/HER2/neu
Molecules of adhesion and invasion
E-selectin
Soluble urokinase plasminogen activator receptor (SuPAR)
Intercellular adhesion molecule-1 (ICAM-1)
Molecules associated with angiogenesis
Vascular endothelial growth factor (VEGF)
Basic fibroblast growth factor (bFGF)
Hepatocyte growth factor (HGF)
HUVEC assay

Antibody response against TAAs
c-erbB-2/HER2/neu
P53

There are several tumor markers correlate with the incidence of breast cancer, but the most important markers are:

1.4.5.1. CEA

Carcinoembryonic antigen is a marker for breast carcinoma (59), lung, gastrointestinal and colorectal (60). CEA is one of the older oncofetal protenis in use. CEA is a large family of related cell-surface glycoproteins with a high molecular mass of 150 to 300 KD, it contains 45 to 55% carbohydrate with increase expression found in a variety of malignancies, including breast cancer(61) .CEA is not recommended for screening, diagnosis, staging or routine surveillance of breast cancer patients following primary therapy (62).

1.4.5.2. TPA

Tissue polypeptide antigen is not a specific tumor marker (63). Antibodies that react with cytokeratin 8,18 and 19 identify it. TPA is a heterogeneous group of molecules with molecular weight range 20-45 KD (64). Both normal and cancerous cells produce TPA; it is useful in the monitoring of metastic diseases(54).

1.4.5.3. TPS

Tissue polypeptide-specific antigen (TPS) is a new tumor marker defined by monoclonal antibody against the soluble tissue polypeptide antigen (TPA) (65). First described as specific tumor marker by Bjorklund in 1957 (63). In breast cancer patients TPS was especially useful in monitoring response to treatment and effectiveness of therapy in metastatic disease (66).

1.4.5.4. CA 549

CA 549 is an acidic glycoprotein and it is a marker for breast carcinoma. CA 549 is not useful in detecting early breast carcinoma but it is useful is detecting recurrence of breast cancer in patients after initial therapy followed by adjuvant therapy (67).

1.4.5.5. CA 27.29

CA 27.29 is detected by a monoclonal antibody B 27.29 (68), this is produced against antigen in ascites of patients with metastatic breast carcinoma. CA 27.29 test above 37.7 KU.L-1 were considered positive, its most useful in monitoring metastatic breast carcinoma (69).

1.4.5.6. CA 125

CA 125 is a high-molecular mass (>200 KD) glycoprotein recognized by the monoclonal antibody OC 125. The level of CA 125 is measured quantitatively by using immunoradiometric assay (70). In healthy population, the upper limit of CA 125 level is 35 KU.L-1. CA 125 is elevated in ovarian carcinoma, endometerial, pancreatic, Lung, breast, colorectal and other gastrointestinal tumors (71,72). CA 125 is useful to detecting residual disease in cancer patients following initial therapy (73).

1.4.5.7. Mammary Antigen

Several new antigens have been recognized by monoclonal antibodies. Which have been identified in patients with breast cancer (74). They have been proposed as "tumor markers":

- MCA

Mucin-like carcinoma associated antigen (MCA) is a mucin glycoprotein with a molecular mass of 350 KD. MCA was identified on the surface of a breast carcinoma cell line by the monoclonal antibody b-12 (54). MCA level is elevated in 60% of metastatic breast cancer patients (75) .

- MAM-6

MAM-6 an epithelial membrane antigen present on ductal and alveoli epithelial cells that is detected by monoclonal antibody raised against human milk-fat globule membranes (76). Partial characterization of the antigen by

SDS-PAGE showed that the antigen is a polymorphic epithelial sialomucin with a molecular mass over 400 KD (77).

- MSA

Mammary serum antigen (MSA) was detected by an antibody raised against a whole cell suspension of intraductal breast cancer (74).

1.4.5.8. Galectin-4

A protein Galectin-4 is expressed in non-invasive and invasive breast cancer but not in normal cell. An anti-Galectin-4 antibody was able to detect the presence of Galectin-4 very specifically. Galectin-4 is specific diagnostic marker of breast cancer whose patterns of expression at early stages of disease could identify those patients with a high risk of progression to aggressive cancer (78).

1.4.5.9. Cathepsin-D

Cathepsin-D is a glycoprotein with molecular weight M.wt: 52 KD. It was discovered in 1979 in the culture medium of hormone dependent human breast cancer. It is a precursor to lysosomal acidic protease. This proteolytic enzyme can react against basement membranes (79).

Cathapsin-D may facilitate cellular actions such as migration, metastasis, and an invasion of other tissues. Estrogen has been shown to stimulate secretion of this tumor marker in certain hormone-dependent breast cancer cell lines. This antigen has been found to have potential application in breast cancer prognosis as its concentration appears to be related to the patients overall change for survival (80,81).

1.5. Carbohydrate Antigen 15-3 (CA 15-3)

CA 15-3 is a breast-associated antigen identified on the apical side of alveoli and ducts of mammary glands and as a circulating antigen (82). Distinct epitopes of this high molecular-weight mucin-like glycoprotein of 300-400 KD(83-85), which carbohydrate side chain account for about 50% (86).

Also known as polymorphic epithelial mucin (87) (PEM), epithelial membrane antigen (88) (EMA) or episialin (89). CA 15-3 can be identified by two monoclonal antibodies DF3 and 115 D8, in a double-determinate or sandwich-type immunoassay (90). The 115 D8 antibody was prepared against human milk-fat globulin membrane (91) while the DF3 antibody was raised against a membrane-enriched fraction of a human breast carcinoma (92).

1.5.1. Structure of CA 15-3

CA 15-3 (Episialin) is synthesized as transmembrane molecule with a relatively large extracellular domain and cytoplasmic domain of 69 amino acids(93). The extracellular domain mainly consists of region of nearly identical repeats population, leading to substantial differences in molecular weights of the CA 15-3 molecules from different individuals (94).

The repeats together with adjacent degenerated repeats contain many serins and threonines that are potential attachment sites for O-liked glycans and constitute the mucin-like domain, which comprises more than half of the polypeptide backbone. The mucin domain of CA15-3 contains many prolines and other helix-breaking amino acids, resulting in a molecule with an extended structure and many β-turns (95). The extended structure is very rigid as aresult of the numerous O-linked glycans attached to the molecule (96). The CA15-3 extends 200 to 500 nm above the cell membrane (97).

1.5.2. CA 15-3 Expression

- CA 15-3 Expression in Normal Tissues

CA 15-3 is predominatly found at the apical side of epithelial cells lining the acini alveoli, or lumens in various organs, i.e. in the mammary glands, salivary glands, sebacious glands, sweat glands, esophagus, stomach, pancreas, bile ducts, lungs, kidney, bladder, prostate, uterus, and rete testis (98-100).

- CA 15-3 Expression In Malignant Tissues

Relative to the expression levels of CA 15-3 found in normal tissues, CA15-3 is often overexpressed several-fold in many types of carcinomas derived from these tissues (101). In these tumors, polarization of the cells is often lost, resulting in the presence of CA 15-3 at the entire cell surface. High levels of CA 15-3 are also detected on carcinoma cells present in pleural effusions on ascites from patients with breast or ovary carcinoma and on many breast carcinoma cell lines.

1.5.3. Biosynthesis of CA 15-3

CA15-3 is synthesized as a large single polypeptide, in most cell lines approximately 200 KD or more (102,103). This precursor is rapidly cleaved by proteolysis in a small moiety, which contains the transmembrane and cytoplasmic domains, and a larger part, which comprises most of the extra cellular domain. Both moieties remain non-covalently associated (104). This proteolytic processing step occurs in the endoplasmic reticulum and may be essential for further maturation. CA 15-3 is mainly processed by adding numerous O-linked glycans, which increases the apparent molecular weight on SDS-polyacrylamide gels to more than 400 KD. The extensive glycosylation protects the molecule against proteolytic degradation, since the precursors without O-linked sugars are degarded rapidly, while the mature molecule is extremely resistant to the action of proteases. The glycosylation also determines the rigidity of the molecule. The last step in the processing of CA15-3 is the addition of sialic acid to the glycans, which increases the mobility of the molecule on SDS-gels (96).

The early proteolytic cleavage step is not directly responsible for the release of CA15-3 for the membrane, which suggests that CA15-3 is most likely released from the membrane by a second proteolytic cleavage step after arrival at the cell surface. The second proteolytic cleavage seems to be a slow and probably a random process, allowing the mucin to remain associated with the cell surface with a half-life of 16-24 hrs (96).

1.5.4. Methodology

The CA 15-3 test from all sources uses both DF3 and 115-D8 antibodies. Serum is initially incubated with a polystyrene bead to which 115-D8 antibody has been attached. This antibody binds to antigenic sites on the glycoprotein, pulling it out of solution. The beads are then washed to remove unbound meterial and incubated with the radioiodine (125I)-labeled DF3 antibody. The radiolabeled DF3 antibody binds its antigenic sites and then the amount of radioactivity is quantitated (105). This is called Immunoradiometric Assay (IRMA).

1.5.5. Biology of CA15-3

- CA15-3 and Cell Adhesion

Similar to mucins in mucus, membran-associated mucins might act as barrier molecules to protect cells against toxic substances, as in pancreatic and bladder ducts. The high densities of CA15-3, due to its extended and relatively rigid structure, might also interfere with the function of the adhesion molecules. In this way, CA15-3 might prevent interactions between opposing apical membrane of polarized normal cells and facilitate the formation and maintenance of the ducts during development (106,107).

In carcinomas, the combination of overproduction and loss of polarization of CA15-3 expression might reduce cell adhesion and facilitate the invasion of tumor cells because CA15-3 might now interfere with the function of molecules required from tissue integrity (104).

- CA15-3 and Immune System

The putative function of CA15-3 in tumor progression may not only be restricted to inhibition of adhesion which will probably result in an increased invasive potential of cells, but CA15-3 overexpression may well be critical to the survival of tumor cells during dissemination (108). A completely different aspect of CA15-3 is its ability to act as a tumor-specific antigen. The underglycosylation of CA15-3 in various tumor cells exposes the protein backbone, leading to the generation of novel epitopes. This could elicit an immune response (109-112).

1.5.6. Clinical Application

In healthy subjects, the upper limit of CA 15-3 concentration is 25 (KU.L-1). At this level, (5.5%) of 1050 normal subjects, (23%) of patients with primary breast cancer, and (69%) of those with metastatic breast cancer show elevated CA 15-3 levels (113).

Elevated CA15-3 levels are also found in other malignancies, including pancreatic (80%), lung (71%), breast (69%), ovarian (64%), colorectal (63%) and liver (28%) cancer. It is also reported to be elevated in benign diseases, although with less frequency (e.g., in benign liver [42%] and benign breast diseases [16%]).

CA15-3 should be used to diagnose primary breast cancer, because the incidence of elevation (23%) is fairly low. CA 15-3 is most useful in monitoring therapy and disease progression in metastatic breast cancer patients. A significant change must be at least (25%) and correlates with disease progression in (90%) of patients, with its regression in (78%). No change correlates with disease stability in (60%). CA 15-3 could replace CEA in metastatic breast cancer owing to its sensitivity and specificity.

1.6. Carbohydrate Antigen 19-9 (CA 19-9)

1.6.1. Marker Definition

CA 19-9 is a carbohydrate antigen identified as a glycolipid-that is, sialylated lacto-N-fucopentose II ganglioside, which is a sialylated derivative of the Lewis a blood group antigen and is denoted as Le a (114). CA19-9 is synthesized by normal human pancreatic and biliary ductular cells and by gastric, colonic, endometerial, kidny, salivary gland, sweat gland and present in ductal epithelium of breast (115-117). In serum it exists as a mucin, a high-molecular weight (200-1000 KD) glycoprotein complete (54). The monoclonal antibody against CA19-9 was developed from a human colon carcinoma cell line, SW-1116 by Koprowski and associates (118).

Monoclonal antibody 19-9 derived from spleen cells of a mouse immunized with human colon adenocarcinoma cell line SW-1116 (119). The epitope of this antibody is carbohydrate with the sugar sequence

NeuNAcα 2-3 Gal β 1-3 GlcNAc β 1-3 Gal…

4

|

Fuc α 1

As described by Magnani et.al. (119).

1.6.2. Methodology

CA 19-9 is measured with a double monoclonal immunoradiometric assay, using monoclonal antibodies raised against the SW-1116 cell line (120). The antibody reacts with CA19-9 found at low concentrations in sera from healthy individuals, but frequently increased in sera from patients with adenocarcinomas (120). The upper limit of normal for healthy subjects has been defined by the cutoff value of 37.0 (U.mL-1) (121). CA 19-9 has become an established marker for pancreatic cancer (121-123), but it must still be regarded as a research test for colorectal cancer.

Another methods to determinate CA 19-9 were enzyme-linked immunosorbent assay. Both the capture and the enzyme-conjugated antibody use the CA 19-9 monoclonal antibody. It should be noted that this antibody is useless for cancer diagnosis when a patient is lacking the enzyme for the synthesis of sialyl Le a. In Japanese, about 5-10% of the population lacks this enzyme. Determination carbohydrate antigen CA 19-9 levels in serum were also measured by radioimmunoassay (RIA) (125). Immunohistochemical technique used for the distribution of CA19-9 in tissues using an immunoperoxidase assay (126).By this technique the CA 19-9 can be detected not only in cancerous tissues but also in non cancerous normal tissues.

1.6.3. Screening

Numerous studies have addressed the potential utility of CA 19-9 in adenocarcinoma of the colon and rectum.

The reported incidence of elevated serum CA 19-9 in colorectal cancer ranges from 20% to 40% (127,128). The incidence of elevated CA 19-9 in stage-related, with the highest sensitivity occurring in patients with metastases (129-131). However, the sensitivity of CA 19-9 was always less than that of the CEA test for all stages of disease (127-130). The false-positive rate (>37.0 U.mL-1) is 15% to 30% in patients with non-neoplastic diseases of the pancreas, liver and biliary tract (131). Consequently, CA 19-9 cannot be used for screening asymptomatic populations.

1.6.4. Monitoring Response to Treatment

Kouri et.al.(132) compared CEA and CA 19-9 for predicting response to chemotherapy in 85 patients. Decreases in CEA more accurately reflect the response to therapy than did the decreases of CA 19-9. The pretreatment CA 19-9 value was, however, an important prognostic factor. Median survival was 30 months for patients with normal CA 19-9 values and 10.3 months for patients with elevated CA 19-9 values. CA 19-9 used to examined the serum levels and immunohistochemistry during the clinical course of female patient treatment with idiopathic interstitial pneumonia (IIp) that had elevated serum levels of   CA 19-9 (133).

1.6.5. Clinical Application

Elevated levels (>37 U.mL-1) were seen in patients with pancreatic (80%), hepatobiliary (67%), gastric (40-50%), hepatocellular (30-50%), colorectal (30%), and breast (15%) cancer. Pancreatits and other benign gastrointestinal diseases show a 10 to 20% elevation; however, the levels are usually lower than 120 (U.mL-1). CA 19-9 levels correlate with pancreatic cancer staging (54). CA19-9 is useful in monitoring pancreatic and colorectal cancer. Elevated levels can indicate the recurrence before clinical finding by 1 to 7 months (134). Unfortunately, early detection of relapse may not be useful because of the lack of effective therapy for pancreatic cancer.

ntroduction

The role of tumor markers in breast cancer is to enhance the clinicians, ability to provide more effective management of the disease (135). Serum CA15-3 concentration was determined by using sandwich enzyme immunoassay of a double monoclonal antibody (136,137), automated chemiluminescent immunoanalyzer (138), immunoradiometric assay (139) and radioimmunoassay(140), in women with benign breast tumor and breast cancer.

CA 15-3 has been used in management of patients, with breast cancer. CA 15-3 has been evaluated for its ability to determine diagnosis, prognosis, monitor therapy and predict recurrence of breast cancer following curative surgery and radiation therapy (141,142). Low incidences of CA 15-3 elevation in early stage cancer (stage I and stage II) have been observed (143).

Incidence of abnormal values of CA 15-3 in stage III and stage IV, and a very high CA 15-3 level have been correlated with metastases of breast cancer(144).

Therefore the development of immunoradiometric was planned to carry out the determination of the optimum conditions of 125I-anti CA 15-3 antibody binding with CA 15-3 in breast tumor tissue homogenate, hence determination of CA 15-3.

Chapter Two

CA15-3 in Breast Tumors

Materials and Methods

2.1. Materials

2.1.1. Chemicals

All chemicals and reagents used in this study were of analytical grade, tabulated in the following table.

Table (2.1): Chemicals used and Companies.

Chemicals	Company
1. Immunoradiometric assay kit for CA 15-3 level	Diasorin Inc. (USA)
2. Bovine serum albumin (BSA) , urea , $ZnCl_2$,$CaCl_2$,NH_4Cl, NaBr, ethylendiamine-tetraaceticdisodium salt (EDTA).	Fluka: (Switzerland)
3. $CuSO_4.H_2O$, NaK-tartarate glycine, NaOH,HCl, $NaCO_3$,NaF,NaCl,NaI,Na_2HPO_4, NaH_2PO_4.	BDH,limited,Poole (UK)
4.Folin-Ciolteau	E.Merck AG. Dastmstapt
5.Blue dextran (2000),sepharose CL-4B.	Pharmacia fine chemicals (Sweden)

2.1.2. Instruments

Table (2-2): Instruments used and Companies.

Instruments	Company
1.Gamma counter type 1270-rack gamma II 2. Spectrophotometer ultraspace type 4050	LKB
3. UV-210 a double beam spectrophotometer	Shimadzu
4.pH-meter	Pye-Unicam
5.Cooling centrifuge; with a maximum speed 5000 r.p.m.	Hettich
Cooling centrifuge type 202-MK; with a maximum speed 13500 r.p.m.	Sigma
7.Memmert water bath, memmert incubator	West Germany

8. SM-shaker	England
9. Combicold rack	LKB

2.1.3. Patients

Three groups of breast tumors patients were included in this study.

Group I : Consisted of 40 patients with benign breast tumors

Group II : Consisted of 32 premenopausal patients with breast cancer.

Group III : Consisted of 15 postmenopausal patients with breast cancer.

Group IV : Consisted of 10 controls.

All patients were admitted for treatment to (Saddam Medical City, Baghdad Teaching Hospital),(University Hospital, Saddam College of Medicine), (Nursing Home Private Hospital) and (Al-Arabi Private Hospital).

Patients suffered from any disease that may interfere with this study were excluded. All surgical operation of breast tumors were carried out under the supervision of the following surgeons:

Dr. Saab Sedeq, Dr.Munthir Al-Aubaidi, Dr.Azam Qanbar Agha, Dr. Abd Al-Salam Al-Tai, Dr. Zuhair Abid Al-Hadi.

The host information of all patients and normal healthy subjects is summarized in table (2-3).

Table (2.3): The host information of breast tumors patients and healthy subjects studied.

Group	Patients	No.	Age	Type of tumor	Metastases
I	Benign breast tumor	40	18-42	23 fibroepithelial tumor (fibroadenoma) 17 fibrocystic changes (adenosis)	– –
II	Premenopausal malignant	32	34-52	22 Infiltrative Ductal	2 lymph

49

	breast tumor			carcinoma 10 Ductal carcinoma	nodes
III	Postmenopau sal malignant breast tumor	1 5	55- 73	Infiltrative Ductal carcinoma	4 lumph nodes
IV	Control	1 0	25- 40		

2.1.4. Preparation of Blood Samples

Five milliliters of blood samples were obtained from patients by venipuncture just before surgery. Ten physically normal age volunteers were used as controls. Blood samples were left for 20 min. at room temperature. After coagulation, sera were separated centrifugation at 3000r.p.m for 10 min., and then sera were aspirated and stored at $-20oC$ until time analysis. The samples were not thawed and refrozen before testing.

2.1.5. Collections of Specimens

The tumors tissues were surgically removed from breast tumor patients by either mastectomy (cancer patients) or lumpectomy (benign tumor patients). The specimens were cut off and immediately rinsed with ice-cold isotonic saline solution. They were collected individually in plastic receptacle and stored at -20 oC until homogenization.

2.1.6. Preparation of Phosphate–Buffered Saline

Phosphate –buffered saline (PBS) 0.15 M, pH 7.2 was prepared as following:

A: Disodium basic phosphate (0.15M); 21.2940g Na_2HPO_4 and 9.0g of NaCl were dissolved in a final volume 1L deionized distilled water.

B:.Monobasic sodium phosphate (0.15M) 17.9970g of NaH_2PO_4 and 9.0g NaCl were dissolved in a final volume 1L deionized distilled water.

Phosphate buffer saline pH 7.2 was prepared by mixing a volume of solution A with appropriate amounts of solution B to obtain the required pH.

2.1.7. Preparation of Breast Tumors Tissues Homogenates

The frozen tissue were weighed, sliced finely and scalped in petri dish standing on ice bath, and then homogenized with fivefold volumes of PBS buffer pH7.2, using manual homogenizer (145). The homogenate was filtered through four layers of nylon gauze in order to eliminate fibers connective tissues, and then centrifuged at 4000 r.p.m for 45 min. at 4 oC in order to precipitate the remaining intact cells and the intact nucleus. The supernatant fraction at this speed was separated, divided in aliquots and freezed-20 oC until use.

2.1.8. Statistical Analyses

Students' t-test was used to determine if the mean values of studied parameters were significant different in the individual groups included in this work. $P < 0.05$ were considered significant (146).

2.2. Methods
2.2.1. Protein Determinations

Total homogenate protein content was determined by the method of Lowry (147), using bovine serum albumin (BSA) as the standard.

Figure (1.1) represents the standard curve of protein, which was constructed by plotting the absorbance at 600 nm against standard protein concentrations

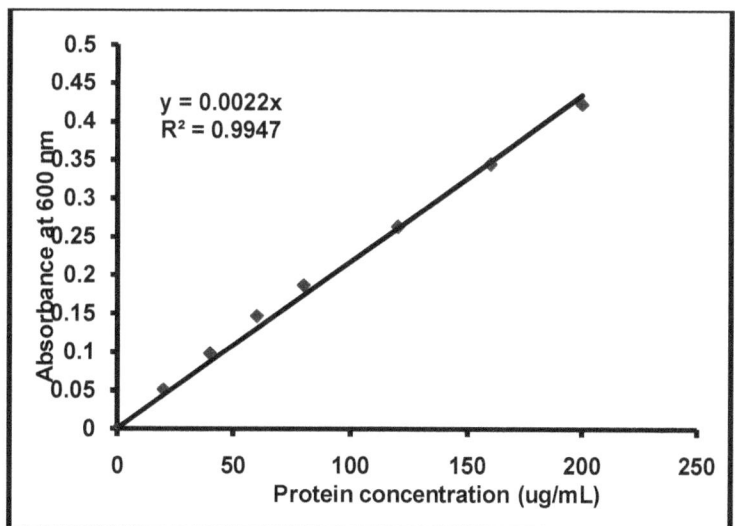

Figure (2.1): Standard curve of protein concentration. (All other details are explained in the text).

2.2.2. Determination of CA 15-3 Levels in Sera of Breast Tumors Patients

Reagents

The following reagents provided in the CA15-3 IRMA kit from Dia–Sorin-U.S.A. were used:

- Tracer: two vials each one contained 1.0 µ Ci/mL (37.1 KBa /mL). CA15-3 antibody labeled with 125I in 10 mL / Tris buffer with protein stabilizer and preservative.

- CA15-3 standards: The vial contained 100 mL, which represented 0 U.mL-1. There are four vials, 1.2 mL in each vial with different concentrations of human CA15-3 (25, 50, 100, 200) U.mL-1 in Tris. buffer with protein stabilizer and preservative.

- One bottle contained 100-coated beads, Anti-CA15-3-mouse, monoclonal.

- One vial contained 0.5mL CA15-3 control, CA15-3 in re-calcified human plasma with preservative.

Procedure

The assay protocol is described in table (2.4).

Table (2.4): IRMA protocol of serum CA 15-3 (U.mL-1) (All other details are explained in the text).

	CA 15-3 standard (U.mL-1)					Control	Unknown samples	
	0	25	50	100	200		1	2-etc.
Reaction trays no.	1.2	3.4	5.6	7.8	9.10	11.12	13.14	15.16-etc.
Standard (µL)	200	200	200	200	200	–	–	–
Control serum or samples (µL)	–	–	–	–	–	200	200	200
125I-anti CA15-3	200	200	200	200	200	200	200	200

The specimens and reagents must be brought to room temperature (20-30 oC) before opening. The reaction trays and data sheets were marked.

First Incubation

The specimens and the control were diluted to (1:15) prepared by adding 20 μL of specimen or control to 1000 μL CA15-3 standard, 0 U.mL-1 in a tube marked proper identification of specimen. Two hundred microliters of diluted specimen and control were pipette to their assigned wells. Two hundred microliters of each standard was pipette to its assigned well (standards are not to be diluted). One bead was dispensed into each well and the adhesive cover sealer was applied. After incubation for 2hrs at room temperature, the adhesive cover sealer was removed and the liquid was aspirated, then each bead was washed three times with 5 mL distilled water.

Second Incubation

Two hundred microliters of 125I-antiCA15-3 was pipetted on each bead. The adhesive cover sealer was applied again. After incubation time for 3hrs at room temperature, the cover was removed and the liquid was aspirated from wells, then the beads were washed as it is above. The beads were transferred to the counting tubes, and then the tubes were counted for 1 min.

Calculations

The standard curve was constructed by plotting counts per min. (Y axis) versus concentration for CA15-3 standard (X axis), figure (2-2). Then the points were connected with straight-line segments.

The CA15-3 concentration of specimens and control were determined directly from the standard curve.

2.2.3. Preliminary Test of CA 15-3 Binding to 125I -Anti CA15-3 Antibody in Breast Tumor Homogenate

Reagents

Phosphate buffered saline 0.15 M; pH 7.2 was prepared as described in section (2.1.6).

Procedure

The supernatant and pellet were centrifuged and detected by using ordinary tubes. In order to detect CA 15-3, 100 µL of the supernatant breast homogenate having (900µg protein) were incubated with 50 µL (0.35 mg.mL-1) of 125I -anti CA15-3.The volume of reaction was completed to 500 µL with PBS buffer pH7.2, then incubated at 37 oC for 2hrs. The assay tubes were centrifuged at 4000 r.p.m. for 45 min. at 4 oC.

The supernatant was discarded, the rim at each tube was swabbed with cotton, and then gamma counter counted the complex formed for one minute. The pellet of CA 15-3 was estimated by dissolving the sediment in PBS-buffer pH 7.2 with the ratio 1:5 (weight: volume) shaking was then carried out. Hundred microliters of the supernatant fraction of the sediment having (540 µg.mL-1 protein) was added to 50 µL (0.35 mg.mL-1) of 125I -anti CA 15-3 antibody. The same steps mentioned in this experiment were followed to determine the radioactivity of the complex formed. For total count two additional tubes with 50µL of 125I –anti CA 15-3 antibody were counted in gamma counter.

Calculations

1. The counted radioactivity in each tube (expressed in c.p.m.) represents the bound fraction (B), (i.e., 125I antiCA15-3 antibody/CA 15-3 complex).

2. The counted radioactivity in the tubes containing 125I-anti CA15-3 antibody only represents the total count (T).

3. The (B/T) ratio for each tube counted as follows:

$$(B/T)\% = \frac{Sample\ Counts\ (B)}{Total\ Counts\ (T)} \times 100$$

2.2.4. Factors Effecting of 125I-Anti CA-3 Antibody Binding to CA 15-3 in Breast Tumors Homogenates

2.2.4.1. The Effect of Different Amounts of Protein Concentration of the Tumor Homogenate on the Binding with 125I-Anti CA 15-3 Antibody

Reagents

All reagents prepared as described previously in sections (2.1.6) and (2.2.3).

Procedure

1. Fifty microliters (0.35 mg.mL-1) of 125I -anti CA 15-3 antibody were added to 100µL of the supernatant (benign Fibroadenoma, pre-and post-menopausal malignant breast tumors (IDC) respectively) containing increasing amounts of protein (50, 100, 150, 200, 250 µg.mL-1) then completed to a final volume of reaction to 500 µL with 0.15 M PBS pH 7.2.

2. The assay tubes were then incubated for 2 hrs at 37oC.

3. Two additional tubes, containing 50µL (0.35 mg.mL-1) of 125I –anti CAB-3 antibody only, for total counts were set-aside until counting.

4. At the end of incubation, the assay tubes were centrifuged at 4000 r.p.m for 45 min at 4oC.

5. The supernatant were decanted, the rims at the tube were swabbed with cotton piece.

6. The radioactivity of the complex were counted using gamma counter.

Calculations

1. The B/T percent were determined according to section (2.2.3).

2. The percent of binding values B/T were plotted versus the increasing amount of protein of the breast tissue homogenate.

2.2.4.2. The Effect of 125I -Anti CA15-3 Antibody Concentration on the Binding

Reagents

All reagents prepared as described previously in section (2.1.6) and (2.2.3).

Procedure

1. Fifty microliters of increasing concentration (0.070,0.140,0.175,0.350, 0.701 mg.mL-1) of 125I -anti CA 15-3 antibody were added to 100µL of

homogenate (benign breast tumor (Fibroadenoma), pre-and post-menopausal malignant breast tumors) (IDC) containing (100, 100, 200 µg.mL-1 protein) respectively.

2. The volume of reaction was made up to 500 µL with PBS pH 7.2.
3. Steps 2,3,4,5 and 6 of the experiment (2.2.4.1) were repeated.

Calculations

1. The same mathematical equation mentioned in section (2.2.4.1) was used to calculate (B/T)%.

2. Values of (B/T)% were plotted versus concentration of labeled antibody (125I -anti CA15-3 antibody).

2.2.4.3. The Effect of pH on the Binding

Reagents

All reagents prepared as described previously in section (2.1.6) and (2.2.3).

Procedure

1. One hundred microliters of human homogenate (benign breast tumor (Fibroadenoma), pre-and post-menopausal malignant breast tumors (IDC)) containing (100,100,200, µg.mL-1 protein) were added to 50µL (0.175, 0.175,0.140mg.mL-1) of 125I -anti CA15-3 antibody respectively.

2. Each mixture was completed to 500 µL with PBS of different pH ranging (6.8-8.0).

3. Step 2,3,4,5 and 6 of the experiment (2.2.4.1) were repeated.

Calculations

1.Values of (B/T) % were calculated as described in section (2.2.4.1).

2. (B/T)% were plotted against their corresponding pH.

2.2.4.4. Time Course of the Binding of 125I -Anti CA15-3 Antibody to CA 15-3 in Breast Tumors Homogenates

Reagents

All reagents prepared as described previously in sections (2.1.6) and (2.2.3).

Procedure

1. One hundred microliters of homogenate (benign breast tumor (Fibroadenoma), pre-and post-menopausal malignant breast tumors (IDC)) containing (100,100,and 200 µg.mL-1 protein) were incubated with 50µL of 125I -anti CA15-3 antibody concentration (0.175,0.175 and 0.140 gm.mL-1).

2. The volume of reaction was made up to 500 µL with PBS pH (7.0,7.6 and 7.8).

3. All tubes were incubated at 37oC at different time intervals (30, 60, 90, 120, 150 and 180) min.

4. Step 3,4,5,6 of the experiment (2.2.4.1) was repeated.

5. To determine the time course of CA 15-3 binding to 125I –anti CA 15-3 antibody at different temperatures. Steps 1, 2, 3 and 4 in the same experiment were repeated at different temperatures (5, 15, 25, 45°C).

Calculations

1. The values of (B/T)% were calculated as described in section (2.2.4.1) at each time and temperature used.

2. The values (B/T)% was plotted against the time of incubation at different temperatures.

2.2.4.5. The Effect of Different Halides on the Binding

Reagents

1. Phosphate buffer (PB) were prepared as described in section (2.1.6) without the addition of NaCl .

2. Halid reagents were prepared in concentration of 0.01M PB at pH (7.0, 7.6 and 7.8) individually, by dissolving each of 0.021gm of NaF, 0.0292gm of NaCl, 0.0515 gm of NaBr, and 0.075gm of NaI in a final volume 50µL of PB and the pH was adjusted.

3. The breast tumors homogenates (benign breast tumor (Fibroadenoma)) were prepared as described in section (2.1.7), except using PB-buffer instead of PBS at the same pH and concentration carried out the homogenization.

Procedure

1. One hundered microliters of each group homogenate (benign breast tumors (Fibroadenoma) and pre-post menopausal malignant breast tumors (IDC)) containing (100,100 and 200 $\mu g.mL^{-1}$ protein) were incubated with 50 μL of 125I -anti CA 15-3 antibody concentration (0.175,0.175 and 0.140 $gm.mL^{-1}$). The volume was made up to 500 μL with PB pH (7.0, 7.6 and 7.8) containing 0.01 M of the following halides: NaF, NaCl, NaBr and NaI in each assay tube. (A sample without the addition of any salt was used as a control).

2. The assay tubes were then incubated for 90min. at 45, 15 and 45°C for the three groups individually.

3. Steps 3, 4, 5 and 6 of the experiment (2.2.4.1) were repeated.

Calculations

1. The values of (B/T) % were calculated as described in section (2.2.4.1).

2. (B/T)% was plotted against halides concentrations.

2.2.4.6. The Effect of Monovalent and Divalent Cations on the Binding

Reagents

1. PB was prepared as described in section (2.1.6) without addition of NaCl.

2. Monovalent and divalent cations salts were prepared in concentration of (0.025 M) PB at pH (7.0,7.6 and 7.8) individually, by dissolving each of 0.0931gm of KCl, 0.0668gm of NH_4Cl, 0.2541 gm of $MgCl_2.6H_2O$, 0.1388 gm of $CaCl_2.2H_2O$, 0.2474gm of $MnCl_2.4H_2O$, 0.3150 gm of $CuSO_4.5H_2O$ and 0.1703gm of $ZnCl_2$, in a final volume 50 mL of PB and the pH was adjusted.

Procedure

1. The same steps mentioned in section (2.2.4.5) were followed to determinate the effect of monovalent and divalent of CA 15-3 in the tissues homogenates of (benign breast tumors (Fibroadenoma) and pre-and postmenopausal malignant breast tumors (IDC)) with 125I -anti CA 15-3 antibody, except the PB buffer containing (0.025M) of the following salts:

KCl, NH4.Cl, MgCl2.6H2O, CaCl2.2H2O, MnCl2.4H2O, CuSO4.5H2O, ZnCl2.

2. A sample without the addition of any salt was used as control.

Calculations

1. The values of (B/T)% were calculated as described in section (2.2.4.1)

2. (B/T)%was plotted against monovalent and divalent cations salts concentrations.

2.2.4.7. Recovery of CA 15-3

Reagents

1. All reagents are described previously in section (2.1.6) and (2.2.3).

2. Standard concentration of CA 15-3 200 U.mL-1 was used.

Procedure

Known concentration of CA15-3 (200 U.mL-1) was added to the three groups of tissues homogenates (benign breast tumors (Fibroadenoma), and pre-and post-menopausal malignant breast tumors (IDC)). The experiment was carried out at optimum conditions that were obtained in experiments of (2.2.4). The CA15-3 was determined according to the experiment in section (2.2.3).

Calculations

1. The bound (c.p.m) of the reaction mixture (standard CA 15-3 was added to tissue homogenate) with 125I -antiCA15-3 antibody, represent the measured value.

2. The bound (c.p.m.) of CA 15-3 in tissue homogenate with 125I – antibody CA 15-3 antibody only , represent the expected value.

3. The recovery % (yield%) was calculated as follows:

$$\mathrm{Re\,cov\,ery\%} = \frac{\mathrm{Measured\,values}}{\mathrm{Expected\,values}} \times 100$$

2.3. Results and Discussion

Human breast tissues in this study were classified according to type of breast tumors (benign and malignant) and the malignant breast tumors were again classified into sub groups (premenopausal and postmenopausal). Each type was examined histologically according to WHO classification. Homogenization was carried out in a cold medium (i.e.4°C) to avoid protein denaturation (148,149), by proteolytic enzymes (150). The filtration of the tissue homogenate through several layers of nylon gauze was used to remove any suspended pieces unhomogenized fragments and blood vessels.

Determination of CA 15-3 Levels in Sera of Breast Tumors Patients

CA 15-3 levels in sera of patients with benign breast tumors (group I) and pre and post-menopausal malignant breast tumors (group II and group III) were measured by IRMA method. These three groups were matched with a group of control subjects.

Table (2.5) summarizes the groups and the mean concentrations of CA15-3 for the control women and patients with benign breast tumors and pre-and post-menopausal malignant breast tumors.

Table (2.5) showes that CA15-3 levels in three different groups (benign breast tumors and pre-and post-menopausal malignant breast tumors) were significantly elevated ($p < 0.05$) for benign breast tumors and highly significantly elevation ($p < 0.0001$) for pre-and post menopausal malignant breast tumors respectively, as compared with the control.

The mean serum CA15-3 level of the control was found to be (17.26 ± 4.06 U.mL-1) as shown in table (2-5), and the cutoff values was (25 U.mL-1) that obtained from (mean +2 SD). This cutoff value is in agreement with Geraghty J.G(151), other study obtained that cutoff value of 40 U.mL-1 (152), 22 U.mL-1 (153), 30 U.mL-1 (154).

It has shown that widely different cutoff value which was described ranging from 20-40 U.mL-1 in different reference (155-159).

According to Bon et al (160) the upper limit of CA 15-3 of normal may be method-dependent. No association between the CA 15-3 and either age or menopausal status was found in the control group. Therefore , the cutoff values do not require adjustments related to these variables. These results were in agreements with Gion M.et.al. (68). Figure (2.3) shows the distribution of the individual values of CA15-3 in sera of patients with benign breast tumors and pre-and post menopausal malignant breast tumors and control, were determined by using the standared curve in figure (2-2).

It was found that the mean of serum CA 15-3 concentration in 20 patients with benign breast tumors was 21.9 ± 6.6 U.mL-1 (mean ± SD). These results are in agreement with Hayes D.F. et.al (161). The results show there was highly significant correlation between serum connections of CA15-3 in both groups pre-and post-menopausal status with control, while it was significantly lower in benign breast tumors status.

This is in agreement with Ichihara S. et.al (162). Therefore all of the cases used in the binding studies were concentrated to this type of carcinoma (IDC) and this type is the common type of breast cancer. In Iraq very high levels of CA15-3 advanced disease and the value 5 to 10 times of normal suggest the presence of metastasis. Increasing numbers of metastatic sites correlate with increasing CA15-3 levels (163,164).

These findings suggest that higher levels of CA15-3 represent the breast cancer extent and reflect the cell differentiation and aggressiveness of the tumor. Therefore, it could be concluded that the determination of CA15-3 before surgical operative may be useful as a prognostic factor in breast cancer.

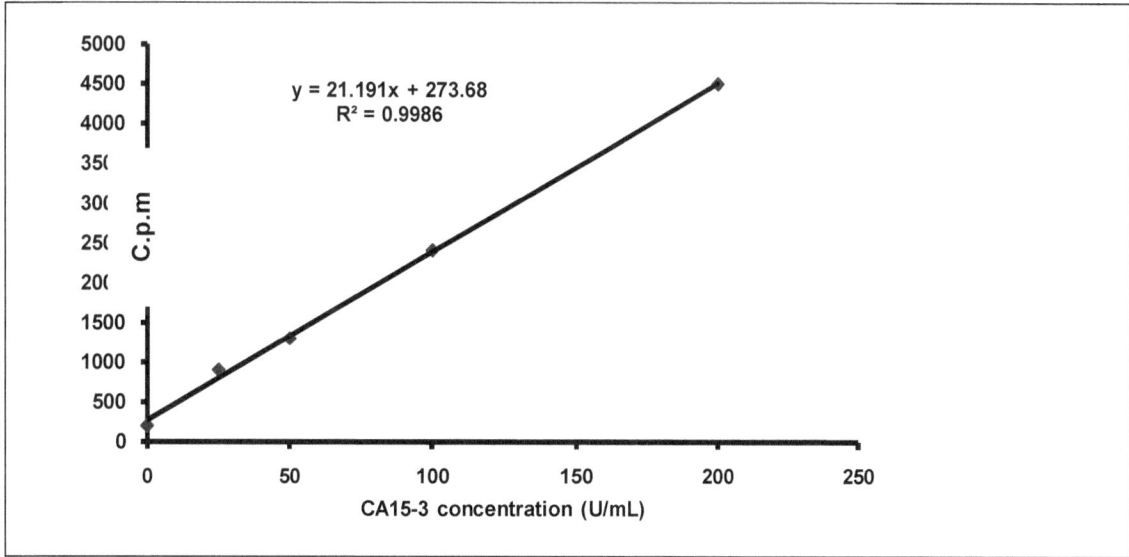

Figure (2.2): Standard curve of CA 15-3 determination in human sera by IRMA method.(All other details are explained in the text).

Table (2.5): Sera CA15-3 levels (U.mL-1) in patients with benign and malignant breast tumors. (All details are explained in the text).

Group	Patients	No. of cases	Age range (Year)	Sera CA15-3 U.mL-1 (mean ± SD)	P values
I	Benign breast tumors	20	18-42	21.9 ± 6.6	P<0.05
II	Premenopausal malignant breast tumors	16	34-52	37.3 ± 6.8	P<0.0001
III	Postmenopausal malignant breast tumors	12	55-73	60.3 ± 10.9	P<0.0001
IV	Control	10	25-40	17.3 ± 4.06	--

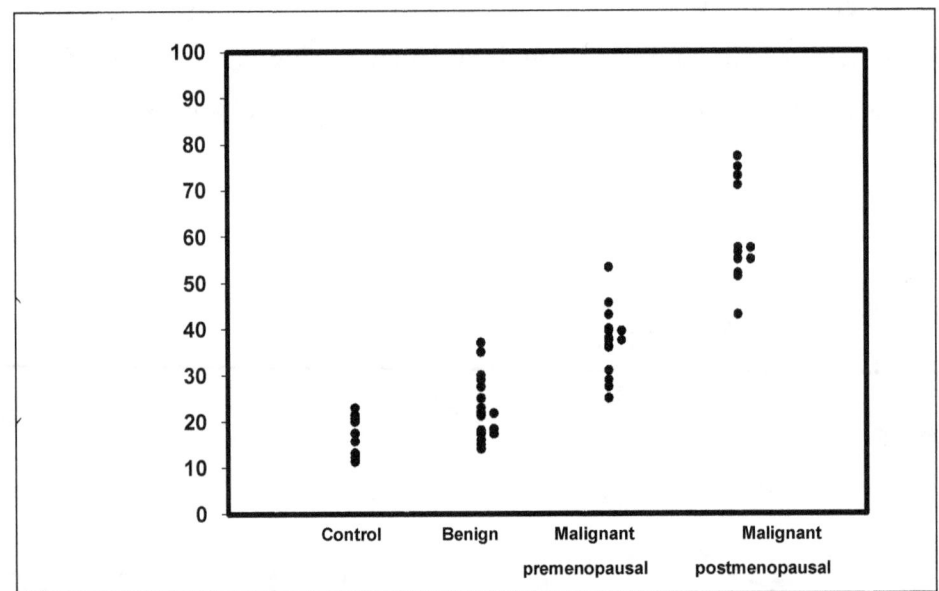

Figure (2.3): Distribution of the individual value of CA15-3 U.mL-1 in the sera of benign and malignant breast tumors patients. (All other details are explained in the text).

Binding Studies of 125I-Anti CA15-3 Antibody with CA15-3 in Benign and Malignant Breast Tumors Homogenates

Preliminary Test of the Binding of 125I-Anti CA15-3 Antibody with CA

Supernatant and pellet formed at speed (4000 r.p.m.) in three groups of human breast tumor homogenate (benign breast tumors, pre-and post-menopausal malignant breast tumors) were used in this experiment. In each fraction CA 15-3 was detected through the incubation of 125I-anti CA15-3 antibody with supernatant fraction and pellet individually for 2hrs at 37°C in PBS buffer as a medium to complete the reaction. The separation of the bound antibody from the unbound was carried out at 4000 r.p.m. for 45 min. to precipitate the (125I-anti CA15-3 antibody/CA15-3) complex formed. Preliminary experimental conditions used in Table (2.6), which is show, the amount of binding B/T% in both fractions. The data revealed that CA15-3 was higher in incidence according to B/T%.

Table (2.6): Incidence of CA15-3 in supernatant and pellet fractions in three different breast homogenate.

Groups	(B/T)%		CA15-3 U.mL-1 in supernatant fraction kit
	Supernatant Fraction	Pellet Fraction	
Benign	6.20	3.40	90
Premenopausal	8.04	5.53	356
Postmenopausal	6.31	4.64	144

B/T% in supernatant is more than in pellet fractions of this speed (4000 r.p.m.). According to these results supernatant fractions was collected and the pellet was then discarded. The CA15-3 levels in the supernatant of breast tumors homogenate were determined according to IRMA method.

In general, results show that CA15-3 concentration in pre-and post-menopausal malignant breast tumors homogenates is more than benign breast tumors homogenates. These results are in agreement with the result obtained from B/T% from IRMA developed method.

From these results, it can be said that developed method was useful for determination CA15-3 in breast tumors homogenate using 125I-anti CA15-3 antibody.

Factors Effecting of 125I-Anti CA15-3 Antibody Binding to CA15-3 in Breast Tumors Homogenates

The Effect of Different Amounts of Protein Concentration of the Tumor Homogenate on the Binding with 125I-Anti CA 15-3 Antibody

To obtain the optimum protein of homogenate for the binding of CA15-3 with 125I-anti CA15-3 antibody, the supernatant homogenate containing increasing amount of CA15-3 in the presence of fixed amount of 125I-anti CA15-3 antibody as it was mentioned in section of (2.2.4.1).

Figure (2.4) represents the quantitative precipitation curve in which the amount of (125I-anti CA15-3 antibody/CA15-3) complex in three groups (benign breast tumors and pre-and-post menopausal malignant breast tumors) was plotted as a function of CA15-3 concentration.

As shown in this figure, in the first phase of the reaction no precipitate was formed. The amount of precipitate increased until a point of maximum binding was reached. After this point as the amount of CA15-3 increased the amount of precipitate diminished; thus the increase in protein concentration which would increase the number of binding site and hence increase the percent of binding until the saturation state at (100, 100, and 200 µg.mL-1) homogenate concentration for (benign breast tumors, pre-and post menopausal malignant breast tumors respectively).

The complex precipitate out of solution because of the multivalent nature of both molecules (165). The radioactive antibody has two binding sites, it can cross-link antigenic sites of two different CA15-3 molecules and can produce maximum complex formation and therefore maximum precipitate will occur. When CA15-3 is in greater excess, large complex are again less probable.

In all subsequent experiments the amonts of (100, 100 and 200 µg.mL-1 protein) of tissue homogenate in benign breast tumors and pre-and post menopausal malignant breast tumors were used according to the result obtained in this experiment.

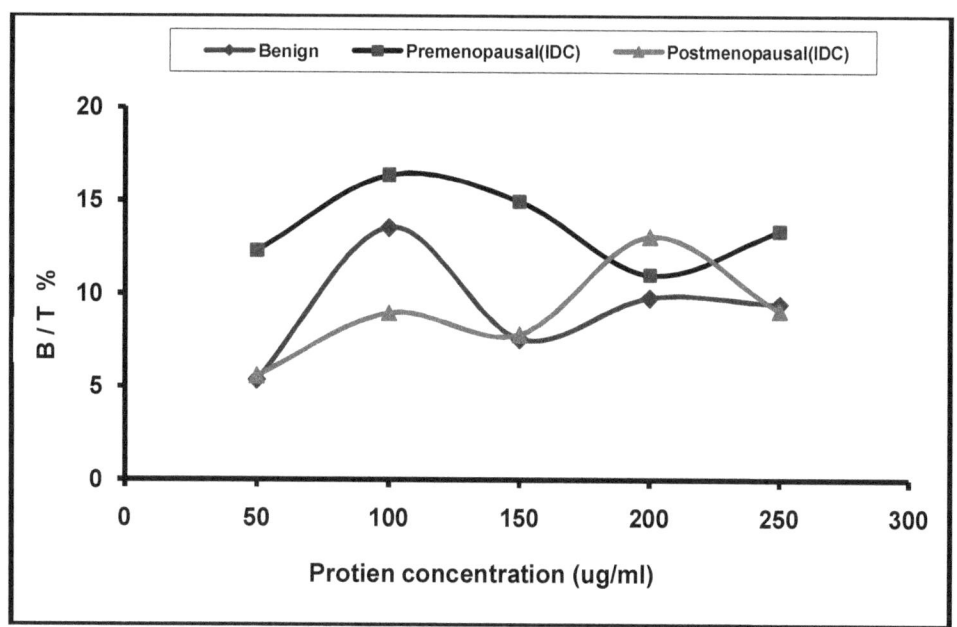

Figure (2.4): Influence of increasing protein concentration on the binding with 125I-anti CA15-3 antibody. (All other details are explained in the text).

The Effect of 125I-Anti CA15-3 Antibody Concentration on the Binding

The experiment was carried out in the presence of fixed amount of protein concentration of the homogenate and increasing concentration of 125I-anti CA15-3 antibody.

The results are illustrated in figure (2.5). Which represent 125I-anti CA15-3 antibody binding curve with supernatant fraction of benign breast tumor, pre-and post-menopausal malignant breast tumors. As shown in figure (2.5) it is obvious that the amount of (125I-anti CA15-3 antibody/CA15-3) complex rises gradually, and then the breast tumor protein was saturated with 125I-anti CA15-3 antibody. When the amount of antibody is in moderate excess, the probability of cross-linking of Ag by Ab in the incubation mixture is more likely, and hence large complex formation is favored. Then the maximum B/T percent was detected. The presence of (0.175, 0.175, 0.14 mg.mL-1) of 125I-anti CA15-3 antibody in benign, pre-and post-menopausal breast tumors homogenates give the optimum concentration of 125I-anti CA15-3 antibody in three groups. Then the binding percent decreased as the amount of 125I-anti CA15-3 antibody increased.

This is because all antigenic sites are covered with antibody and complex formation is inhibited (166). These results indicate that the binding is principally dependent on the amount of the antibody in the reaction mixture (167).

According to the results of this experiment the above concentration of 125I-anti CA15-3 antibody was used in the subsequent experiments.

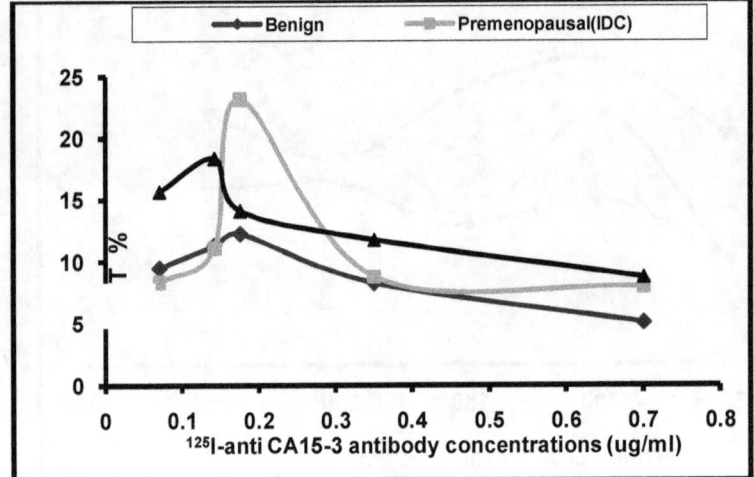

Figure (2.5): Effect of different concentrations of 125I-anti CA15-3 antibody on the binding of with CA15-3. (All other details are explained in the text).

The Effect of pH on the Binding

Figure (2.6) shows the values of the binding of 125I-anti CA 15-3 antibody to CA 15-3 in benign breast tumor, pre- and post-menopausal malignant breast tumors, at different pH values. Maximum value of the binding occurs at (pH 7, pH 7.6, pH 7.8) for benign breast tumor, pre-and post-menopausal malignant breast tumors respectively.

The formation of (125I-anti CA 15-3 antibody/CA 15-3) complex is usually performed at pH between 6.8-8.0; the results indicate that the shift in the pH of the environment may affect the properties of CA 15-13 molecules involved in the binding. This effect may include the protonation deprotonation processes occurring within the possible ionizable groups of the amino acids present in the binding domain of these molecules (168).

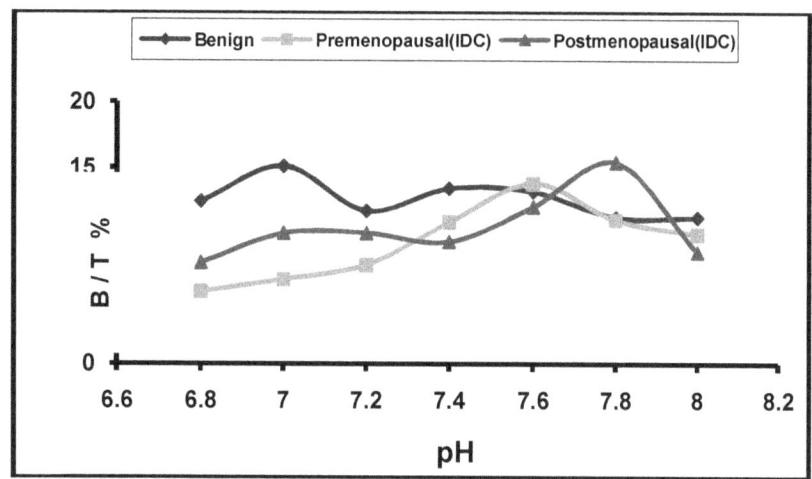

Figure (2.6): Effect of pH on the binding of 125I-anti CA 15-3 antibody with CA 15-3 in breast tumors homogenates. (All other details are explained in the text).

Time Course of the Binding of 125I -Anti CA15-3 Antibody to CA 15-3 in Breast Tumors Homogenates

The results of time course pattern at different temperatures (5, 15, 25, 37, 45oC) indicate the 125I-anti CA 15-3 antibody binding to crude fractions of CA 15-3 is temperature and time dependent process, as shown in figures (2.7), (2.8) and (2.9). The maximum binding was obtained at 45oC after incubation for 90 min. in crude fractions of benign breast tumors and postmenopausal malignant breast tumors respectively, whereas the binding in crude fractions of premenopausal malignant breast tumors occurs at 15oC after incubation for 30 min.

The decrease of the binding activity may be due to reversible dissociation of (125I-anti CA 15-3 antibody/CA 15-3) complex after reaching the equilibrium state.

At 45oC the CA 15-3 molecule preserve the nature of protein structure and gave the maximum binding, but at higher temperature than 45oC denaturation may occur.

In the premenopausal malignant breast tumors the maximum binding occurs at 15oC for 30 min., in this state the energy is enough to overcome the energy barrier and give the maximum binding (169), the decrease in binding after 15oC may be due to proteolytic enzyme.

The difference in incubation time to give the maximum binding may be due to the different source of CA 15-3. According to this results, the binding studies of the subsequent experiments were carried out a 45oC for 90 min incubation for benign and postmenopausal breast tumors homogenate, whereas 15oC for 30 min. incubation for premenopausal malignant breast tumors homogenate.

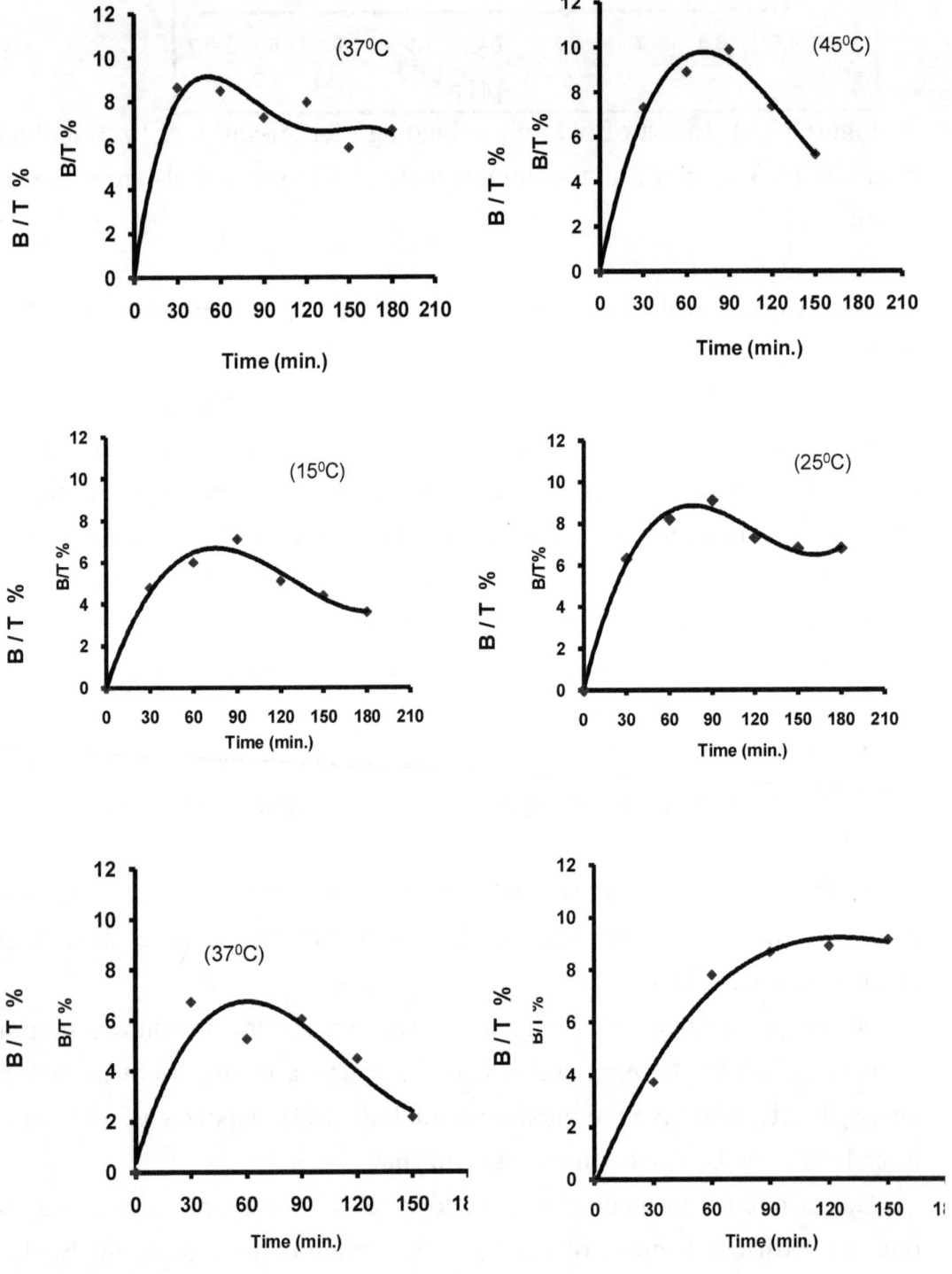

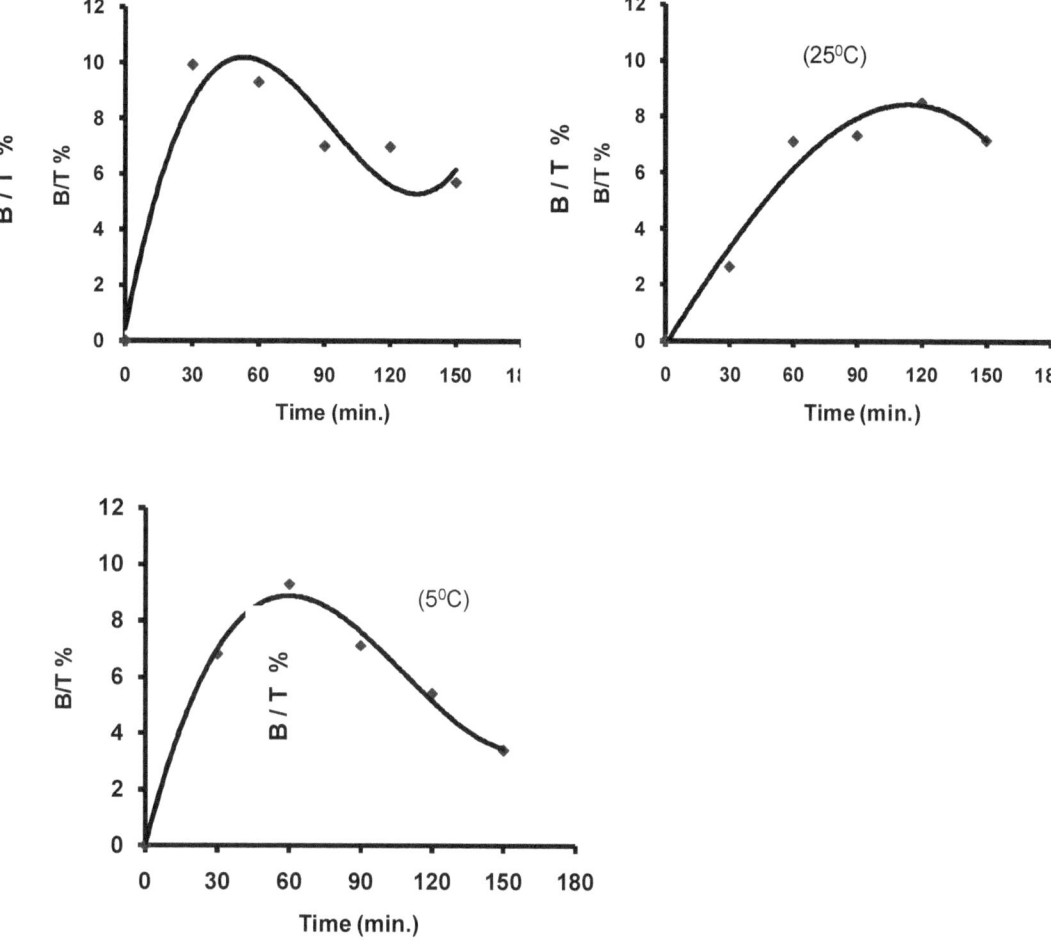

Figure (2.8): Time course of the binding of 125I-antiCA15-3 antibody with CA15-3 in premenopausal malignant breast tumor. (All other details are explained in the text).

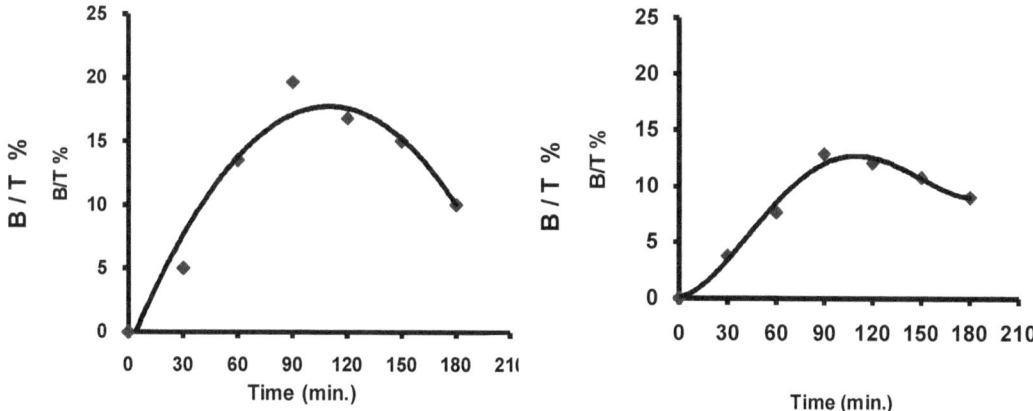

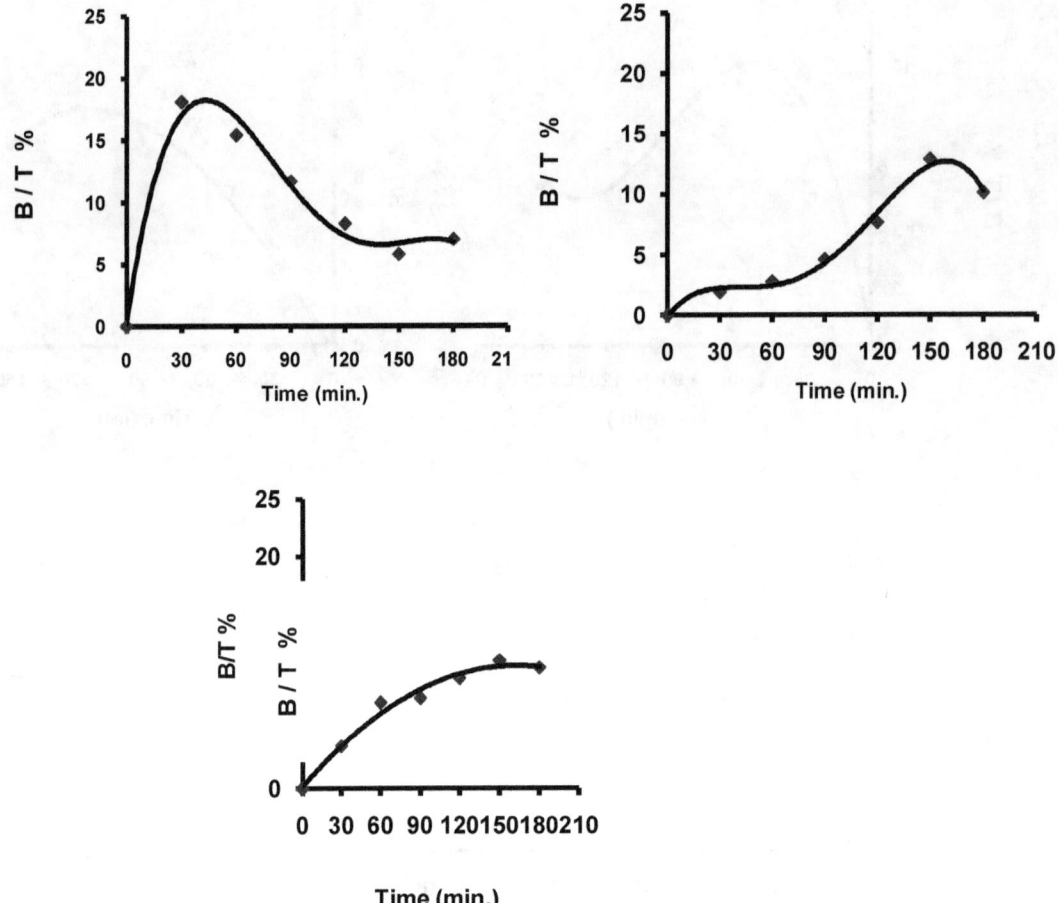

Figure (2.9): Time course of the binding of 125I-antiCA15-3 antibody with CA15-3 in postmenopausal malignant breast tumor. (All other details are explained in the text).

The Effect of Different Halides on the Binding

Different sodium halides at 0.01 M concentration were investigated to study their action on the binding 125I-anti CA 15-3 antibody with CA 15-3 in the three groups (benign breast tumors, Pre-and postmenopausal malignant breast tumors), as shown in figure (2.10).

The presence of the sodium halides in the incubation medium tends to promote the binding of 125I-anti CA 15-3 antibody to CA 15-3 in these groups, the following sequence of effects have occurred.

1. Benign breast tumor tissue homogenate

NaI > NaBr > NaCl > NaF

2. Premenopausal breast cancer tissue homogenate

NaCl > NaI > NaBr > NaF

70

3.Postostmenopausal breast cancer tissue homogenate

NaCl > NaBr > NaF > NaI

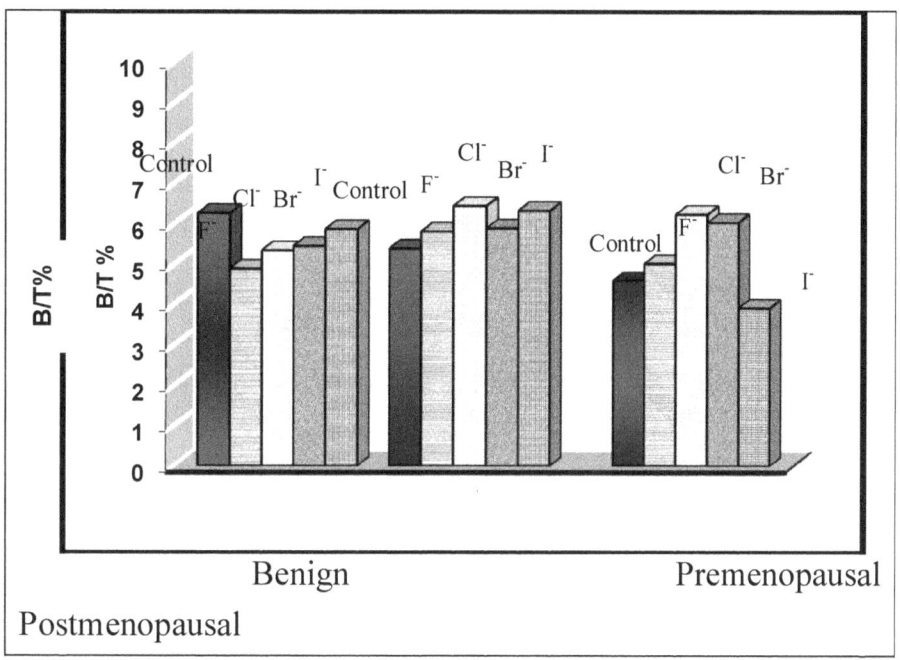

Figure (2.10): Effect of different halides on the binding of 125I-anti CA 15-3 antibody with CA 15-3. (All other details are explained in the test).

As shown in figure (2.10), the sodium halides inhibited the binding in benign breast tumors, according to the decreasing ionic radius and increasing radius of hydration. It seemed that fluoride ion causes lower binding, this could be due to higher electro negativity of fluoride ion that tend to interact with the positive residue in the binding site of the antibody and/or the antigen which lead to decrease the interaction between CA 15-3 and its antibody (170).

Melander and Horvath (1977) reported that the effect of halide salt type on hydrophobic interactions is quantified by its molar surface tension increment (MSTI) that is a measure of the increase in surface tension by the salt (171). On the other hand, figure (2.10) shows the effect of different halides salts at 0.01 M on the extent binding of 125I-anti CA 15-3 antibody to pre-and postmenopausal malignant breast tumors homogenate. It seems that halides salts increased the binding, especially NaCl, this could be due to that NaCl in lower concentration (0.15M) or in physiological concentration, increased the binding between CA 15-3 and its antibody (172).

The Effect of Monovalent and Divalent Cations on the Binding

The effect of different salts on the extent of binding of 125I-anti CA 15-3 antibody to CA 15-3 in benign and malignant breast tumors are shown in figure (2.11).

The results indicate that the binding process is sensitive to the presence of cation metal ions. CuSO4.5H2O at concentration (25mM) was shown to increase the binding more than other divalent cations, while ZnCl2 increased the binding less than other divalent cations. One hypothesis assumes that salts may alter the nature of the hydrophobic forces controlling stabilization of the complex formed and these vary depending on the nature of the interacting groups (172). From the results illustrated in figure (2.11), it is suggested that these salts maybe provide some conformational changes in the CA 15-3 and the charged groups of the binding domain of the antibody and antigen molecule (173,174), that hinder maximal binding are shielded. If the interaction is dominated by ionic strength, high salt concentration lowers the affinity.

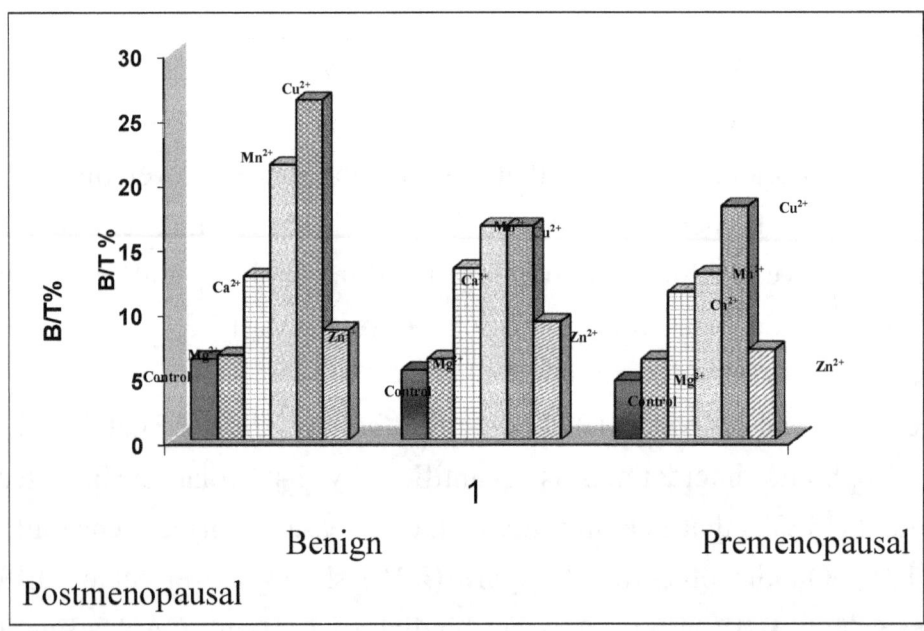

Figure (2.11): Effect of different divalent cations on the binding of 125I-anti CA 15-3 antibody with CA 15-3. (All other details are explained in the text)

Figure (2.12) shows the effect of monovalent cations (KCl and NH4Cl) on the extent of the binding of CA 15-3 to its antibody 125I-anti CA 15-3 in benign and malignant breast tumors. KCl at 25mM was shown to increase the binding in benign and premenopausal malignant breast tumors as compared with the control value, while KCl at the same concentration slightly inhibiting the binding in postmenopausal malignant breast tumors. These results may be due to conformational changes. NH4Cl at 25 mM was shown to inhibit the binding but to a lesser extent.

This result shows that NH4Cl effect on the binding is nearly unremarkable. Presumably, the lesser degree of hydration permits greater interaction of the salt with an anionic group located in the antibody-combining site and then inhibits the complex formation.

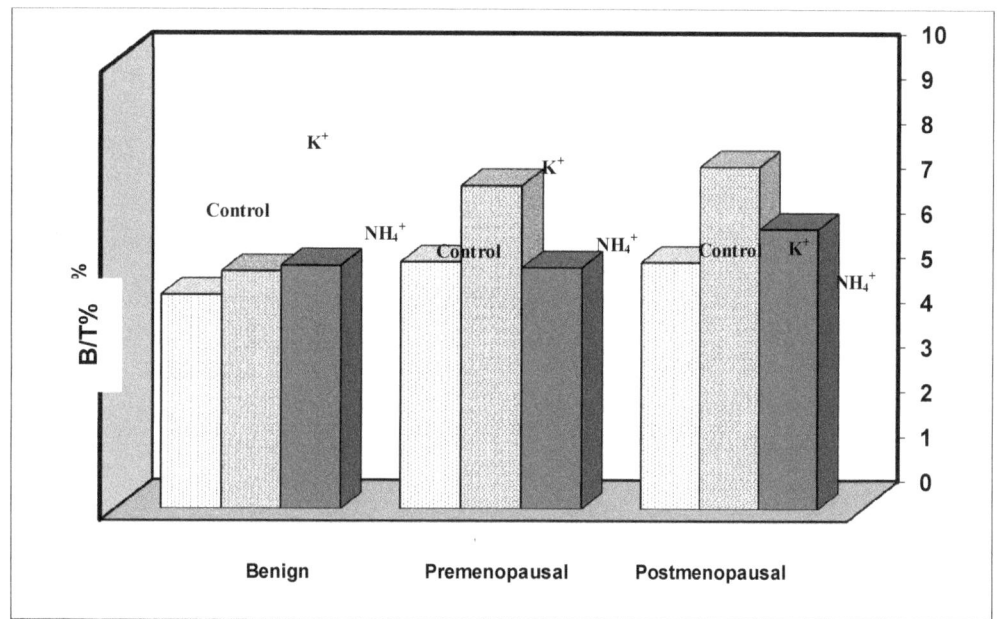

Figure (2.12): Effect of different monovalent cations on the binding of 125I-anti CA 15-3 antibody with CA 15-3. (All details are explained in the text).

Recovery of CA 15-3

This method was used to estimate the percent recovery of CA 15-3 in supernatant fractions of benign and malignant breast tumors homogenates. The results are summarized in table (2.7) and indicate that the CA 15-3 extracted from benign breast tumors, and CA15-3 extracted from malignant breast tissues homogenate were recovered more than CA 15-3 extracted from postmenopausal malignant breast tumors homogenates were recovered more than CA 15-3 extracted from premenopausal malignant breast tumors homogenates. Also the results indicate that total CA 15-3 can be determined through the developed method of immunoradiometric assay, as well as the percent of recovery indicates the precision of the used method.

Table (2.7): Recovery of CA 15-3. (All other details are explained in the text).

Type of CA 15-3	Measured B/T	Expected B/T	Recovery% Measured / Expected
Benign (Fibroadenoma)	16.23	24.84	65.34
Premenopausal (IDC)	20.04	27.64	72.50
Postmenopausal (IDC)	24.20	26.80	90.30

s witAbstract

Gel filtration chromatography technique was used for partial purification of CA15-3 from breast tumor homogenates.

The results revealed the presence of one form of CA15-3 with a high molecular weight (440 KD). This type possesses a high affinity for the binding to its antibody 125I-anti CA15-3 at the same conditions performed in section (2.2.4) of chapter two.

The elution volume (Ve) and the Kav value for elution of CA15-3 from sepharose CL-4B column were calculated. The experiments of optimum conditions of the binding between the partially purified CA15-3 and 125I-anti CA15-3 antibody were determined, in benign breast tumor and premenopausal malignant breast cancer homogenates.

Studies on the stability of both partial purified CA15-3 and crude CA15-3, show that the crude CA15-3 was more stable than the purified CA15-3.

Chapter Three

Purification of CA-15-3

CA15-3 is high molecular weight glycoprotein (>400 KD) identified at the apical side of alevoli and duct of mamary glands (83). Several authors have isolated, purified and characterized CA15-3 from different sources;either by the

isolation of CA15-3 from a breast cancer patient's sera, using affinity chromatography, gel filtration, and then characterized by SDS-PAGE(175,176) ,or by purifing a high molecular weight glycoprotein from human milk and breast carcinoma by using gel filtration, affinity chromatography and then PAGE (177).   In the present study , benign breast tumors and premenopausal malignant breast cancer were used as a source for partial purification of CA15-3 and then determination its yield. The factors effect the binding of partial purified CA15-3 to its antibody 125I-anti CA15-3 antibody were also studied.

Materials and Methods

3.1. Materials
3.1.1. Chemicals
All chemical and reagents mentioned in section (2.1.1) and (2.1.6) were used in the experiments of this chapter.

3.1.2. Instruments
All instruments mentioned in section (2.1.2) were also used in the experiments of this chapter.

3.1.3. Patients
The same patients tissues mentioned in section (2.1.5) were used in the following experiments. Benign breast tumor and premenopausal malignant breast cancer homogenates that showed maximal binding in the preliminary test in section (2.2.3) were used for the purification of CA15-3.

3.2. Methods
3.2.1. Isolation of CA15-3 by Sepharose CL-4B Column
3.2.1.1. Preparation of the Column

The dimensions of the column were chosen according to the following equation (150).

$$\text{Diameter} = \sqrt{\frac{m}{10}}$$

Where:

m= amount of protein in mg.

L = 30 x diameter

Where:

L : length of the column

3.2.1.2. Preparation of Phosphate Buffered Saline

PBS buffer pH 7.0 containing 0.02% sodium azide was prepared as described previously in section (2.1.6).

3.2.1.3. Preparation of the Gel

The gel was prepared by allowing the pre-swollen gel to swell again in PBS buffer (0.05 M) pH 7.0, then left to settle and the excess of buffer was decanted. The step was repeated several times. Suction was then used to degas the gel and slurry was left for 24 hrs to equilibrate with buffer.

The swollen gel was suspended and carefully poured into a vertical glass-column (0.7 x 30 cm) down the wall using a glass rod. After the gel has settled, the column was equilibrated with PBS for 24 hrs.

3.2.1.4. Void Volume Determination

The void volume of the column was determined by using blue dextran 2000 at concentration of 2 mg.mL-1 dissolved in PBS buffer pH 7.0 , then the elution was carried out with the same buffer at a flow rate of 20 mL.hrs-1.

Fractions of 2 mL were collected and their absorbance was measured at 600 nm. Figure (3-1) shows the elution profile of blue dextran 2000. The volume of the buffer required to elute the blue dextran which represents the void volume was (6 mL).

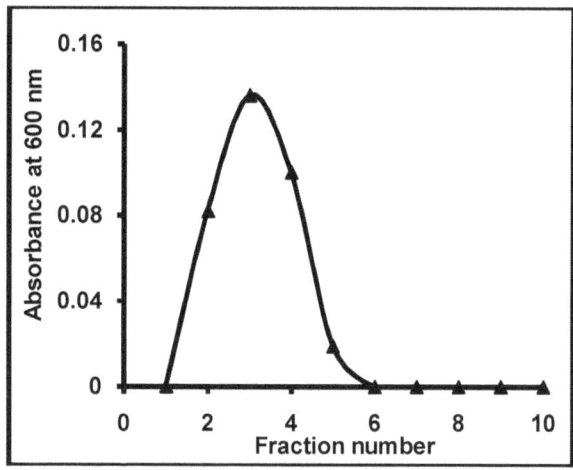

Figure (3.1): The elution profile of blue dextran 2000. (All other details are explained in the text).

3.2.1.5. Column-Calibration

The column was calibrated by gel filtration kit, purchased from pharmacia fine chemicals which contained standard proteins. Standard protein solutions were prepared according to the manufacturers instructions, then applied through two 0.5 mL portions, proteins 1 and 3 in the first portion, protien 2 and 4 in the second portion. Elution was carried out with PBS buffer at a flow rate of 20 ml.hrs-1. the absorbance of the fractions collected was measured at 280 nm to evaluated the elution volume (Ve) of the standard protein.

Standard Proteins

Pharmacia calibration kit for determination of M.wt by gel filtration was used. The kit comprises the highly purified proteins and their high M.wt are detailed in table (3.1).

Table (3.1): Standard proteins and their molecular weights (All other details are explain in text).

Protein	M.wt (KD)	Conc. mg.mL-1
Thyroglobulin	669	4.0
Ferritin	440	1.0
Catalase	322	6.0
Aldolase	158	6.0

Calculations

The Kav values of the proteins eluted were determined using the following equation:

$$K_{av} = \frac{V_e - V_o}{V_t - V_o}$$

Where:

Vo= Void volume

Ve=Elution volume of each protein

Vt=Total gel - bed volume.

The calibration curve of Kav values vs. log M.wt. of the proteins were plotted.

3.2.1.6. Separation Procedure

PBS buffer pH (7.0, 7.2 and 7.6) containing 0.02% sodium azide was prepared as described previously in section (2.1.6).

Procedure

The sample of tissue homogenate (0.5 mL) containing approximately 3.43 mg protein was applied to the surface of gel , equilibrated with 0.15 M PBS buffer pH 7.2 for benign and premenopausal malignant breast tumor respectively. The sample was eluted by using the same buffer pH (7.0 and 7.6) for (benign and premenopausal malignant breast tumors respectively) with a flow rate of 20 mL.hrs-1 and fractions volume 2 mL were collected, gel filtration was carried out at 10 oC. The protein content of each fraction was determined using Lowry.et.al method (147).

The fractions contained CA15-3 were identified by the assay method. The binding of each fraction was calculated and plotted against the elution volume. The degree of purification (folds) of CA15-3 was calculated from the following formula.

$$\text{Purification fold of CA15-3} = \frac{\text{Specific binding of purified CA15-3}}{\text{Specific binding of crude CA15-3}}$$

Then yield % was determined as follows:

$$\text{Yield \%} = \frac{\text{Total protein content of purified CA15-3}}{\text{Total protein content of crude CA15-3}} \times 100$$

3.2.1.7. Dialysis for Concentration

After preparing dialysis tube, the fractions that contained high levels of the binding activity were pooled and concentrated by dialyzing against sucrose at 4 oC for 2hrs to get the required concentration to be used in the next experiments.

3.3. The Choice of the Optimum Conditions for the Binding of the Partially Purified CA15-3 to 125I-Anti CA15-3 Antibody

3.3.1. Optimum Protein Concentration

Reagents

PBS buffer pH 7.0 and 7.6 was prepared as described previously in section (2.1.6).

Procedure

One hundred microliters of increasing amount (50,100,150,200 and 250) µg.mL-1 protein of the dialyzable fractions of the partially purified CA15-3 from benign breast tumor was incubated with 50 µL of 125I-anti CA15-3 antibody (0.35 mg.mL-1) and completed to a final volume of 500 µL with 0.15 M PBS pH 7.0. The assay tubes were incubated for 90 min. at 45 oC. Two additional tubes, containing 50 µL (0.35 mg.mL-1) of 125I-anti CA15-3 antibody only, for total radioactivity computation, were set a side until counting.

Steps 4,5 and 6 of the experiment (2.2.4.1) were repeated. The same experiment was repeated on premenopausal malignant breast tissues homogenates (100 µg.mL-1 protein) with PBS buffer pH 7.6 and incubation time for 90 min at 15 oC.

Calculations

The (B/T) % was calculated as mentioned in experiment (2.2.3) and plotted against increasing amounts of protein concentration.

3.3.2. Influence of 125I-Anti CA15-3 Antibody on the Binding

Reagents

PBS buffer pH 7.0 and 7.6 was prepared as described previously in section (2.1.6).

Procedure

Fifty microliters of increasing concentration (0.070, 0.140, 0.175, 0.210, 0.245, 0.280 mg.mL-1) of 125I-anti CA15-3 antibody were added to 100 µL (150 µg.mL-1 protein) of partially purified CA15-3 from benign breast tumors. The reaction was completed to 500 µL with PBS pH 7.0. The assay tubes were incubated for 90 min at 45 oC. Two additional tubes containing increased concentration of 125I-anti CA15-3 antibody only, for total counts were counted. Steps 4,5 and 6 of the experiment (2.2.4.1) were repeated. The same experiment was repeated on premenopausal malignant breast tissues homogenate (100 µg.mL-1 protein) with PBS pH 7.6 and incubation time for 90 min at 15 oC.

Calculations

The (B/T) % was calculated as mentioned in experiment (2.2.3) and plotted against increasing concentration of 125I-anti CA15-3 antibody.

3.3.3. Optimum pH

Reagents

PBS buffer pH (6.8, 7.0, 7.2, 7.4, 7.6, 7.8, and 8.0) was prepared as described previously in section (2.1.6).

Procedure

To determine the optimum pH, 100 µL of a dialyzable fractions of partially purified CA15-3 from benign breast tumors (150 µg.mL-1 protein) were added to 20 µL of 125I-anti CA15-3 antibody (0.140 mg.mL-1). The volume of each fraction was completed to 500 µL with 0.15 M PBS of different pH (6.8 , 7.0 ,7.2 , 7.4 , 7.6 , 7.8 , 8.0). The assay tubes were incubated for 90 min at 45 oC. Two additional tubes, containing 20 µL (0.140 mg.mL-1) of 125I-anti CA15-3 antibody only , for total count , were set aside until counting. Steps 4,5 and 6 of experiment (2.2.4.1) were repeated. The same experiment was repeated on premenopausal malignant breast tissues homogenates (100 µg,mL-1 protein) and 25 µL (0.175 mg.mL-1) of 125I-anti CA15-3 antibody was incubated for 90 min at 15 oC.

Calculations

The (B/T) % was calculated as mentioned in experiment (2.2.3) and plotted against their corresponding pH values.

3.3.4. Optimum Temperature

PBS buffer pH 7.0 was prepared as described previously in section (2.1.6).

Twenty microliters (0.140 mg.mL-1) of 125I-anti CA15-3 antibody was added to 100 μL dialyzable fractions of partially purified CA15-3 from benign breast tumors (150 μg.mL-1 protein).

The volume of reaction was completed to 500 μL with 0.15 M PBS buffer pH 7.0. The assay tubes were incubated for 90 min at 45 oC. The same steps were repeated at (37, 25, 15, 5oC). Two additional tubes containing 20 μL (0.140 mg.mL-1) of 125I-anti CA15-3 antibody only, for total count, were set aside until counting. Steps 4,5 and 6 of experiment (2.2.4.1) were repeated.

The same experiment was repeated on the premenopausal malignant breast tissues homogenates (100 μg.mL-1 protein) and 25 μL (0.175 mg.mL-1) of 125I-anti CA15-3 antibody in PBS buffer pH 7.0, with incubation time 90 min at 15 oC. The experiment was repeated at different temperatures (45, 37, 25 and 5 oC).

Calculations

The (B/T) % was calculated as mentioned in experiment (2.2.3) and plotted versus temperatures of incubation.

3.3.5. The Effect of Incubation Time

PBS buffer pH 7.0 was prepared as described previously in section (2.1.6).

Twenty microliters (0.140 mg.mL-1) of 125I-anti CA15-3 antibody were added to 100 μL of dialyzable fractions of partially purified CA15-3 from benign breast tumors containing (150 μg.mL-1 protein). The reaction volume was completed to 500 μL with 0.15 M PBS buffer pH 7.0 , then incubated at 37 oC for (30, 60, 90, 120, 150, 180 min). Two additional tubes counting 20 μL (0.140 mg.mL-1) of 125I-anti CA15-3 antibody for total counts , were set aside until counting. Steps 4,5 and 6 of the experiment (2.2.4.1) were repeated. The same experiment was repeated on the premenopausal malignant breast tissues homogenates (100 µg.mL-1 protein) and 25 µL (0.175 mg.mL-1) of 125I-anti CA15-3 antibody with 0.15 M PBS buffer pH 7.0 and incubated at 15 oC for (30, 60, 90, 120, 150 and 180 min).

Calculations

The (B/T) % was calculated as metioned in experiment (2.2.3) and plotted versus the time of incubation for each group.

3.3.6. Stability of CA15-3 at –20 oC

Reagents

PBS buffer pH 7.0 was prepared as described previously in section (2.1.6).

Procedure

Crude and purified CA15-3 were stored at –20 oC for several time intervals. The frozen specimen was thawed at the end of each interval and the binding activity was measured at optimum conditions as described in section (2.2.4) and (3.6). The remaining activity was calculated and plotted against storage periods.

Calculations

The (B/T) % was calculated as mentioned in experiment (2.2.11) and plotted versus time storage for each group.

3.4. Results and Discussion

Partial Purification of CA15-3

Isolation of CA15-3 was performed by gel exclusion chromatography technique. CA15-3 was found to be separated from aggregates and other protein having smaller molecular weight by sepharose CL-4B. Figure (3-2, A & B) shows the elution profile of CA15-3 from benign breast tumors and premenopausal malignant breast cancer homogenates. The homogenate was loaded on the column as described in section (3.2.1). The void volume (Vo) of column was (6 mL) as predicted from the elution profile of the blue dextran. The elution was performed with PBS buffer. The resultant fractions containing the binding activity of CA15-3 were collected, pooled and concentrated, then subjected to protein determination as in section (2.2.1).

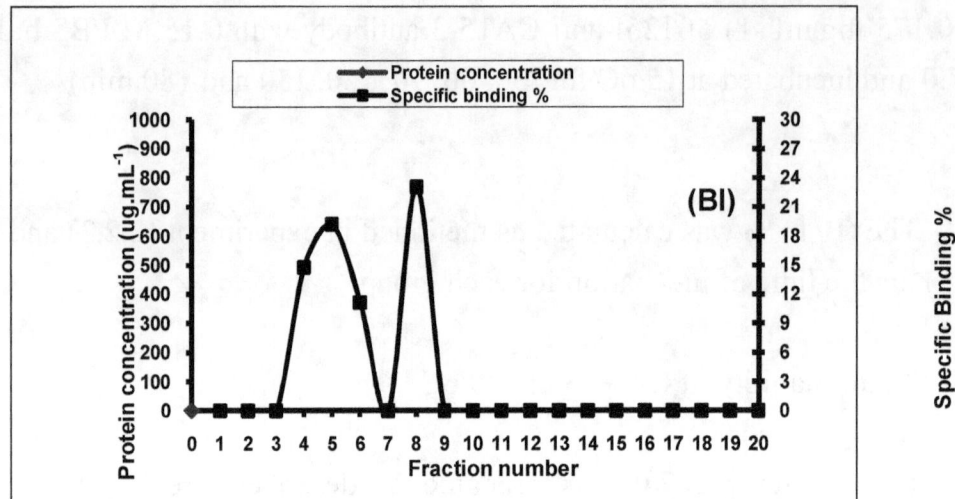

Figure (3.2A): The elution profile of human CA15-3 from benign breast tumors (BI). (All other details are explained in the text).

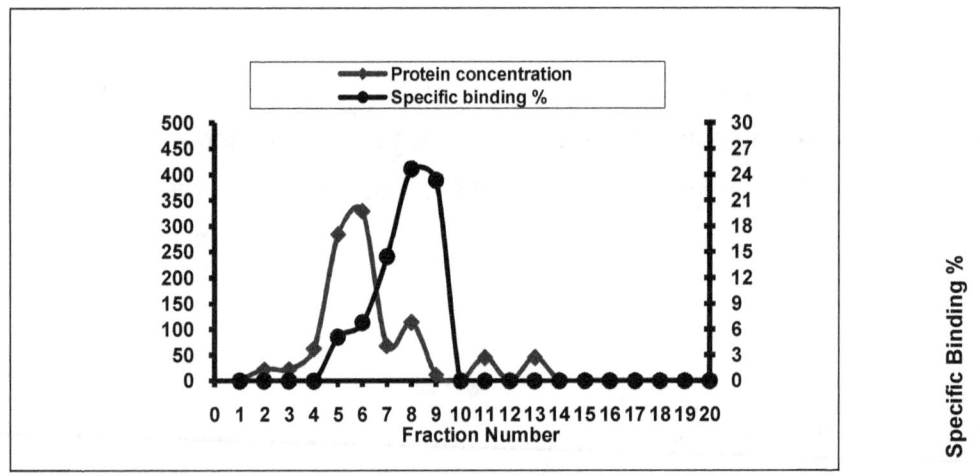

Figure (3.2B): The elution profile of human CA15-3 from premenopausal malignant breast cancer (MI). (All other details are explained in the text).

The elution volume Ve and then Kav values for the two peaks of CA15-3 (BI & MI) from benign breast tumors and malignant breast cancer respectively were calculated. The molecular weight of the partially purified CA15-3 obtained from figure (3-3) was 440 KD for peak (BI) and peak (MI) in two cases.

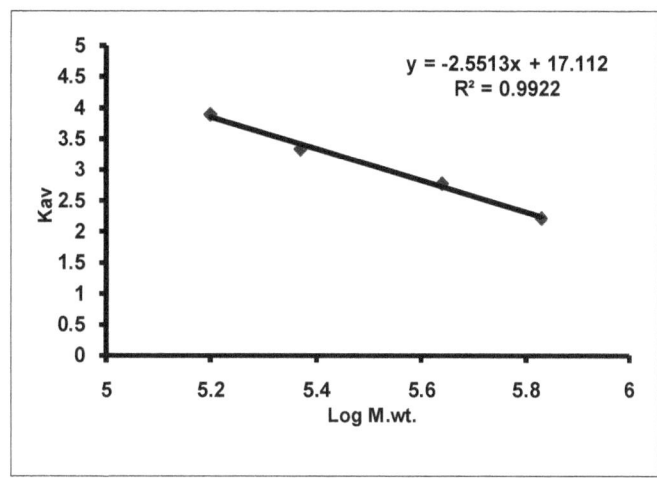

Figure (3.3): Calibration curve for determination of M.wt by gel filtration chromatography. (All other details are explained in the text).

The values ranged from 300-450 KD (178-180). Peaks of partially purified CA15-3 may be heavily aggregated, CA15-3 was obtained near the void volume of the column under separation conditions. From these results it was concluded that these components are capable of binding to the 125I-anti CA15-3 antibody with different affinities and in general CA15-3 type (BI) have lower binding affinities than CA15-3 type (MI), the isolation of CA15-3 from benign breast tumors on gel filtration column showed 3.02 folds of purification for peak (BI), while the isolation of CA15-3 from premenopausal malignant breast cancer showed 5.0 folds of purification. Table (3-2) illustrates the purification parameters for the different purified CA15-3 forms isolated by gel exclusion chromatography technique. The glycosylation of the protein backbone may differ in carcinoma cells from normal epithelial cells causing a wide range of molecular weight for this mucin (181).

Table (3.2): Partial purification of CA15-3 by gel filtration. (All other details are explained in the text).

CA15-3 Source	Total protein mg .mL-1	Specifically bound 125I-anti CA15-3	Specifically binding 125I-anti CA15-3/mg protein	Yield %	Purification fold

Benign	Crude extract	3.43	10.17	2.97	100	1.00
	Gel filtration on sepharose CL-4B	2.91	30.70	10.55	84.84	3.02
Malignant	Crude extract	3.43	8.18	2.39	100	1.00
	Gel filtration on sepharose CL-4B	2.21	40.93	18.52	64.43	5.00

The Choice of Optimum Conditions for the Binding of Partially Purified CA15-3 with 125I-Anti CA15-3 Antibody

Optimum Protein Concentration

Figure (3-4) shows the effect of increasing amounts of partially purified CA15-3 to a fixed amount of 125I-anti CA15-3 antibody to produce (125I-anti CA15-3 antibody/CA15-3) complex, that grow in size until they formed a precipitate. Above this zone an equivalence between CA15-3 and its antibody concentration is obtained, and amount of complex shows no further increases. A further addition of CA15-3 give rise to a solubilization of complex. The results revealed that 150 µg protein was the most appropriate concentration for the binding of (BI) and 100 µg protein for (MI). From these results, it could be concluded that the binding of 125I-anti CA15-3 antibody with its partially purified CA15-3 (BI) needed a higher amount of protein concentration than partially purified CA15-3 (MI). This is may be due to lower concentration of CA15-3 in benign breast tumor as compared with malignant breast tumors. According to these results, in all subsequent experiments, (150 µg.mL-1 protein) in benign breast tumors and (100 µg.mL-1 protein) in malignant breast tumors were used, since they give the highest binding.

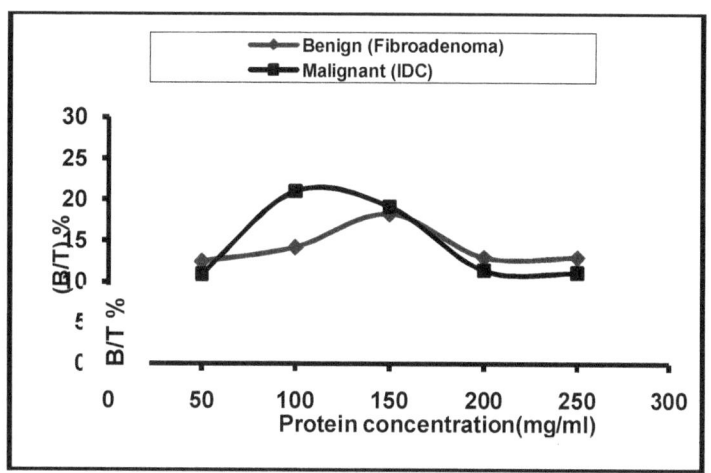

Figure (3.4): Influence of protein concentration on the binding of 125I-anti CA15-3 antibody with partially purified CA15-3 from breast tumors. (All other details are explained in the text).

Influence of 125I-anti CA15-3 Antibody on the Binding

Figure (3.5) illustrate the effect of 125I-anti CA15-3 antibody concentration on the binding with partial purified CA15-3 from benign breast tumors and premenopausal malignant breast cancer.

The maximum binding obtained when 0.140 mg.mL-1 of antibody in benign breast tumors and 0.175 mg.mL-1 of antibody in malignant breast tumors were used. From these results, it was found that (BI) purified fraction was saturated with small concentration of 125I-anti CA15-3 antibody than those required for (MI). This is may be due to the increasement of the epitope (is the part of an antigen molecule that binds to any single antigen-combining site) (182) in partially purified CA15-3 in malignant breast tumors as compared to benign breast tumors.

According to these results, in all subsequent experiments (0.140 mg.mL-1) and (0.175 mg.mL-1) of 125I-anti CA15-3 antibody in benign and malignant breast tumors were used, since they give the highest binding.

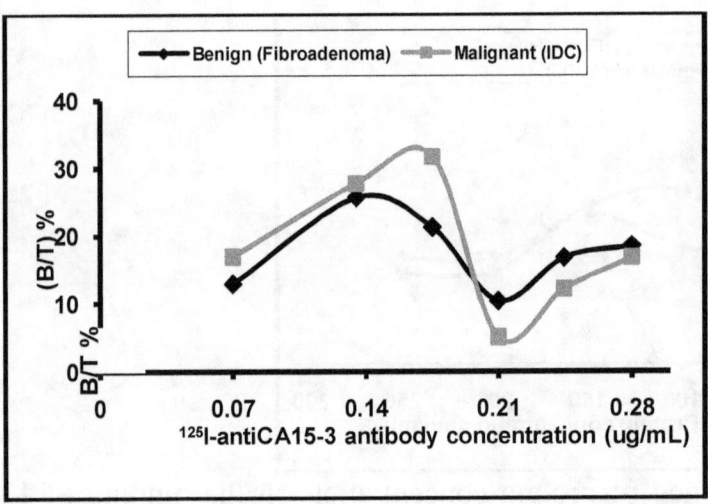

Figure (3.5): Effect of 125I-anti CA15-3 antibody concentration on its binding with partially purified CA15-3 from breast tumors. (All other details are explained in the text).

Optimum pH

Figure (3-6) shows the effect of increasing pH on the binding of 125I-anti CA15-3 antibody to its purified antigen. The results revealed that the optimum pH for (BI) and (MI) purified fractions for the binding with its antibody was 7.0. These results indicate that the binding was pH dependent.

The similarity in pH (7.0) suggests that the CA15-3 isolated from different sources of tissues either benign or malignant breast tissues homogenates possesses the same epitopes in both cases. That means the induction of protonation-deprotonation process (183) occurs within the same changed polar groups on the amino acid residues present in the binding domain. According to the results obtained, the pH of the buffer used in all subsequent experiments was adjusted to pH 7.0.

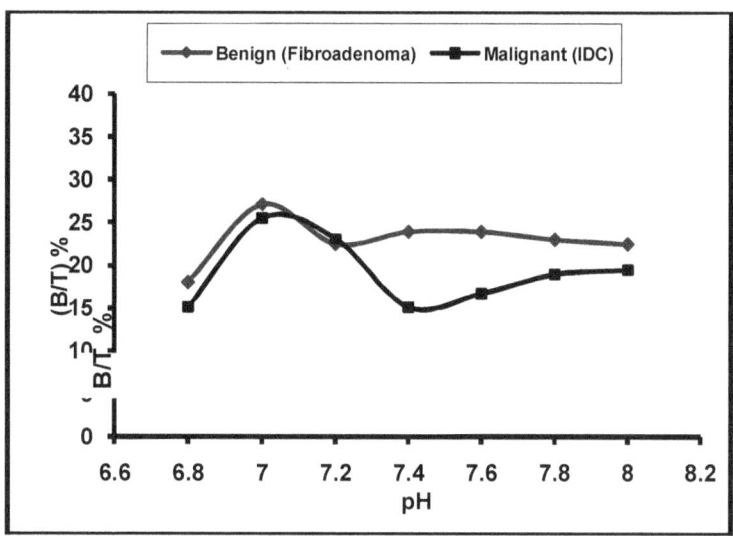

Figure (3.6): pH effect on the binding of 125I-anti CA15-3 antibody with partially purified CA15-3 from breast tumors. (All other details are explained in the text)

Optimum Temperature

The temperature dependency of the isolated CA15-3 binding to its antibody 125I-anti CA15-3 was investigated.

Figure (3.7) show the optimum temperatures on the binding of 125I-anti CA15-3 antibody was 37oC with partially purified CA15-3 (BI) and 15oC with partially purified CA15-3 (MI).

The difference of the temperature between crude and purified CA15-3 occurs in benign breast tumors, i.e. the optimum temperature was 45oC of the binding of 125I-anti CA15-3 antibody to crude CA15-3 while in the purified CA15-3 (BI) was 37oC. On the other hand, the optimum temperature in both crude and partially purified CA15-3 from premenopausal malignant breast tumors was 15oC.

The temperature dependency of the binding suggests that the whole process is controlled by diffusion of the interacting of 125I-anti CA15-3 antibody to CA15-3 in benign and malignant breast tumors (184).

In view of these results, the temperatures (37 & 15 oC) for both benign and malignant breast tumors were used in all subsequent experiments.

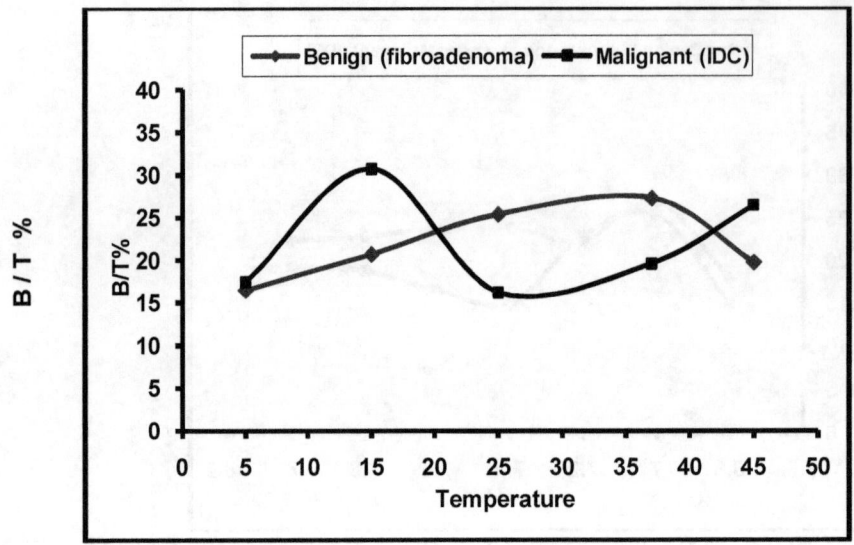

Figure (3.7): Effect of the temperature on the binding of 125I-anti CA15-3 antibody with partially purified CA15-3 from breast tumors. (All other are explained in the text)

The Effect of Incubation Time

Figure (3.8) shows the time required for the highest binding of 125I-anti CA15-3 antibody to partially purified CA15-3 in (BI) and (MI) was 90 min at 37 and 15 oC respectively.

In view of these results, the incubation time used in all subsequent experiments was 90 min.

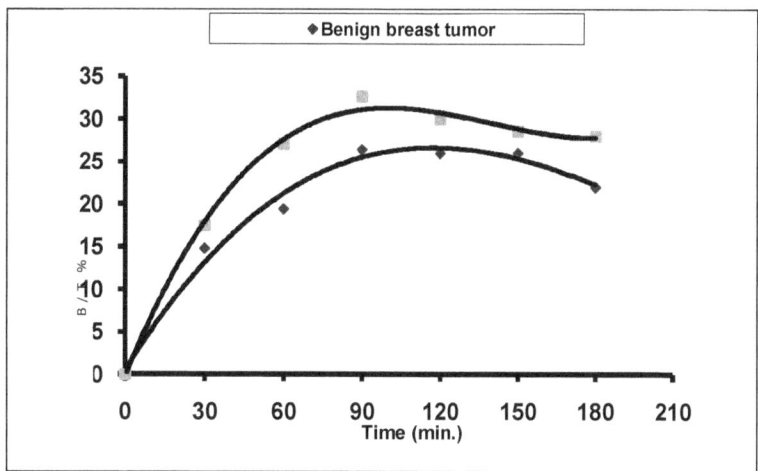

Figure (3.8): Time dependence of 125I-anti CA15-3 antibody binding with partially purified CA15-3 from breast tumors. (All other details are explained in the text)

Stability of CA15-3 at –20 oC

The crude and isolated fractions of CA15-3 from malignant breast tumors were stored at –20 oC during the experiments. It was carried out in order to study the stability of CA15-3 and check their efficiencies of the binding through out the storage period. The results showed that CA15-3 of crude fraction was more stable than the isolated fractions as shows in figure (3-9). This result is in agreement with Al-Atrakchi observations (185).

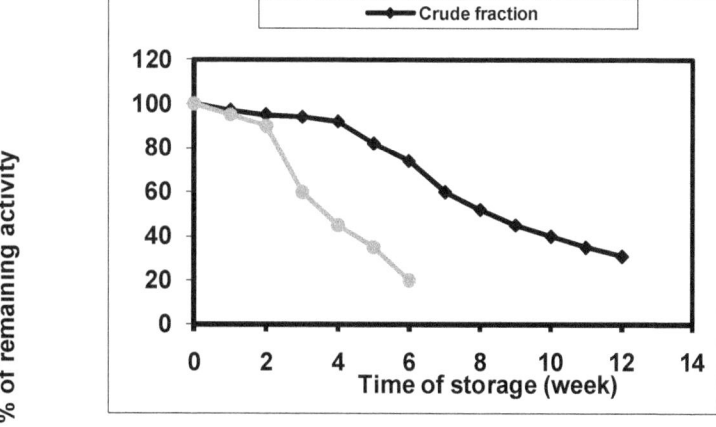

Figure (3.9): Stability of partially purified and crude CA15-3 upon storage at Abstract

Kinetic and thermodynamic parameter associated with the binding of 125I-anti CA15-3 antibody to partially purified CA15-3 in both cases, benign and malignant breast tumors were investigated.

It was shown that the reaction in all studied cases follow pseudo-first order reaction kinetics. The maximum binding (Bmax) of partially purified CA15-3 in benign breast tumors (Fibroadenoma) was 10.48×10^{-3} mg.mL-1 after 90 minutes incubation at 37oC, while the (Bmax) of partially purified CA15-3 in malignant breast tumors (IDC) was 13.38×10^{-3} mg.mL-1. The (Bmax) was decreased with increasing temperature. The values of affinity constant (Ka) were dependent on the temperature, Ka increased from 14.18 mg-1.mL at 5oC to 31.65 mg-1.mL at 45oC in benign breast tumors (Fibroadenoma), while Ka was increased from 13.87 mg-1.mL at 5oC to 23.81 mg-1.mL at 45oC in premenopausal malignant breast tumors (IDC). The association constant K+1 increased with temperature in benign breast tumors (Fibroadenoma). On the other hand, K+1 was independent of temperatures in premenopausal malignant breast tumors (IDC). The Van't Hoff plot demonstrated a linear relationship between Ka and 1/T, using the partially purified CA15-3 in benign and malignant tumor homogenate. Arrhenius plot indicate that there was a linear-relationship between log K+1 and 1/T. The transition state thermodynamic parameters (Ea, ΔH^*, ΔG^*, ΔS^*) for the formation of (125I-antiCA15-3 antibody /CA15-3) were determined.

Chapter Four

Kinetic and thermodynamics of Binding CA-15-3 to its antibody

Introduction

The specific reaction between an antibody (Ab) and an antigen (Ag) is usually driven by electrostatic forces between oppositely charged amino acids, hydrogen bonding, and hydrophobic interactions. The equilibrium reaction, termed "biospecific interaction", is characterized by the affinity of reactants to form Ag-Ab complex (186).

Kinetic studies supplement the information for differences between the initial, final states of each reactant and an intermediate activated complex, (i.e, the pathway taken by the reactants reach the final product) (187). On the other hand, thermodynamics of the binding describes the system in its initial, final states. Using kinetic and equilibrium data also determined thermodynamic formation constant.

Al-Mudhuffar et.al, have many studies on the kinetic and thermodynamic of protein-protein interaction in human breast tissue, like kinetic and thermodynamic of purified steroid receptor of malignant breast tumors with hormone (188), also kinetic and thermodynamic studies on the binding of lectin in human malignant breast to glycoprotein (189).

In this chapter, the basic mathematical analysis was described and used to explain the mechanism through kinetics of binding of CA15-3 from both breast tumor homogenates (fibroadenoma and Infiltrating ductalcarcinoma) to its antibody to form (125I-anti CA15-3 antibody / CA15-3) complex in partially purified fraction.

Materials and Methods

4.1. Materials
4.1.1. Chemicals
All chemical and reagents mentioned in section (2.1.1) in chapter two were used in the experiments of this chapter.

4.1.2. Instruments
All instruments that were described in section (2.1.2) in chapter two were used in the experiments of this chapter.

4.2. Methods

4.2.1. Kinetic Studies

4.2.1.1. The Time-Course of the Binding of 125I-anti CA15-3 Antibody with CA15-3 in Breast Tumor Homogenate

1. One hundred microliters of partially purified CA15-3 from benign breast tumor (fibroadenoma) and premenopausal malignant breast tumor (Infitrating ductal carcinoma, IDC) containing (150 and 100 □g.mL-1 protein) respectively, were added to (20 and 25 µL) of 125I-anti CA15-3 antibody containing (0.140 and 0.175 mg.mL-1) respectively.

2. The volume of reaction were completed to 500 µL with PBS buffer pH 7.0.

3. All tubes were incubated at 37oC at different time intervals (30, 60, 90, 120, 150, 180) min.

4. Steps 3, 4, 5 and 6 of experiment (2-4-2-1) were repeated.

5. To determine the time-course of partially purified CA15-3 binding to 125I-anti CA15-3 antibody at different temperatures, step 1,2,3 and 4 in the same experiment were repeated at different temperatures 5, 15, 25 and 45Co.

Calculation

The values of (B/T)% were calculated as described in section (2.4.1) and plotted against incubation time at each temperature for both types of homogenates.

4.2.1.2. Determination of Kinetic Parameters of 125I-Anti CA 15-3 Antibody Binding with Partially Purified CA 15-3 in Benign and Malignant Breast Tumors

Determination of the affinity constant (Ka) and the maximal binding capacity (Bmax) of:

A. Partially Purified CA15-3 in Benign Breast Tumor Homogenate Binding with 125I-Anti CA15-3 Antibody

1. One hundred microliters of partially purified CA15-3 from benign breast tumor (Fibroadenoma) containing (150 □g.mL-1 protein) were added to increasing volumes (4, 8, 12, 16, 20 and 24 µL) of 125I-anti CA15-3 antibody containing (0.0280, 0.0560, 0.0841, 0.1121, 0.1402 and 0.1684 mg.mL-1) to each assay tube. The final volume of each assay tube was completed to 500 µL with PBS buffer pH 7.0.

2. All tubes were incubated for 90 min at 37oC.

3. Steps 3, 4, 5 and 6 in experiment (2.4.2.1) were repeated at different temperatures (5, 15, 25 and 45oC).

4. The time of incubation required to reach the equilibrium state are reported in table (4-1) according to the following:Table (4.1): The time of incubation for benign and malignant breast tumor homogenate at different temperatures

Temp. oC	Time (min.)	
	Benign breast tumor homogenate (Fibroadenoma)	Malignant breast tumor homogenate (IDC)
5	180	180
15	60	90
25	90	150
37	90	90
45	180	90

Calculations

1- The B/F ratio was computed for each tube, where:

B: is the bound radioactivity (mean counts in c.p.m), which represent the formation of (125I-anti CA15-3 /CA15-3) complex.

F: is the free radioactivity (mean counts in c.p.m.), which represents the (unbound or unreacted), 125I-anti CA15-3 antibody.

T: is the total activity (mean counts in c.p.m.)

F = T (total counts) - B (bound radioactivity)

2- The concentration of (125I-anti CA15-3/CA15-3) complex in mg.mL-1 which found after time (t) was calculated from the following equation:

$$B(\text{mg.mL}^{-1}) = \frac{B(\text{c.p.m})}{T(\text{c.p.m})} \times \text{Concentration of } ^{125}I - \text{anti CA15} - 3 \text{ antibody in}$$

the incubation medium in mg.mL^{-1}

3- The affinity constant and maximal binding capacity were determined according to Scatchard equation (190,191).

$$\frac{B}{F} = \frac{1}{K_d} \times (B_{max} - B)$$

$$K_a = \frac{1}{K_d} = \frac{K_{+1}}{K_{-1}}$$

Where: Ka = affinity constant

Kd = dissociation constant

Bmax = maximal binding capacity

The value of affinity constant of the binding Ka at each temperature can be calculated from the slop of the straight line in figure (4.2), while the value of the total concentration of CA15-3 (Bmax) in breast tumor homogenate for each group was calculated from the intercept of the x-axis.

B. Partially Purified CA15-3 in Human Malignant Breast Tumor Homogenate Binding with 125I-Anti CA15-3 Antibody

1. One hundred microliters of partially purified CA15-3 from premenopausal malignant breast tumor (IDC) containing (100 µg.mL-1 protein) were added to increasing volumes (5, 10, 15, 20, 25 and 30 µL) of 125I-anti CA15-3 antibody containing (0.035, 0.070, 0.105, 0.140, 0.175 and 0.210 mg.mL-1) to each assay tube. The final volume of each assay tube was completed to 500 µL with PBS buffer pH 7.0.

2. All tubes were incubated for 90 min at 15oC

3. Steps 3, 4, 5 and 6 in experiment (2.4.2.1) were repeated at different temperatures (5, 25, 37 and 45 oC).

4. The times of incubation required to reach the equilibrium state are reported in table (4.1).

Calculations

The method outlined in experiment (4.3.2.A) was followed exactly to obtain the values of Ka and Bmax at each temperature as shown in figure (4.3).

4.3. The Thermodynamic Studies of 125I-Anti CA15-3 Antibody Binding to Partially Purified CA15-3 in Benign and Malignant Breast Tumors

The same steps mentioned in section (4.2.1.1) and (4.2.1.2) were performed using the dialyzable protein fraction of benign and malignant breast tumor homogenate from fibroadenoma and (IDC) as the partially purified
CA15-3 source.

Calculation

1. The thermodynamic parameters of standard state were obtained from Van't Hoff plot, the values of the natural logarithm of equilibrium constant (affinity constant Ka) obtained at different temperatures were plotted against the reciprocal values of the absolute temperature in Kelvin (1/T), according to the following equation:

$$\ln K_a = \frac{\Delta S^\circ}{R} - \frac{\Delta H^\circ}{RT}$$

Where:

ΔHo = the enthalpy change of the standard state.

ΔSo = the entropy change of the standard state.

R   = the gas constant (8.314 J.K-1.mol-1).

ΔHo value obtained from the slop of a linear relationship of the plot.

The change in Gibbs free energy of the standard state ΔGo was obtained from the following equation:

ΔGo = -RT Ln Ka

Where Ka is the affinity constant, while the standard state entropy change was obtained from (192):

$$\Delta S^\circ = \frac{\Delta H^\circ - \Delta G^\circ}{T}$$

2. The thermodynamic parameters of the transition state were obtained from Arrhenius plot of Ln K+1 values against (1/T) values, that given a linear relationship according to the following equation:

$$\text{Ln K}_{+1} = \text{Ln A} - \left(\frac{E_a}{RT} \right)$$

Where:

A: Arrhenius constant .

The values of activation energy (Ea) of the binding reaction can be determined from the slop of the straight line.

The enthalpy of transition state ΔH^* was obtained from:

$\Delta H^* = Ea - RT$

Transition state of free energy change ΔG^* is calculated from the following equation:

$$\Delta G^* = -RT\, \text{LnK}_{+1} + RT\, \text{Ln} \frac{KT}{h}$$

where K and h were Boltzmann and Plank's constant which equal (1.38x10-23 J.K-1), (6.62x10-34 J.sec-1) respectively.

The change in entropy of the transition state ΔS^* is calculated from the following equation:

$$\Delta S^* = \frac{\Delta H^* - \Delta G^*}{T}$$

Results and Discussion

Kinetic Studies

The Time-Course of the Binding of 125I-anti CA15-3 Antibody with CA15-3 in Breast Tumor Homogenate

Figure (4.1.A & B) shows the time – course of the formation of (125I-anti CA15-3 /CA15-3) complex at five different temperatures (5, 15, 25, 37 and 45oC) of partially purified CA15-3 from benign and malignant breast tumors homogenates samples.

The concentration of (125I-anti CA15-3/CA15-3) complex formed after time (t) was calculated from the following equation:

$$[\text{125I-antiCA15-3/CA15-3}]\text{ in mg.mL-1 after time (t)} = \frac{\text{Count (c.p.m.) of 125I-antiCA15-3 specifically bound after time (t)}}{\text{Total counts (c.p.m.) of 125I-anti CA15-3 used in the incubation}} \times \text{Concentration of 125I-antiCA15-3 in the incubation (mg.mL-1)}$$

The results of time-course pattern at different temperatures indicated that the equilibrium binding studies is temperature and time dependent process. In case premenopausal malignant breast tumor (IDC) the maximum binding occurs at 15oC (after incubation for 90 minutes), while in benign breast tumors (fibroadenoma) the maximum binding occurs at 37oC at the same incubation time. This is may be due to the different source of CA15-3. Several authors studied the time – course of purified steroid receptors of malignant breast tumors(188), others studied the time – course on the binding of lectin in human malignant breast to glycoprotein (189), these studies revealed that the time-course must be done to find the maximum binding at different incubation time as a step to prepare the kinetic and thermodynamic studies.

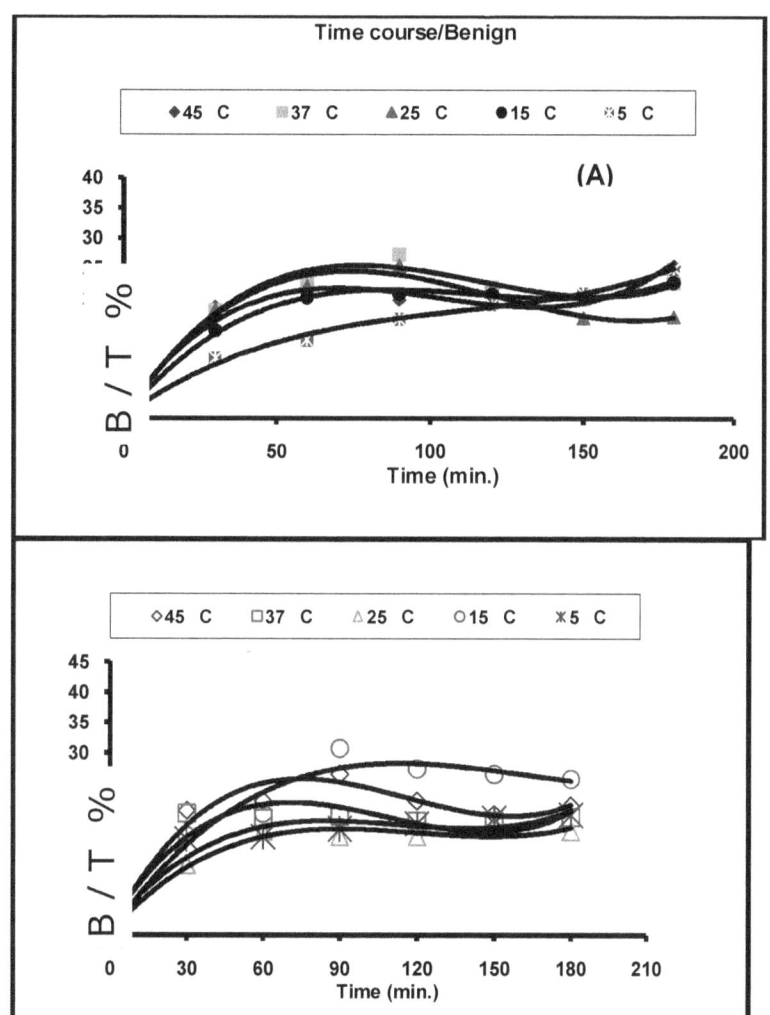

Figure (4.1): Time-Course of 125I-anti CA15-3 binding to partially purified CA15-3 in:

(A)   Benign tumor (Fibroadenoma) tissue homogenate.

(B)   Malignant tumor (IDC) tissue homogenate.

(All other details are explained in the text).

Determination of Kinetic Parameters of 125I-Anti CA15-3 Antibody Binding with Partially Purified CA15-3 from Benign and Malignant Breast Tumors

The time course of (125I-anti CA15-3/CA15-3) complex formation was carried out to describe the kinetic parameters of the binding. The simplest proposed model representing this interaction is:

$$125\text{I-antiCA15-3} + \text{CA15-3} \underset{K_{-1}}{\overset{K_{+1}}{\rightleftharpoons}} [125\text{I-antiCA15-3/CA15-3}]$$

Where:

K+1: is the association rate of 125I-anti CA15-3 to /or CA15-3.

K-1: is the dissociation rate of (125I-anti CA15-3/CA15-3) complex formed.

At equilibrium:

$$K_a = \frac{[^{125}I - antiCA15 - 3/CA15 - 3]}{[^{125}I - antiCA15 - 3][CA15 - 3]} \ldots\ldots\ldots\ldots(2)$$

$$K_d = \frac{[^{125}I - antiCA15 - 3][CA15 - 3]}{[^{125}I - antiCA15 - 3/CA15 - 3]} \ldots\ldots\ldots\ldots(3)$$

Thus:

$$K_a = \frac{1}{K_d} = \frac{K_{+1}}{K_{-1}} \ldots\ldots\ldots\ldots\ldots\ldots\ldots\ldots\ldots\ldots\ldots(4)$$

Where:

The value Ka and maximal binding capacity (Bmax). Were calculated from Scatchard plot at five different temperatures at incubation time of 90 minutes, figure (4-2) and (4-3).

It is clear from table (4-2), that the affinity constant (Ka) is depended on the type of the tumor (i.e., benign or malignant) and on the temperature. Ka increased with increased temperature for the same tumor (Fibroadenoma), Ka increased from 14.18 mg-1.mL at 5oC to 31.65 mg-1.mL at 45oC. Whereas the values of dissociation constant (Kd) was calculated by using equation (4), which show that the lowest Kd value of (125I-anti CA15-3/CA15-3) complex occurs at 45oC at time of incubation 180 minutes.

The concentration of CA15-3 in partially purified fractions of (Fibroadenoma) was determined to be 10.48x10-3 mg.mL-1 and the maximum binding (Bmax) occurred after 90 minutes incubation at 37 oC. While in the same table the maximum Ka value for the binding 125I-anti CA15-3 antibody with CA15-3 present in partially purified fraction of (IDC) occurred at 15oC and it was increased with temperature in the following order: 5 >15 >25 >37 > 45 oC.

The lowest Kd value of (125I-anti-CA15-3 /CA15-3) complex occurs at 45 oC at the time of incubation.

Scatchard plot analysis gave straight line as shown in figure (4.2) and (4.3) indicating that the (125I-anti CA15-3/CA15-3) complex is directed against the same epitopes on CA15-3 molecules. On the other hand, the maximum binding occurred at 15oC and was 13.38x10-3 mg.mL-1 also shows that the (Bmax) decreased with increasing temperatures of incubation.

Table (4-2): The Kinetic parameter of 125I-anti CA15-3 antibody binding to partially purified CA15-3 in breast tumor homogenate. (All other details are explained in the text).

Temp oC	Benign breast tumors (Fibroadenoma)			Malignant breast tumors (IDC)		
	Binding Capacity Bmaxx10-3 (mg.mL-1)	Ka (mg-1.mL)	Kdx10-2 (mg.mL-1)	Binding Capacity Bmax x10-3 (mg.mL-1)	Ka (mg-1.mL)	Kdx 10-2 (mg.mL-1)
5	9.22	14.18	7.05	10.82	13.87	7.21
15	8.05	16.73	5.98	13.38	20.78	4.81
25	9.02	16.38	6.10	12.57	20.84	4.79
37	10.48	18.66	5.36	9.63	22.22	4.50
45	6.67	31.65	3.16	11.67	23.81	4.20

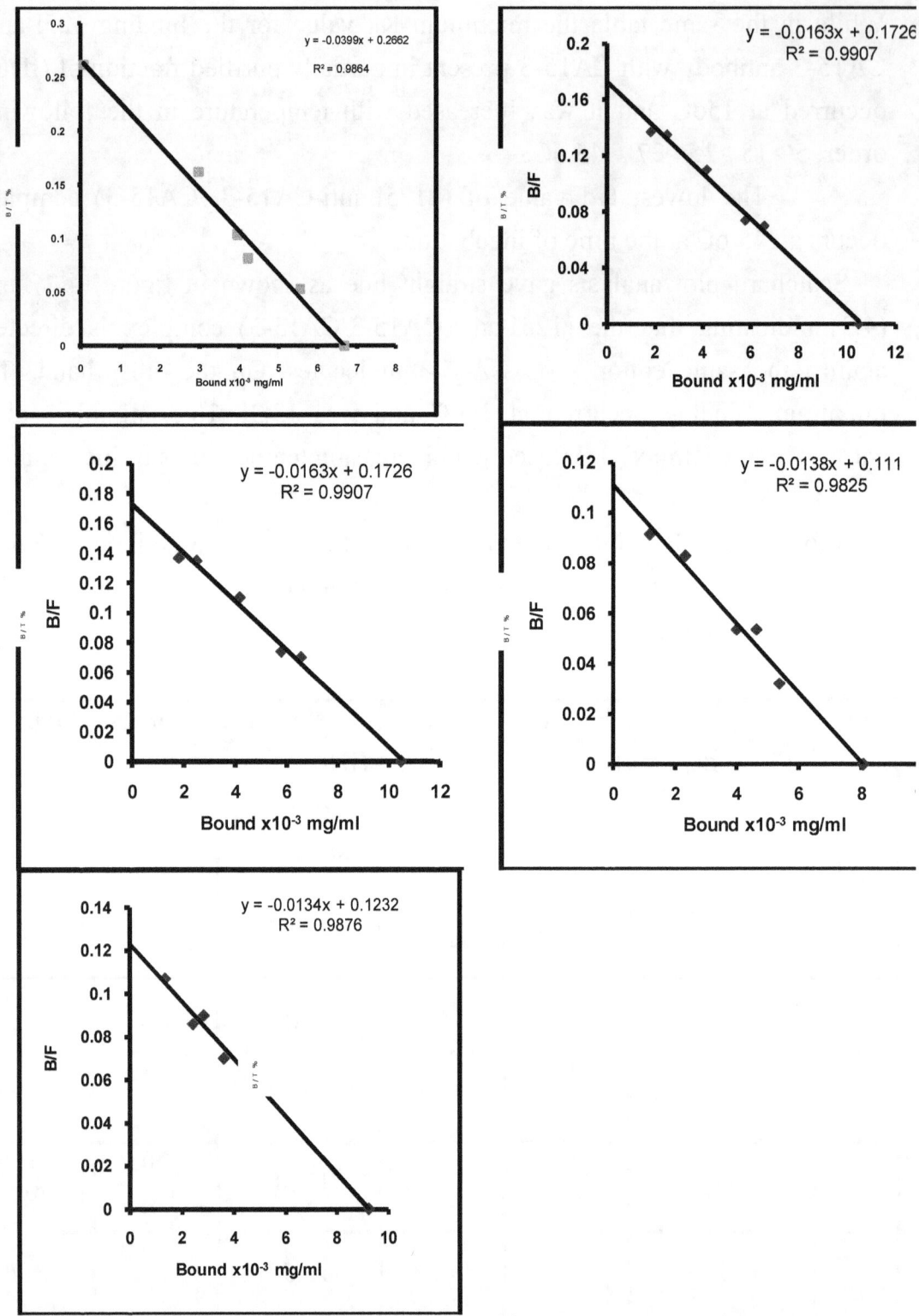

Figure (4-2): Scatchard plot of 125I-anti CA15-3 antibody binding to the partially purified CA15-3 in benign breast tumors (Fibroadenoma) at five different temperatures. All details are explained in the text.

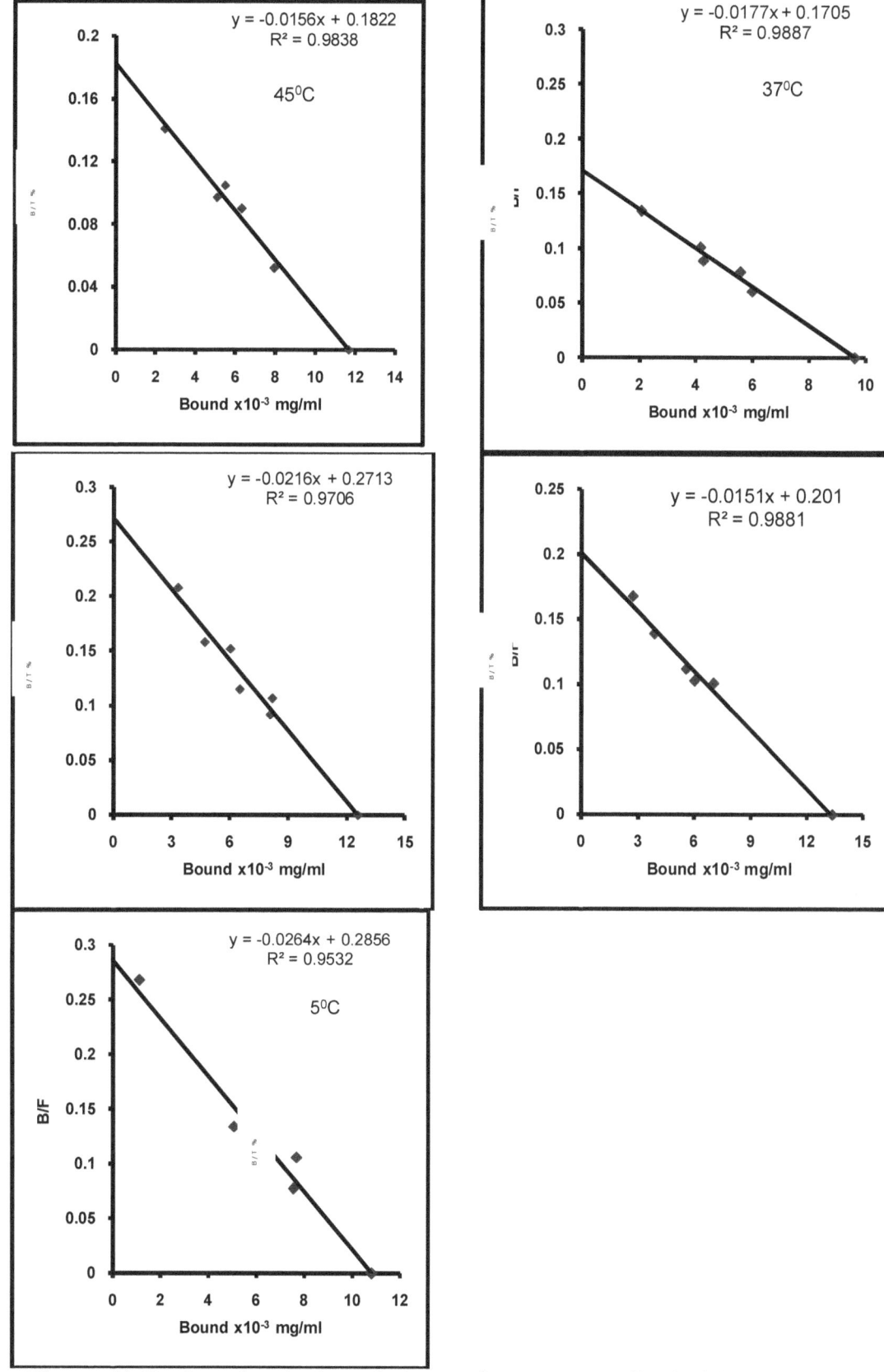

Figure (4.3): Scatchard plot of 125I-anti CA15-3 antibody binding to the partially purified CA15-3 in Malignant breast tumors (IDC) at five different temperatures. All details are explained in the text.

However, the time-course data shown in figure (4-1) could be used to determine the reaction order of CA15-3 binding to its specifically 125I-anti CA15-3 using the following equation (193):

$$Ln[AbAg]_e\left[\frac{[Ab]_t - [AbAg]_t [AbAg]_e /[Ag]_t]}{[Ab]_t [AbAg]_e - [AbAg]_e}\right] = K_{+1}t\left[\frac{[Ab]+[Ag]_t - [AbAg]_e}{[AbAg]_e}\right] \quad ......(5)$$

Where:

k+1 : is the kinetic association constant in mg-1. min-1. mL.

[AbAg]e : is the concentration of (125I-antiCA15-3/CA15-3)complex formed at equilibrium.

[AbAg]t : is the concentration of (125I-antiCA15-3/CA15-3) complex after time (t).

[Ab]t : is the total concentration of 125I-anti CA15-3 antibody in mg. mL-1.

[Ag]t : is the total concentration of CA15-3 in mg. mL-1.

Equation (5) represents the second order kinetics, but the percent of binding was in some cases, small and most labeled antibody remains free and only small fraction binds even at equilibrium, i.e , [Ab]t >> [AbAg]e

Thus :

$$[Ab]_t >> \frac{[AbAg]_t [AbAg]_e}{[Ag]_t}$$

So that the following equation (187) could be used in order to fit the pseudo-first order kinetics:

$$Ln\frac{[AbAg]_e}{[AbAg]_e - [AbAg]_t} = K_{+1}t\frac{[Ab]_t [Ag]_t}{[AbAg]_e} \quad ....(6)$$

On the other hand, figure (4-4) and (4-5) show the plot of $ln\frac{[AbAg]_e}{[AbAg]_e - [AbAg]_t}$ Against time (t) in both benign and malignant breast tumors, which give a straight line with a slope equal to the observed value of first rate constant Kbos in min-1. The rate constant (k+1) in mg-1. mL. min was calculated at five different temperatures by using the following equation (194)

$$K_{obs} = K_{+1} \frac{[^{125}I - antiCA15 - 3]_t [CA15 - 3]_t}{[^{125}I - antiCA15 - 3 / CA15 - 3]_e} \quad............(7)$$

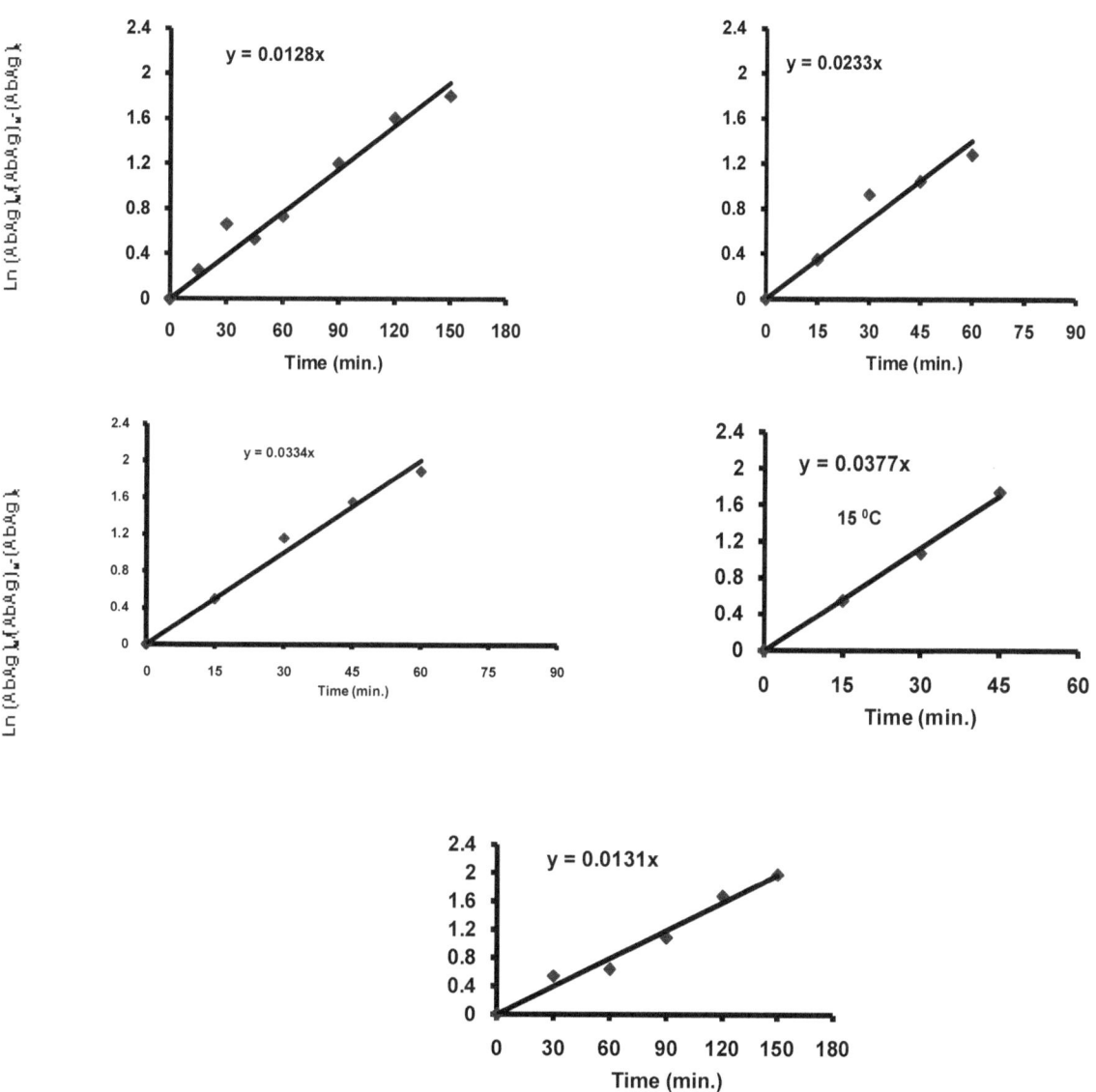

Figure (4.4): Kinetics of 125I-anti CA15-3 antibody binding to partially purified CA15-3 in benign breast tumors (Fibroadenoma). All details are explained in the text.

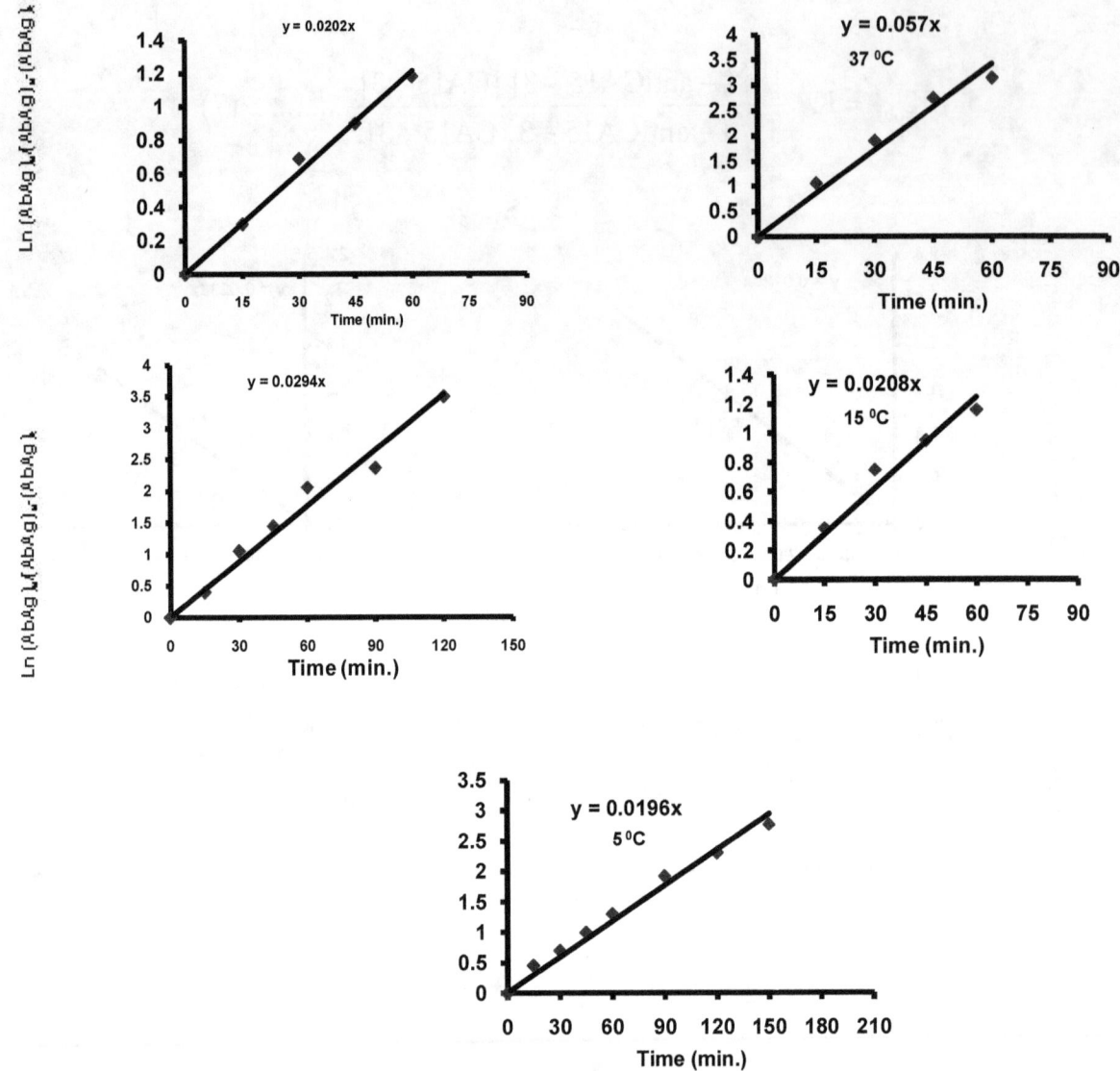

Figure (4.5): Kinetics of 125I-anti CA15-3 binding to partially purified CA15-3 in malignant breast tumors (IDC). (All details are explained in the text).

The value of k-1 at five temperatures was calculated by using equation (4). Whereas, the half-life time of association (t ½)ass. , Which represented the time needed for the formation of half amount of the complex at equilibrium was determined from the concentration of the complex at equilibrium and the time-course curve. The half-life time of dissociation (t ½) diss. , was calculated from the following relation:

$$(t_{1/2})_{diss.} = \frac{\ln 2}{k_{-1}} = \frac{0.693}{k_{-1}}$$

$$(t_{1/2})_{ass.} = \frac{\ln 2}{k_{obs}} = \frac{0.693}{k_{+1}}$$

The value of kobs. , k+1, k-1, (t ½)ass. ,(t ½)diss. at five different temperatures are summarized in table (4.3). Data analysis in this table shows that highest rate for the association reaction k+1 , in benign breast tumors (Fibroadenoma) and malignant breast tumors (IDC) occurs at 37°C and 15°C respectively , while the lowest rate occurs at 45°C. This means the dependence of reaction rate on temperature (Table 4.3) that also shows the values of the rate constant for the reverse reaction k-1 calculated from equation (4). Results show that the rate of dissociation of 125I-anti CA15-3 antibody, from its CA15-3 is temperature independent.

Table (4.3): The effect of temperature on the kinetic parameters of 125I-anti CA15-3 binding to partially purified CA15-3 in benign and malignant breast tumors at five different temperature.

Temp. °C	$k_{obs} \times 10^{-3}$ (min^{-1})		K_{+1} mg^{-1}.ml.min^{-1}		$k_{-1} \times 10^{-1}$ (min^{-1})		$(t_{1/2})_{ass}$ (min)		$(t_{1/2})_{diss}$ (min)	
	Benign (Fibroadenom)	Malignant (IDC)	Benign (Fibroadenoma)	Malignant (IDC)	Benign (Fibroadenoma)	Malignant (IDC)	Benign (Fibroadenoma)	Malignant (IDC)	Benign (Fibroadenoma)	Malignant (IDC)
5	12.8	20.20	48.69	45.81	15.38	34.68	54	34	45	20
15	23.3	57.00	60.65	116.16	32.50	73.75	30	12	21	9
25	33.4	24.9	93.98	35.48	57.37	18.07	21	28	12	38
37	37.7	20.30	103.54	46.61	61.89	22.43	18	34	11	31
45	13.1	19.60	35.40	21.34	24.96	10.58	53	35	28	66

The Thermodynamic Studies of 125I-Anti CA15-3 Antibody to the Partially Purified CA15-3 in Benign and Malignant Tumors

Thermodynamic Parameters of Standard State

Figure (4.6) and (4.7) show Van't Hoff plot of the binding of 125I-anti CA15-3 antibody to the partially purified CA15-3 in benign breast tumors (Fibroadenoma) and malignant breast tumors (IDC) respectively, at different temperatures (5 , 15 , 25 , 37 and 45 °C).

These figures revealed that the equilibrium binding constant (affinity constant) for CA15-3 to its antibody is a temperature dependent. The results indicated that $\Delta H°$, in general, had small values and their positive sign ascertains that the reaction was nearly endothermic. The $\Delta H°$ value in the case of the binding of 125I-anti CA15-3 antibody to partially purified CA15-3 in benign breast tumors 12.71 KJ.mol-1 was higher than that in case of binding in malignant breast tumors (IDC) 6.7 KJ.mol-1, so more energy is needed in case of benign breast tumor for the reaction (binding) to occur. The small positive value of $\Delta H°$ may indicate a favorable interaction between 125I-anti CA15-3 antibody with partially purified CA15-3 in both cases.

These include the non-covalent interaction, which are fundamentally electrostatic in nature such as charge-charge, charge-dipole, dipole-dipole, charge-induced dipole, dipole-induced dipole interactions, and hydrogen bonds. The sum of these types of interactions can yield some stabilization to the folded structure of the complex (195).

The other values of thermodynamic parameters of standard state at five temperatures, such as $\Delta G°$ values and $\Delta S°$ values are summarized in table (4.4) and (4.5).

Table (4.4): Thermodynamic parameters at standard state of 125I-anti CA15-3 to the partially CA15-3 in benign breast tumors (Fibroadenoma). (All other details are explained in the text).

Temp. °C	$\Delta H°$ KJ .moL-1	$\Delta G°$ KJ .moL-1	$\Delta S°$ J .mol-1.K-1
5	12.71	-36.87	137.20
15	12.71	-38.59	138.42
25	12.71	-39.88	138.10

37	12.71	-41.82	139.01
45	12.71	-44.30	143.30

Table (4.5): Thermodynamic parameters at standard state of 125I-anti CA15-3 to the partially purified CA15-3 in malignant breast tumors (IDC). (All other details are explained in the text).

Temp. °C	$\Delta H°$ KJ .moL-1	$\Delta G°$ KJ .moL-1	$\Delta S°$ J .mol-1.K-1
5	6.70	-36.82	156.55
15	6.70	-39.11	159.06
25	6.70	-40.48	158.32
37	6.70	-42.27	157.97
45	6.70	-43.54	157.99

The negative values of $\Delta G°$ reflects the stability of the complex hence. The high affinity of the reactants. The high negative values of $\Delta G°$ for the binding reaction are controlled by high positive $\Delta S°$ values of the complex formed. So, our system is characterized by the sole contribution of $\Delta S°$ to the stability of the complex formed, which $\Delta H°$ has little or no effect (196). Whereas, the negative values of $\Delta G°$ indicates that the reaction is spontaneous at the standard condition. On the other hand, the high positive of $\Delta S°$ suggest that the binding was entropically driven. Entropy has a driven force for the occurrence of the binding reaction, this indicates that the hydrophobic interactions played an important role in the stability of complex formation (197).

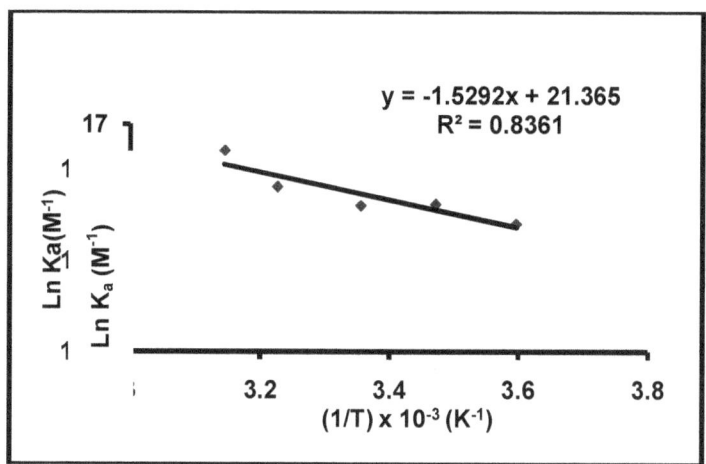

Figure (4.6): Van't Hoff plot for the binding of 125I-anti CA15-3 antibody to the partially purified CA15-3 in benign breast tumors (Fibroadenoma). All details are explained in the text.

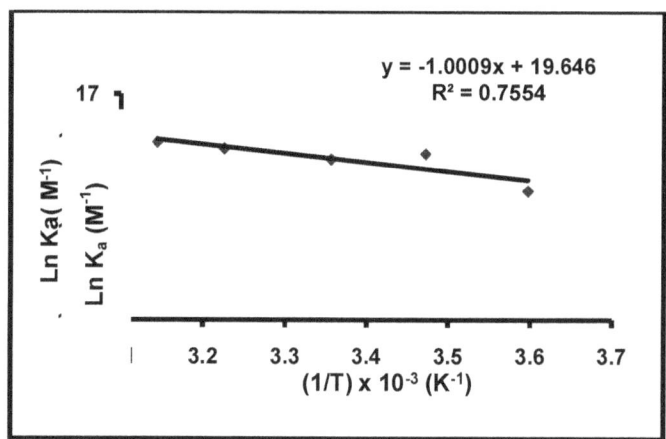

Figure (4.7): Van't Hoff plot for the binding of 125I-anti CA15-3 antibody to the partially purified CA15-3 in malignant breast tumors (IDC). All details are explained in the text.

B. Thermodynamic Parameters of Transition State

Transition state theory postulated that the interaction of two substances to form the final product proceeds through the formation of an activated complex (transition state).

Consequently, the association of 125I-anti CA15-3 antibody with its CA15-3 can be represented as follows:

$$^{125}I - antiCA\,15 - 3 + CA15 - 3 \rightarrow \left[^{125}I - antiCA\,15 - 3/CA15 - 3\right]^{*} \rightarrow \left[^{125}I - antiCA\,15 - 3/CA15 - 3\right]$$

State (A)   An Activated Complex   Final Pr oduct

               Transition State   State (B)

Thermodynamic parameters (ΔH^*, ΔG^* and ΔS^*) of the transition state were determined from the application of Arrhenius equation to the kinetic data. Figure (4.8) and (4.9) show Arrhenius plots for the binding of CA15-3 to its antibody, the slope of the line represents the activation energy (Ea) of the binding reaction, the linear relationship indicates the dependency of the association rate constant of the binding of CA15-3 to its antibody for benign and malignant breast tumors homogenate on temperature.

Table (4.6) and (4.7) show the values of thermodynamic parameters of the transition state (Ea, ΔH^*, ΔG^* and ΔS^*).

The high values of activation energy 9.96 KJ.mol-1 and 41.76 KJ.mol-1 of CA15-3 partially purified from benign and malignant breast tumors respectively, represents the required energy to overcome the energy barrier of the transition state for the formation of (125I-anti CA15-3 antibody / CA15-3) complex. Also the value of activation energy is in accordance with the high positive values of ΔG^*, which indicates that the formation of the activated complex is a non-spontaneous process and requires a lot of energy (equal to Ea) to overcome the transition state energy barrier and giving the final product, whereas the high negative ΔS^* revealed that the activated complex had a more order structure than the reactants.

From the result obtained of the thermodynamic parameters in the transition state, it can be concluded that the positive values of ΔH^* and high positive values of ΔG^* are favorable to overcome the energy barrier of the transition state, the high negative values of ΔG^* is mainly attributed to the decrease in entropy of the transition state ($\Delta S^* < 0$).

In addition the positive values of ΔH^* show that the heat content of the activated complex is more than that in isolated species (193,198).

It is proposed that the formation of a complex occurs in the two steps. The first is the stabilization of the complex by hydrophobic interactions and second is the stabilization by short range interactions , such as electrostatic interaction, hydrogen bonding and Van der Waals interactions (199).

Hydrophobic interactions contribute to the complex stability via high positive entropy change ($\Delta S^* > 0$), while electrostatic interactions, hydrogen bonding and Van der Waals interactions contribute to the stability of the complex via negative entropy change ($\Delta S^* > 0$) (199,200).

The thermodynamic data indicate that the binding of 125I-anti CA15-3 antibody to partially purified CA15-3 are entropy driven and in agreement with the concept that hydrophobic interaction play an important rote in the formation of (125I-anti CA15-3 antibody / CA15-3) complex.

Table (4.6): Thermodynamic parameters at transition state of 125I-anti CA15-3 antibody to the partially purified CA15-3 in benign breast tumors (Fibroadenoma). (All other details are explained in the text).

Temp. °C	Ea KJ . mol-1	ΔH* KJ . mol-1	ΔG* KJ . mol-1	ΔS* J .mol-1. K-1
5	9.96	7.65	58.94	-184.50
15	9.96	7.57	60.62	-184.20
25	9.96	7.48	61.72	-182.01
37	9.96	7.38	64.06	-182.84
45	9.96	7.32	68.62	-192.77

Table (4.7): Thermodynamic parameters at transition state of 125I-anti CA15-3 antibody to the partially purified CA15-3 in malignant breast tumors (IDC). (All other details are explained in the text).

Temp. °C	Ea KJ . mol-1	ΔH* KJ . mol-1	ΔG* KJ . mol-1	ΔS* J .mol-1. K-1
5	41.76	39.45	59.08	-70.61
15	41.76	39.37	59.09	-68.47
25	41.76	39.28	64.14	-83.42
37	41.76	39.18	66.12	-86.90
45	41.76	39.12	70.00	-97.11

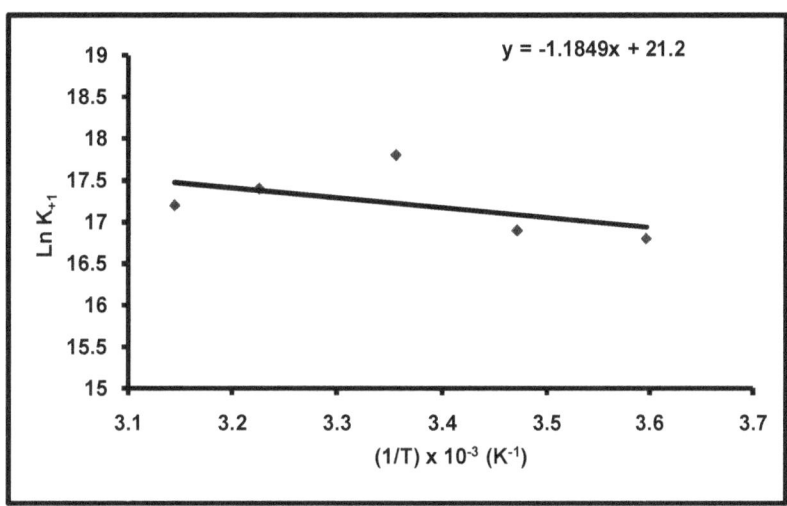

Figure (4-8): Arrhenius plot for the binding of 125I-anti CA15-3 to the partially purified CA15-3 in benign breast tumor (Fibroadnoma). All details are explained in the text.

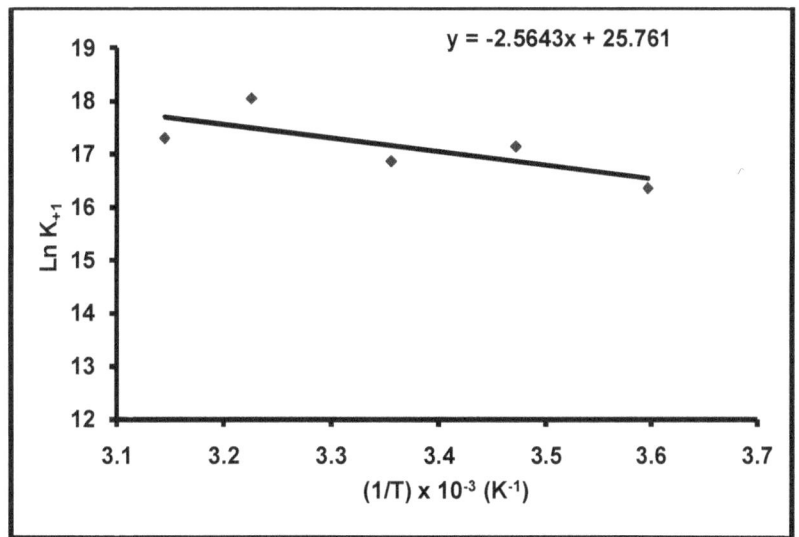

Figure (4.9): Arrhenius plot for the binding of 125I-anti CA15-3 to the partially purified CA15-3 in malignant breast tumor (IDC). All details are explained in the text. Gel filtration technique was used to separate 125I-anti CA 15-3 antibody bound to partially purified CA 15-3 using benign (Fibroadenoma) and malignant (IDC) breast tissue homogenate (as CA 15-3 source) from unbound (Free) 125I-anti CA 15-3 antibody.

Chapter Five

characterization
of complexes of CA 15-3

Introduction

The characterization of the complexes (125I-anti CA 15-3 antibody/CA15-3) from both benign and malignant breast tumors was carried out through the ultraviolet spectroscopic studies. Factors affecting the absorption properties of the two types of complexes such as pH, solvent polarity (solvent perturbation technique), spectrophotometric pH titration, and thermal stability in the presence of different concentrations of sodium chloride have been studied. pH titration of the two types of the complexes show that about (41.43%) and (44.29%) of histydyl residues are located on the surface of the two types of protein complexes (benign and malignant) respectively, while (40%) and (50%) of tyrosyl residues are buried interiorly in the complexes of (benign and malignant) respectively.

Molecules absorb light; the efficiency of absorption depend on both the structure and environment of the molecule making absorption spectroscopy a useful tool for characterizing both small and large molecule.

The ultraviolet absorption spectra of protein solutions in the region 250 to 310 nm are contributed from phenylalanyl, tyrosyl and tryptophanyl residues. But at the shorter wavelengths the contributions come from other groups such as histidyl residues and the peptide bond (169). Changes in the environment of these chromophores can lead to alteration in the absorption spectrum, and the conformational changes of a protein may also involve environmental changes of its chromophoric groups (1201). A variety of environmental changes (e.g. pH, temperature) can affect the absorption spectrum if the interaction of chromophore and perturbing agent affects the ground and excited states, the altered spectrum of the chromophore can be shifted to longer (red shift) or shorter (blue shift) wavelengths. The shift may or may not be accompanied by a change in intensity of the spectrum (170,202). Saif-Alla, P.H., studied the UV spectra of h-PRL-antibody complex and CA15-3 molecule (203).

Interaction of h-CA 15-3 partially purified from benign (fibroadenoma) and malignant (IDC) tissues homogenate with its antibody is an example of protein-protein association. Although several new immunochemical techniques were developed to study such interactions (204,205), UV spectral remain as one of the most important methods in immunology because it provides a sensitive and quantitative measurements for the study of antibody structure and its specific ligand binding (206,208).

Very limited work concerning the physical properties of CA 15-3 specially those related to UV spectroscopy has been done, also the UV studies on CA 15-3 antibody interaction are not wide spread. Hence, this work is planned to study the association of the partially purified h-CA 15-3 and its antibody at different conditions.

Materials and Methods

5.1. Materials

5.1.1. Chemicals

All chemicals and reagents used in the experiments of this chapter were mentioned in section (2.1.1).

5.1.2. Instruments

The instruments used in this chapter are Shimadzu double beam UV-Visible spectrophotometer type 160, and instruments listed in section (2.1.2).

5.1.3 Buffers and Reagents

Buffers and reagents mentioned in section (2.1.6) are used in this chapter. Other additional solutions are indicators in each experiment.

5.2. Methods

5.2.1. Gel Filtration Technique for Separation of Free and Bound125I - Anti CA 15-3 Antibody

5.2.1.1. Preparation of the Column

The dimensions of the column were (1x30 cm) chosen according to the equation in section (3.2.1.1).

5.2.1.2. Preparation of the Gel and Determination of Void Volume

The sepharose CL-4B was used to separate free and bound 125I - anti CA 15-3 antibody, and was prepared as mentioned in section (3.2.1.3) and (3.2.1.4), the void volume was determined and found to be 10 mL.

5.2.1.3 Separation Procedure of (125I-Anti CA 15-3   Antibody/CA15-3) Complex

A) Partially Purified CA15-3 from Benign Breast Tumor (Fibroadenoma) and its Antibody 125I -Anti CA 15-3

Reagents

Buffer PBS 0.15M, pH 7.0 containing 0.02% sodium azide was prepared as described previously in section (2.1.1.3).

Procedure

1-   Partially purified CA 15-3 (475μL) containing (0.665 mg. mL-1) was incubated with 120 μL of 125I-anti CA 15-3 antibody   (0.8412mg. mL-1) and complete the reaction to a final volume of 700 μL with PBS buffer 0.15 M pH 7.0. The tubes were incubated for 90 min. at 37oC.

2-   At the end of incubation, the mixture was applied to the surface of a sepharose CL-4B (1x30 cm) with a bed volume (23.5 cm3) equilibrated with PBS buffer 0.15M, pH 7.0. Elution was carried out using the same buffer to separate CA 15-3 bound to 125I-anti CA 15-3 antibody from unbound (Free) CA 15-3 and 125I-anti CA 15-3 antibody with a flow rate (1 mL per 7 min), and fraction volumes of 1 mL were collected.

3-   The radioactivity of each fraction was counted by gamma counter for one minute.

4-   Protein concentration was measured at 280 nm.

5-   One hundred and twenty microliters of 125I-anti CA 15-3 antibody (0.84 mg. μL-1) was completed to 700 μL with PBS buffer (0.15M, pH7.0), then this volume was injected to the column as mentioned in step2, then steps 2,3 and 4 were repeated.

Calculations

1. Radioactivity (c.p.m) of each eluted fraction was plotted against the fraction number.

2. The absorbance of each eluted fractions was measured at 280nm, and the absorbance was plotted against the fraction number.

3. The percent radioactivity was calculated by dividing the sum of the radioactivity of the fractions under each peak by the sum of radioactivity of all peaks appeared in the profile:

$$\text{Percent radioactivity of each peak} = \frac{\text{Radioactivity per peak (c.p.m)}}{\text{Sum of radioactivity of all peaks (c.p.m.)}} \times 100$$

B) Partially Purified CA15-3 from Premenopausal Malignant Breast Tumors (IDC) and Its Antibody 125I-anti CA 15-3

Reagents

Buffer PBS 0.15 M, pH 7.0 containing 0.02% sodium azid was prepared as described previously in section (2.1.1.3).

Procedure

1.   Four hundred and twenty four microliters of partially purified CA 15-3 (0.147 mg. mL-1 protein) and incubated with 106 μL of 125I-anti CA 15-3 antibody (0.743 mg.mL-1) in a final volume 700 mL with PBS buffer 0.15M pH 7.0. The tubes were then incubated for 150 min at 15oC.

2.   Steps 2,3,4 and 5 in section (5.2.1.3 A) were repeated.

Calculation

The same calculation that mentioned in section (5.2.1.3 A) was used to calculate the radioactivity; protein was measured at 280nm and the percent of radioactivity of each peak was determined.

5.2.2. The UV Spectrum of (125I-Anti CA 15-3 Antibody/CA15-3) Complex from Benign and Malignant Breast Tumors

The gel filtration profile in section (5.2.1.3 A&B) gave two peaks. The fractions under each peak were pooled and the absorption spectrum was scanned in UV Region against the appropriate blank in the reference beam.

5.2.3. The UV. Spectrum of 125I-Anti CA 15-3 Antibody

Half milliliter of 125I-anti CA 15-3 antibody was placed in a 0.25 cm cuvette in the sample beam and the absorption spectrum was measured immediately against an appropriate blank in the reference beam.

5.2.4. The UV Spectrum of Partially Purified CA 15-3

Half milliliter of partially purified CA 15-3 from benign (Fibroadenoma) and malignant (IDC) breast tumors was placed in a 0.25 cm curette in the sample beam and the absorption spectrum was measured immediately against an appropriate blank in the reference beam.

5.2.5.Factors Affecting the Absorption Properties of (125I-Anti CA 15-3 Antibody/CA 15-3) Complex from Benign and Malignant Breast Tumors

5.2.5.1. The pH Effect on the Complex

Reagents

1. KCl-HCl buffer (pH 2) was prepared as follows:

Solution A: Potassium chloride (0.15M), 1.11825 gm was dissolved in a final volume of 100mL deionized distilled water.

Solution B: Hydrochloric acid (0.15M).

The required pH (2.0) was prepared by mixing a volume of solution A with an appropriate amount of solution B to obtain the required pH.

1. Citrate-phosphate buffer at different pH was prepared as follows:

Solution A: Citric acid (0.15M); 2.8815 gm citric acid dissolved in 100mL deionized distilled water.

Solution B: Dibasic sodium phosphate (0.15M); 2.1294 gm of Na_2HPO_4 was dissolved in a final volume of 100 mL deionized distilled water.

Working buffer pH (4 and 6) was prepared by mixing a volume of solution A with an appropriate amount of solution B to obtain the required pH.

2. Phosphate buffer at different pH values was prepared as follows:

Solution A: Dibasic sodium phosphate (0.15M), 2.1294 gm Na_2HPO_4 was dissolved in a final volume of 100 mL deionized distilled water.

Solution B: Monobasic sodium phosphate (0.15M), 1.7997 gm NaH_2PO_4 was dissolved in a final volume of 100 ml deionized distilled water.

Phosphate buffers at different pH rang (7-8) were prepared by mixing a volume of solution A with an appropriate amount of solution B to obtain the required pH.

3. Glycine - NaOH buffer was prepared as follows:

Solution A: Glycin (0.15M); 1.12575gm $C_2H_5NO_2$ was dissolved in a final volume of 100 mL deionized distilled water.

Solution B: Sodium hydroxide (0.15M); 0.6gm NaOH was dissolved in a final volume of 100 mL deionized distilled water.

Working buffer pH (9-11) was prepared by mixing a volume of solution A with an appropriate a mount of solution B to obtain the required pH.

Procedure

Two hundred and fifty microliters of pooled fractions under the first peak that represent (125I-anti CA 15-3 antibody/CA 15-3) complex, was completed to 500μl with different buffers at different pH values (4 to 11), then each sample beam and the buffer at the adjusted pH in the reference beam. The absorption spectrum was scanned.

Calculations

The molar absorption coefficient (□) for (125I-anti CA 15-3 antibody/CA 15-3) complex at 278 nm was calculated from Lambert-Beer's law.

5.2.5.2. Effect of Solvent Polarity on UV Spectra of the Complex

The effect of 20% ethanol, and the same amount for ethylene glycol, glycerol, sucrose, urea, dimethyl sulphoxide, dioxane, and polyethylene glycol; on the complex. Two hundred and fifty microliters of complex from benign and malignant breast tumors of pooled fractions under the first peak were completed to 500 μL with phosphate buffer containing any of the following solvent at pH 7.4 in the test cell and the 20% ethanol, ethylene glycol, glycerol, sucrose, urea, dimethyl sulphoxide, dioxane, and polyethylene glycol was adjusted and placed in the reference cell using 0.25 cm cuvette (i.e., the experiment was repeated by using solvents individually).

Calculations

The absorption spectrum of each sample was scanned immediately in the area of (200-350 nm).

5.2.5.3. Spectrophotometric pH Titration on the Complex

A series of complex from benign (Fibroadenoma) and Malignant (IDC) breast tumors (250 µL) were completed to 500 µL with buffer at pH ranging from 8 to 11. The maximum absorbance of each sample was measured at 295 nm; the absorbance of λ max at each pH value was plotted versus the corresponding pH. Other series of complexes isolated from benign (Fibroadenoma) and malignant (IDC) breast tumors (250 µL) were completed to 500 µL with buffer at pH ranging 4 to 8. The maximum absorbance of each sample was measured at 211nm. The absorbance of λmax at each pH value was plotted against the corresponding pH.

5.2.5.4. The Effect of NaCl Concentration on the Thermal Stability of the Complex by UV Spectral Studies

Reagents

Twenty percent ethylene glycol buffur was prepared by dissolving 20mL of ethylene glycol in 80mL of phosphate buffer. NaCl (0.01M) in 20% ethylene glycol was prepared by dissolving 0.05844 gm of NaCl in 100mL of 20% ethylene glycol buffur, while NaCl (0.1M) in 20% ethylene glycol was prepared by dissolving 0.5844 gm of NaCl in 100mL of 20% ethylene glycol buffer.

Procedure

Two hundred and fifty microliters of complex from benign (Fibroadenoma) or malignant (IDC) breast tumors were completed to a final volume 500 μL with 20% ethylene glycol buffer pH7.4 containing 0.01 M NaCl Each mixture was placed in 0.25 cm cuvette in the sample beam and the buffer at the adjusted pH in the reference beam. The absorbtion was measured at the wavelength of (292 and 295 nm) at different temperatures 20, 30, 40, 50, 60, 70oC. The experiment was repeated for each complex with another solution (20% ethylene glycol 0.1 M NaCl), at 295 nm.

Calculations

The absorbance of each complex was plotted against the different temperatures at two wavelengths (292 and 295 nm).

5.2.5.5. Effect of Urea, KCl and (Urea, KCl) Mixture on the Spectrum of the Complex

Reagent

1. Eight molar of urea was prepared by dissolving 24.02gm of Urea in a final volume of 50 mL of PBS buffer at pH 7.4.

2. KCl (0.03 M) was prepared by dissolving 0.2737gm of the salt in a final volume of 50mL of corresponding buffer.

3. PB buffer solution was prepared as described in section (5.2.5.1).

Procedure

Two hundred and fifty microliters of complex isolated from benign (fibroadenoma) and malignant (IDC) were pipetted in a set of three tubes. The volume was completed to 500 μL with PBS buffer at pH 7.4 contains (0.03 KCl, 8 M urea and mixture 1:1 of both 0.03 KCl and 8M Urea) respectively, then each sample was placed in 0.25cm cuvette in the sample beam and the buffer at the same pH in the presence of the same salt in the reference beam.

Calculations

The absorption spectrum of each sample was scanned immediately in the area of (200-350 nm).

5.3. Results and Discussion

Protein UV light maximum absorption is at approximately 280nm, caused by tryptophan, tyrosine and (to a lesser extent) phenylalanine residues, and at lower wavelength (215-230 nm) due to polypeptide chain backbone. Absorbance at 280 nm varies for each protein. The absorbance at lower wavelengths is directly related to the amount of polypeptide material and is usually considerably more sensitive than at 280nm.however,many buffers and other molecules also absorb at these lower wavelengths (phosphate and tris buffers are acceptable but the preservative sodium azide absorbs strongly).

Absorbance at 215-230 nm is useful for monitoring peptides that may not contain tryptophan or tyrosine (205).

Gel Filtration Technique for Separation of Free and Bound 125I-Anti CA15-3 Antibody

Figure (5-1) and (5-2) show the results of gel filtration technique to separate 125I-anti CA 15-3 antibody bound to partially purified CA 15-3 from benign (Fibroadenoma) and malignant (IDC) breast tumors respectively. The profile of separation revealed two peaks. The first peak represents (125I-anti CA 15-3 antibody/CA 15-3) complex, the second peak represents the unbound (Free) 125I-anti CA 15-3 antibody. Figure (5-3) show the gel filtration profile of 125I-anti CA 15-3 antibody, the results revealed only one peak in the same position of the second peak of figures (5-1) and (5-2), which represent the unbound 125I-anti CA 15-3 antibody. The percent of 125I-anti CA 15-3 antibody/ CA 15-3) complex was 49.74% in benign (Fibroadenoma) breast tumors patients, while the percent of complex was 56.20% in malignant breast tumors patients (IDC). On the other hand the percent of 125I-anti CA 15-3 antibody was 34.40% in benign breast tumors (Fibroadenoma) and 31.42% in malignant breast tumors (IDC). This is because the epitope of CA 15-3 in malignant breast tumors was higher than in benign breast tumors.

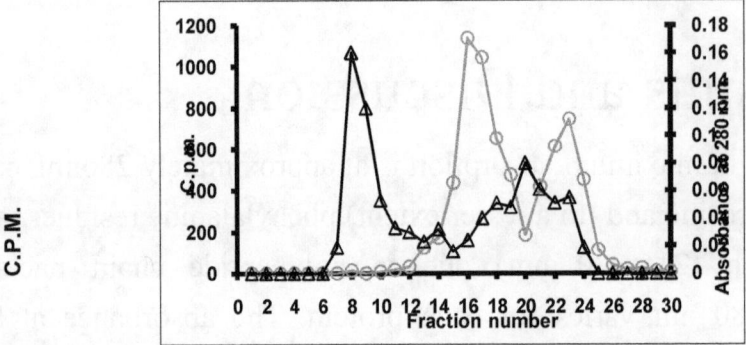

Figure (5.1): The elution profile of the isolated complex (125I-antiCA15-3 antibody/CA15-3) and free antibody in benign breast tumors on Sepharose CL-4B. (O) radioactivity, (△) protein. (All other details are explained in the text).

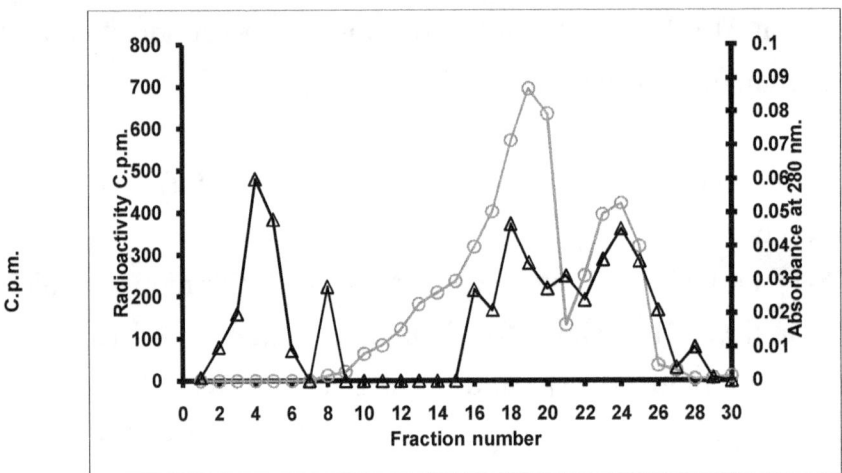

Figure (5.2): The elution profile of the isolated complex (125I-antiCA15-3 antibody/CA15-3) and free antibody in malignant breast tumors (IDC) on Sepharose CL-4B, (O) radioactivity, (△) protein. (All other details are explained in the text).

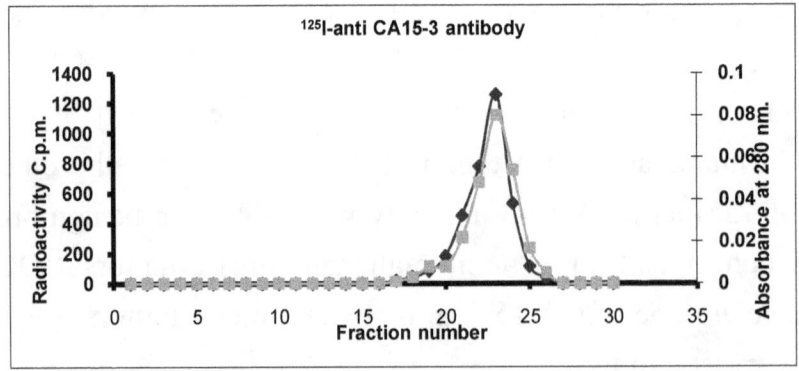

Figure (5.3): The elution profile of the 125I-antiCA15-3 antibody on Sepharose CL-4B, (■) radioactivity, (◆) protein. (All other details are explained in the text).

The UV Spectra of Partially Purified CA 15-3, Anti CA 15-3 Antibody and (125I-Anti CA 15-3 Antibody/CA 15-3) Complex Molecules

The UV spectra of partially purified h-CA 15-3, 125I-anti CA 15-3 antibody and (125I-anti CA 15-3 antibody/CA 15-3) complex were scanned from 200-350 nm to determine the absorption spectra, and the alternation in the UV spectra as a results of their interaction.

The UV Spectrum of Partially Purified CA 15-3

The UV spectra of partially purified h-CA15-3 in benign tumors (Fibroadenoma) and malignant tumors (IDC) at neutral pH shows that the λmax for purified CA15-3 from benign (Fibroadenoma) consisted of two peaks; a large one at 208nm and smaller one at 270nm, while the UV spectra of purified CA 15-3 from malignant tumors (IDC) shows two peaks at 205 and 270nm as shown in table (5.1). Therefore it seemed that each human CA 15-3 has a characteristic spectrum and can be identified by its peaks, the first peak (at 208nm or 205nm) such results could be due to the amide group in polypeptide bond of h-CA 15-3 molecule with contribution of the histidyl residues (207), while the second peak (at 270) is assigned to the side chain chromophore of phenylalanine or tryptophyl residues (208).

The UV spectrum of 125I-Anti CA 15-3 Antibody

The UV spectrum of 125I-anti CA 15-3 antibody at neutral pH shows that the λ max consisted one peak at 203.6nm, which is assigned to the amide groups in the polypeptide bond (207), with contribution of hisidyl residues (209) as shown in table (5.1) .

The UV spectrum of (125I-Anti CA 15-3 Antibody/CA 15-3) Complex

The UV spectra of partially purified CA 15-3 extracted from benign (Fibroadenoma) and malignant (IDC) bound to 125I-anti CA 15-3 antibody at neutral pH show that the λ max is consisted of two peaks at (203.4nm and 274nm) in benign complex, while the λ max is consisted of two peaks at (204.2 nm and 278nm) in malignant complex as shown in table (5-1). The

131

first peak at (274nm or 278nm) is assigned to tyrosyl residues(208), it is very weak band and it seems that the tyrosyl residues in the benign or malignant complexes is located on the surface of protein complex.

The strong absorption of the second peaks (at 203.4 or 204.2nm) arises form electronic transition in the peptide backbone itself and is therefore sensitive to backbone conformation (207).

Table (5-1): The λ max valves of (125I-anti CA 15-3 antibody/CA 15-3) complex, partially purified CA 15-3 and unbound (Free) 125I-anti CA 15-3 antibody in both cases benign and malignant breast tumors. (All other details are explained in the test).

No.	Fractions	Benign λmax (nm)	Malignant λmax (nm)
1	CA 15-3 partially purified	208, 270	205, 270
2	125I-anti CA 15-3 antibody	203.6	203.6
3	125I-anti CA15-3 antibody/ CA15-3) complex	203.4, 274	204.2, 278

Factors Affecting the Absorption Properties of (125I-Anti CA15-3 Antibody/ CA15-3) Complex from Benign and Malignant Breast Tumors

The Effect of pH on the Complex

The pH of the solvent determines the ionization state of ionizable chromophores in the protein molecule (208). The UV spectrum of isolated (125I-anti CA 15-3 antibody/CA15-3) complex from benign (Fibroadenoma) and malignant (IDC) breast tumors was determined at different pH (2, 4, 6, 7, 7.4, 8, 9, 10, and 11). Table (5-2) shows the effect of different pH on both complexes. At an acidic pH 2 and neutral pH (7,7.4) the both complexes benign (Fibroadenoma) and malignant (IDC) have one maximum wavelength near 200 nm as compare to UV spectrum of h-CA 15-3 and the 125I anti CA15-3 antibody.

The λ max of CA 15-3 (270 and 208 nm) in benign (Fibroadenoma) and its antibody λ max (203.6) disappeared. The λ max of CA 15-3 (270 and 205 nm) in malignant (IDC) and its antibody λmax (203.6) also disappeared. The absorption near 200 nm is characteristic of the amide group in the polypeptide bond of the complex (209). The blue shift is due to the increasing of hydrogen bond formed in the presence of highly positively charged state (210). The disappearance of λmax 280 nm of tyrosine and phenylalanine due to conformational changes and chromophore in native complex were buried in the interior of their complexes(211,212). Protein shows a strong absorption in range (180-225 nm), absorption at such wavelength arises from electronic transition in the polypeptide backbone itself and is therefore sensitive to back bone conformation(207).

At pH (4, 6, 9, 10, and 11) no band was observed and all peaks disappeared. The disappearance of the λ max at these pH's may be due to conformational changes of the protein complex.

Table (5-2): The effect of different pH on λ max values of (125I-anti CA 15-3 antibody/CA 15-3) complex. (All other details are explained in the text).

pH	λ max (nm)	
	(125I-anti CA 15-3 antibody/CA15-3) benign complex	(125I-anti CA 15-3 antibody/CA 15-3) malignant complex
2	200	200
4	-	-
6	-	-
7	200	200
7.4	200	200
8	200	200
9	-	-
10	-	-
11	-	-

Effect of Solvent Polarity on UV Spectra of the Complex

The immediate environment of a chromophore affects its absorption. The determination of whether an amino acid is internal or external by measuring the spectra of protein in a polar and non-polar solvent is called the solvent perturbation method (208). In fact, proteins are rarely studied in completely non-polar solvents because most proteins are either insoluble or denatured in these solvents. However, significant solvent effects can be induced by use of a mixture of water and substance of a reduced polarity such as ethanol, ethylene glycol, polyethelyne glycol, sucrose, dioxane and dimethyl sulfoxide (DMSO)(208). Several spectra changes were obtained in the precence of these perturbants, like the alteration of λmax positions and intensities of protein spectrum and the appearance of new chromophores on the surface of the complex. These chromophores on the region of the protein disappeared in the absence of the solvent. One of the main assumptions of the solvent perturbation technique is that solvent alters the peak positions and intensities by altering the energy and probably of electronic transitions. Other considerations include the following (213,214):

a. Polarization effect

b. Change in permanent dipole moment during excitation, which will tend to produce either a short wave or a long wave shift depending on the nature of the electronic transition and wheather the solute is a hydrogen donor or hydrogen acceptor (215).

The effects of different solvents on the (125I-anti CA 15-3 antibody /CA 15-3) complex from benign (Fibroadenoma) and malignant (IDC) breast tumors at pH 7.4 were investigated. The data obtained are illustrated in table (5-3). It was found that one λ max specific for the amide groups of polypeptide bond at pH 7.4, this shift toward the shorter wavelength is due to the n-π^* transitions in the presence of 20% ethanol, ethylene glycol and glycerol. In the presence of polyethylene glycol there was a significant red shift in the λ max (204 nm) in benign (Fibroadenoma) complex and λ max (205nm) in malignant (IDC) complex. When 20% Dioxane was used there were a significant red shift in the λ max (220nm) of the amide bond at pH 7.4, which assigned to tyrosyl residue. The value of λ max is for n-π^* transitions which occur at longer wavelength because the nonbonded electrons in the anion are available for interaction with the π electron system of the ring (215), while in the presence of 20% sucrose the complex has a slight blue shift and show λ max at 202 nm and 201nm in both benign and malignant complexes. Finally the effect of 20% DMSO on the complex, show that the amide bands at pH 7.4 were disappeared, this may be due to the denaturation of protein complex in presence of 20% DMSO.

The application of spectrophotometric solvent perturbation on the complex is to determine the location of tyrosyl residues, whether they are buried and inaccessible or exposed and accessible to the solvent approach (216). Laskowski (201) has listed the major assumptions of solvent perturbation experiments. There are: (1) buried chromophors are unperturbed, that is only the groups located on the surface or near the surface of the protein should experience the perturbing effects of the solvent; groups buried in the interior of the protein, not accessible to the solvent; which should not be affected and consequently could not contribute to the overall spectral shift observed. (2) No conformational changes take place upon addition of perturbant, and (3) the solvation layer around the chromophore contains the same concentration of perturbation experiments when employed at convenient concentrations (often 20%), do not appear to produce conformational changes in most protein studied under reasonable conditions of pH, ionic strength, and temperature. This concentration is large enough to cause measurable shifts in the spectra of chromophoric residues. Conformational changes can be expected if perturbation is carried out under conditions in which the protein structure has marginal stability (low-or high pH for many protein)(201,216). Chromophore may not completely bury. It has been, distinguished between chromophores in crevices and chromophores that are partially buried (201). The former are observed to be fully perturbed by perturbant solvent smaller than a certain critical size (e.g ethanol), but not by larger perturbants (e.g polyethylene glycol). The degree of exposure is thus determined only by the size of perturbant molecule (solvent) or by the size of the crevice in which the chromophore is located (216). Partially buried chromophores on the other hand, show a degree of exposure that depends on the nature of the perturbant, rather than on its size. The observed degree of exposure decrease in the order:

Sucrose ≥ Glycerol ≥ Ethyleneglycol ≥ Methanol, Ethanol > Polyethylene glycol ≥ Dimethyl sulfoxide.

The first perturbant in this series modify the solvent nonspecifically, while the later ones in the series may specifically interact with chromophore(216).

When comparing the effects of the six solvents used, ethanol, ethleneglycol, glycerol, dioxane, polyethylene glycol and dimethylsulfoxide at pH 7.4 on the UV spectrum, especially on the shift of λ max, which is due to tyrosyl residues. It seems that the maximum effect was observed in the presence of 20% dioxane as perturbant solvent, where there was a shift in the λ max about 16nm, while minimum effect was observed in the presence of 20% polyethylene glycol where the λ max remained unchanged. Since the change in the λ max of tyrosyl residues does not depend on the size of the perturbant solvent, the tyrosyl residues showing the changes in λ max ; absorbance must be partially buried.

Table (5-3): The effect of 20% of ethanol, ethyleneglycol, glycerol, polyethylene glycol, sucrose, dioxane, DMSO and on the λ max of (125I-Anti CA 15-3 Antibody/ CA15-3) complex at pH 7.4. (All other details are explained in the text).

Solvent of 20% of	λ max (nm)	
	(125I-anti CA 15-3 antibody/ CA15-3) benign complex	(125I-anti CA 15-3 antibody/ CA 15-3) malignant complex
Ethanol	Near 200	200
Ethylene glycol	Near 200	Near 200
Glycerol	Near 200	Near 200
Polyethylene glycol	204	205
Sucrose	202	201
Dioxane	220	220
DMSO	-	-

Spectrophotometric pH Titration of the Complex from Benign and Malignant Breast Tumors

To study (125I- anti CA 15-3 antibody/ CA15-3) complex structure, this requires the determination of pka values for proton dissiociation from ionizable amino acid side chains, because these values give an indication of the location of amino acid in the protein. This can often be done spectrophotmetrically because dissociation often changes the spectrum of one of the chromopores (tyrosyl)(208). For proteins this usually amounts to the titration of the phenolic groups of tyrosine residues. By the measurement of the absorption at 295 nm (λ max for the ionized form of tyrosine), or observation of histidine dissociation by measurment at 211nm.

The titration curves of (125I- anti CA 15-3 antibody/ CA15-3) complex from benign (Fibroadenoma) and malignant (IDC) for both histidyl and tyrosyl residues are illustrated in figure (5-4 A&B) respectively. Figure (5-4) shows that the pka for histidine is (6.69) for (125I- anti CA 15-3 antibody/ CA15-3) complex from benign breast tumors, while the pka for histidine is (6.65) for (125I- Anti CA 15-3 Antibody/ CA15-3) complex from malignant (IDC) breast tumors. From the same curve it could be concluded that about (41.43%) histidyl residues are located on the surface of the protein complex (217) of benign (Fibroadenoma), while about (44.29%) histidyl residues are located on the surface of the protein complex from malignant (IDC). The other residues are buried interior the benign and malignant complex. Figure (5-4 B) shows that the pka value of the benign complex of tyrosyl residues is (8.9) and it's about (40%) at tyrosine residues are internal and a large arise in the absorbance at very high pH was observed. While in the malignant (IDC) complex the pka value of tyrosyl is (8.4) and it's about (50%) this indicates that the internal tyrosines have become exposed to the solvent, which is the protein complexes in folded (become denatured)(217).

The two curves also illustrated the low content of histidine compared to the high content of tyrosine in the benign and malignant complex.

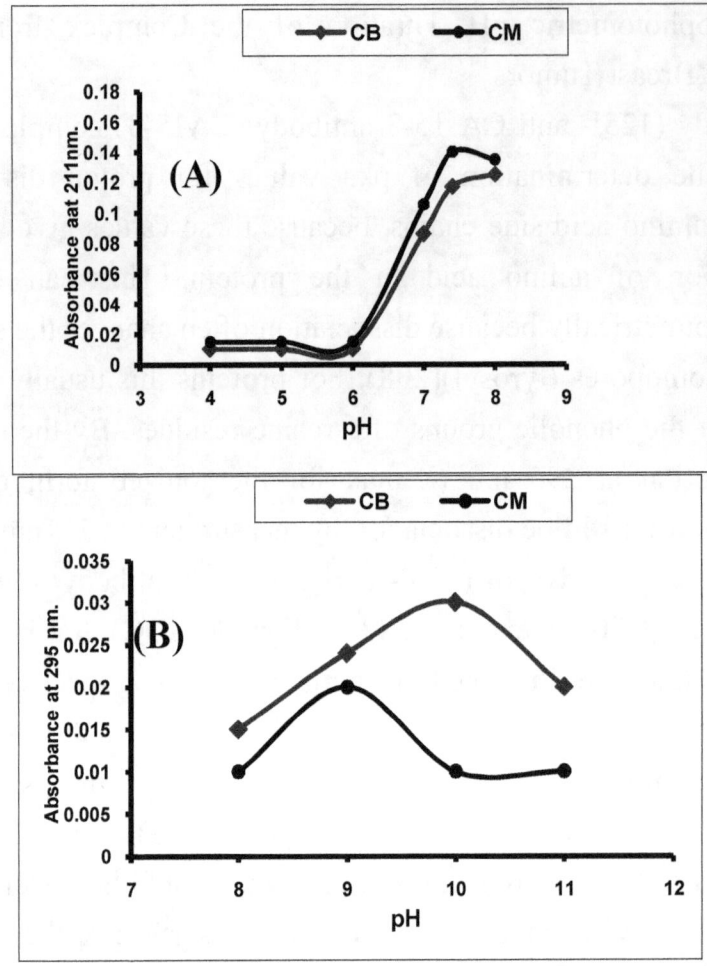

Figure (5.4): Spectrophotometric pH titration of 125I-anti CA15-3 antibody/CA15-3 complex from benign and malignant breast tumors:

(A)   for histidine, (B) for tyrosine.

(CB): Complex of benign breast tumors, (CM): Complex of malignant breast tumors. (All other details are explained in the text).

The Effect of NaCl Concentration on the Thermal Stability of the Complex by UV Spectral Studies

The effect of different concentrations of NaCl on the thermal stability of the protein complex isolated from benign and malignant breast tumors was examined in this experiment. The values of absorbance at λ max (292, 295nm) for tryptophyl and tyrosyl residues respectively, in two different concentrations of NaCl 0.01 M and 0.1 M in 20% ethylene glycol buffer are shown in figure (5.5 A&B) and (5.6 A&B). The λ max was used to examine if the protein contains internal tryptophans and tyrosines .

As shown in figure (5.5 A&B), the absorbance of both tryptophane and tyrosine reach higher absorbance at 60oC, in the presence of 0.01 M NaCl in benign and malignant complex. The increment in the absorbance of both tryptophyl and tyrosyl residues with increasing temperature could be due to that buried chromophores becomes exposed to the solvent during thermal denaturation (209).

Figure (5.6 A) shown the absorbance of tyrosin reach higher absorbance at 70oC in the presence of 0.1 M NaCl in benign and malignant complex. On the other hand figure (5.6 B) shown the absorbance of tryptophane reach higher absorbance at 60oC and 30oC in benign and malignant breast tumors complexes in presence of 0.1 M NaCl respectively. Which means that the complexes were very stable at 70oC in presence of higher concentration of NaCl, 70oC was needed for unfolding benign and malignant complex at λ max 292 nm and benign complex was more stable at 60oC in presence of 0.1 M NaCl,while the temperature is decreased to 30 oC in the presence of 0.1M NaCl at λ max 295. This is due to conformational changes required more energy 70oC in presence of 0.1 M NaCl than in 0.01 M NaCl.

The decreased absorbance in presence of 0.1 M NaCl as compared with that in 0.01 M NaCl could be due to salt concentration. Each protein in solution containing salts will collect around it a counter ion atmosphere enriched in oppositely charged small ion (chloride ion, sodium ion) and such a cloud of ions will tend to screen the protein, the more effective electrostatic screening will be, and decrement in the absorption intensity will be observed (207).

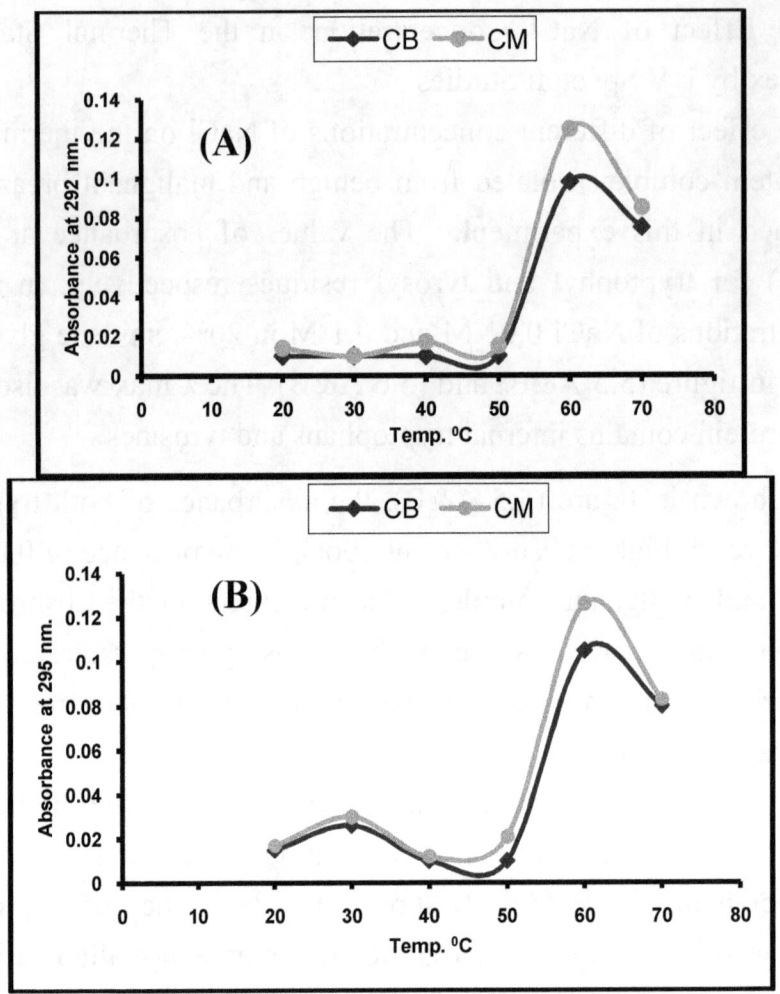

Figure (5.5): Thermal stability curve for benign and malignant: (A) at □max 292 in the presence of 0.01 M NaCl, (B) at □max 295 in the presence of 0.01 M NaCl. (CB): Complex of benign breast tumors, (CM): Complex of malignant breast tumors. (All other details are explained in the text).

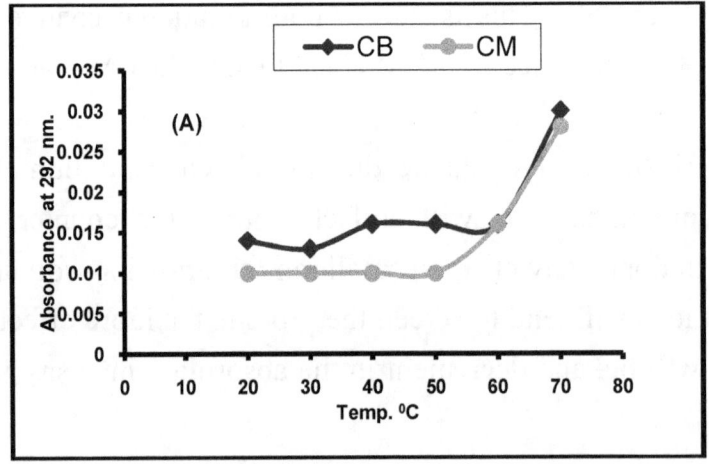

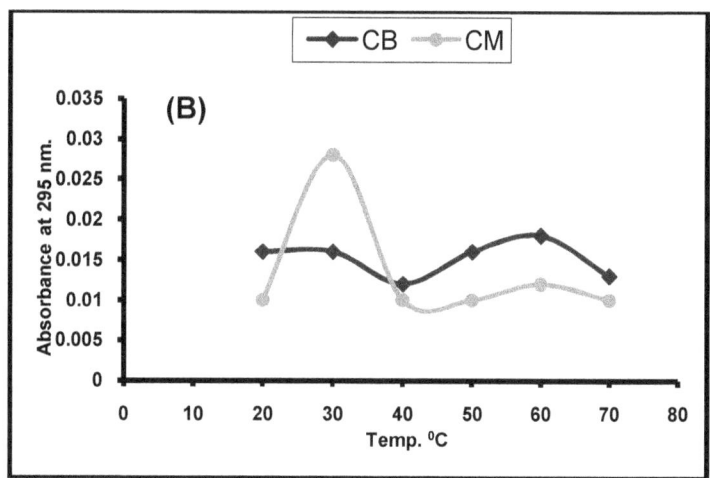

Figure (5.6): Thermal stability curve for benign and malignant: (A) at □max 292 in the presence of 0. 1 M NaCl, (B) at □max 295 in the presence of 0.1 M NaCl. (CB): Complex of benign breast tumors, (CM): Complex of malignant breast tumors. (All other details are explained in the text).

Effect of Urea, KCl and (Urea, KCl) Mixture on the Spectrum of the Complex

The effect of 8 M urea, 0.03 M KCl and a mix of 1:1 of 8 M urea and 0.03 M KCl on the λ max of the benign (fibroadenoma) and malignant (IDC) complexes, were examined. The values of λ max are illustrated in table (5-4). When table (5-4) is compared with table (5-1), it seems that the presence of 8 M urea at pH 7.4, there was a red shift of the λmax1 of polypeptide bond from 200 to 227.4 nm in benign complex and a red shift of λmax1 from 200 to 226 nm in malignant complex respectively. While λmax2 of aromatic amino acid i.e., tyrosine residues in both complexes was disappeared. The red shift is due to intramolecular hydrogen bonding between the oxygen of the amide group and the solvent (218).

When 0.03 M KCl was used, there was no alternation in the position of the λ max2 of the tyrosyl at pH 7.4 in both benign and malignant complexes. There was a slight blue shift (3-4nm) in the λ max1 of the polypeptide bond in the benign and malignant complex spectra respectively. On the other hand the λ max of the aromatic ring of tyrosyl residues at (274 or 278nm) disappeared. Such blue shift can arise by introducing positive (K+) or negative (Cl-) charges near the chromophore (the amid group), which might interact with □-electron system of the amide group (201).

When 8 M urea was mixed with 0.03 M KCl there was significant red shift in λ max (203.4 and 204.2nm) to λ max (221.4 and 219.4nm) in both benign and malignant complexes. The same shift was observed when 8 M urea was used alone with each benign and malignant complexes, this mean that the red shift due to the effect of urea, but not to 0.03 M KCl. On the other hand, there was no alternation in positions of the λ max of the tyrosyl residues near 278nm.As was seen,the changes in absorption were near 230 nm and near 280 nm. This was also observed by Glazer who that solvent perturbation or denaturation of protein poduces may changes in absorption near 230 nm and 280 nm. Some of this change in absorption may be produced by change in the n-□□□absorption of poly peptide bond in protein either because of a change in their geometrical arrangement, or because of an environment changes (219).

Table (5-4): The effect of 8M urea, 0.03M KCl and mixture (urea+KCl) on the λ max of the complex UV spectrum at pH 7.4. (All other details are explained in the text).

Solvent	λ max (nm)	
	(125I-anti CA 15-3 antibody/ CA 15-3) Benign Complex	(125I-anti CA 15-3 antibody/ CA 15-3) Malignant Complex
Urea 8M	227.4	226
KCl 0.03M	200	200
Urea+ KCl mixture 1:1	221.4 278.6	219.4 278

Chapter six

Immunoradiometric assay

Abstract

Asolid-phase Immunoradiomertric Assay sandwich technique (IRMA) was used for the determination of the carbohydrate antigen 19-9 (CA19-9) defined by a monoclonal antibody 125I-anti CA19-9. The antibody 125I-anti CA19-9 reacts with CA19-9 found at low concentrations in sera of healthy women but increased slightly in sera of patients with breast cancer.

The factors affecting the binding of 125I-anti CA19-9 antibody with CA19-9 in the breast tumor homogenate (benign and pre-and post-menopausal malignant) were determined. The results revealed that 100, 75 and 75 μg protein was the most appropriate amount of protein used in each incubation at pH 7.8, 8.0 and 7.0 respectively, with 0.0565 mg. mL-1 of 125I-anti CA19-9 antibody for 4,1 and 6 h incubation time at optimum temperatures 25, 37 and 45 oC respectively. The use of 0.01 M sodium halides and 0.025 M of divalent salts were shown to cause different effects on the binding in the three groups.

The recovery of the method was calculated and found to be 99% , 98% and 95% for binding CA19-9 present in (benign and pre-and post-menopausal malignant) breast tumor homogenates respectively.

Introduction

The Carbohydrate antigen 19-9 (CA19-9) (Koprowski etal.,1979) (118), is specific carbohydrate fraction of a circulating antigen found in sera of normal adults (Koprowski etal.,1981) (220), has sialyl Lewisa structure and is present in individually expressing the Lewisa and /or Lewisb blood group antigen (114). CA19-9 is identified as a glycolipid- that is , sialylated lacto-N-fucopentose II ganglioside (221). In serum, it exists as a mucin , a high molecular mass (200-1000 KD) glycoprotein complex (54). In Normal tissues, sialyl Lewisa antigen is present in ductal epithelium of breast, kidney, salivary gland, and sweatglands(115-117).

CA19-9 is measured with a double monoclonal immuno-radiometric assay (178). Another techniques used for the detection of CA19-9 in tissues and sera were performed by an immunoperoxidase assay (126) and by radioimmunoassay (125) of samples from patients, and enzyme immunoassay (124) for quantitative determination of CA19-9 in human serum. The upper limit of normal value 37.0 U.mL-1 (121,222). The abnormal expression of the sialyl Lewis a is closely correlated with various forms of cancer including pancreatic cancer (223-225), gall bladder (226) and bile duct (227) cancer.

A monoclonal antibody CA19-9 against sialyl Lewis a is a popular diagnostic agent for these tumors. The antibody is useless for cancer diagnosis when a patient is lacking the enzyme for the synthesis of sialyl Lewis a. In Japan, about 5-10% of the population lacks this enzyme leading to false negative results (228). CA19-9 represents the most important and basic carbohydrate tumor marker. The immunohistologic distribution of CA19-9 in tissues is consistent with the quantitative determination of higher CA19-9 concentrations in cancer than in normal of tissues (126,229). Recently reports indicates that serum CA19-9 level is frequently elevated in the serum subjects with pancreatic (80%), hepatobiliary (67%), gastric (40-50%), hepatocellular (30-50%), colorectal (30%) and breast (15%) cancer (51).

Research studies demonstrate that serum CA19-9 values may have utility in monitoring subjects with the above-mentioned diagnosed malignancies(230-232). A declining CA19-9 value may be indicative of a favorable prognosis and good response to treatment (233). Therefore, the development of immunoradiometric assay was planned to carry out the determination of the optimum conditions of 125I-anti CA19-9 antibody.

Materials and Methods

6.1. Materials

6.1.1. Chemicals

All chemical and reagents mentioned in the section (2.1.1) were used in the experiments of this chapter; other reagents used were indicated in each experiment.

6.1.2 Instruments

All instruments described in section (2.1.2) all were used in the experiments of this chapter.

6.1.3 Patients and Blood Samples

Thirty breast patients and specimens mentioned in section (2.1.3) were used in this chapter, classified to three group of patients, one group with benign and two groups with malignant breast tumors. The fourth group is a healthy women used as control.

• Group I: Consisted of 10 patients with benign (Fibroadenoma) breast tumors.

• Group II: Consisted of 10 premenopausal patients with breast cancer (IDC).

• Group III: Consisted of 10 postmenpausal patients with breast cancer (IDC).

• Group IV: Consisted of 10 normal healthy subjects.

Blood samples were prepared as described in section (2.1.4). PBS buffer was prepared as described in section (2.1.6), while homogenization of breast tumor tissues was carried out as described in section (2.1.7). Statistical analysis was determined by student's t-test as mentioned in section (2.1.8).

6.2. Methods

6.2.1. Determination of CA19-9 Levels in Sera of Patients with Benign and Malignant Breast Tumors

Reagents

The reagents IRMA-ELSA CA19-9 Kit was provided from CIS-bio international ORIS Group/France.

1. Anti CA19-9 monoclonal antibody coated on the ELSA fixed in the bottom of the tube.

2. Anti 125I-CA19-9 monoclonal antibody, radioactivity content < 10 µCi (<370 KBq)

3. Six standard ready for use, Human serum, Human CA19-9 in sodium azide (0,14,30,66,130 and 255 U.mL-1).

4. Diluent (0.0 U.mL-1), human serum in sodium azide.

5. Control (35 U.mL-1), human serum, human CA19-9 in sodium azide.

Patients sera and control were used without dilution in this assay.

Procedure

The assay protocol is described in table (6-1).Table (6.1): IRMA protocol of serum CA19-9 (U.mL-1).

	CA19-9 (U.mL-1)						Control		Unknown Samples	
	0	1 4	3 0	6 6	13 0	25 5	Level I	Level II	1	2 etc.
Coated tube no.	1 ,2	3 ,4	5 ,6	7 ,8	9, 10	11, 12	13,14	15,16	17, 18	19, 20
Standards (µL)	←					100 □L			→	
Control serum or samples (µL)	←					100 □L			→	
Buffer (µL)	←					200 □L			→	
	Incubation for 3 h. at 37 oC in water bath									
	The solution was aspirated, and washed the tubes 3 times with 3 mL distilled water									
125I-anti CA19-9 (µL)	←					300 □L			→	
	All tubes were mixed gently with vortex-type mixer and									
	Incubated for 3 hrs. at room temperature (18-25 oC)									
	The solution was aspirated, the tubes were washed 3 times with 3 mL distilled water									
	The remaining bound radioactivity was measured with gamma counter.									

Calculations

1. The mean net count for each group of tubes was counted in gamma counter for 1 min, represents the bound c.p.m.

2. The standard curve was constructed by plotting counts per min. (Y-axis) versus concentration of CA19-9 standard (X-axis) figure (6.1). Then the points were connected with straight-line segments.

6.2.2. Preliminary Test of the Binding of CA19-9 in Breast Tumor Tissues with 125I-Anti CA19-9 Antibody in Breast tumors Homogenates

Reagents

Phosphate buffered saline pH 7.2 was prepared as described in section (2.1.6).

Procedure

The pellet and the cytosol fractions were obtained from the supernatant of breast homogenate were centrifuged at 4000 r.p.m. In order to detect CA19-9, 20 μL of crude cytosol fraction having 1100 μg protein were incubated with 60 μL (0.1356 mg.mL-1) of 125I-anti CA19-9 antibody. The volume of mixture was completed to 500 μL with PBS buffer pH 7.2, and then incubated at 37 oC for 3 hrs. The assay tubes were centrifuged at 4000 r.p.m. for 45 min. at 45 oC. The supernatant was discarded, the rims at tube were swabbed with cotton piece, then the complex formed was counted in gamma counter for 1 min. Pellet CA19-9 were determined by dissolving the sediment in PBS buffer pH 7.2 with ratio 1:5 (weight: volume), then 20 μL of supernatant fraction of pellet breast homogenate having 800 μg protein, was added to 60 μL (0.1356 mg.mL-1) of 125I-anti CA19-9 antibody. The same steps mentioned above were followed to determine the radioactivity of the complex formed. For total radioactivity two additional tubes with 60-μL of 125I-anti CA19-9 antibody were counted in gamma counter.

Calculations

1. The counted radioactivity in each tube (expressed in c.p.m.) represents the bound fraction (B); (i.e., 125I-anti CA19-9 antibody/CA19-9 complex).

2.   The counted radioactivity in the tubes counting 125I-anti CA19-9 antibody only represents the total radioactivity (T).

3.   The (B/T) % ratio for each tube was calculated as follows:

$$(B/T)\% = \frac{\text{Sample counts (B)}}{\text{Total counts(T)}} \times 100$$

6.2.3. Factors Effecting of 125I-Anti CA19-9 Antibody Binding to CA19-9 in Breast Tumors Homogenates

6.2.3.1. Effect of Protein Concentration on the Binding

Reagents

All reagents prepared is described in section (2.1.6) and (2.2.3.1).

Procedure

Sixty microliters (0.1356 mg.mL-1 protein) of 125I-anti CA19-9 antibody were added to 20 µL of cytosolic fraction of benign (Fibroadenoma) and malignant (premenopausal IDC and postmenopausal IDC) breast tumors respectively, containing increasing amounts of protein (50, 75, 100, 150, 200 and 250 µg.mL-1) and were completed to a final volume of 500 µL with 0.15 M PBS pH 7.2. The assay tubes were incubated for 3 hrs. at 37 oC. At the end of incubation, the assay tubes were centrifuged at 4000 r.p.m. for 45 min. at 4 oC. The supernatant was decanted; the rims at the tube were swabbed with cotton piece. The radioactivity of the complex formation was counted using gamma counter.

Calculations

1. The (B/T) % values were determined as in section (6.2.2).

2. Values of (B/T) % were plotted against their corresponding amount of protein of the breast tumor homogenate.

6.2.3.2. Effect of 125I-Anti CA19-9 Antibody Concentration on the Binding

Reagents

All reagents prepared is described in section (2.1.6) and (2.2.3.1).

Procedure

Sixty microliters of increasing amounts (0.0226, 0.0452, 0.0565, 0.113, 0.1356, 0.226 mg.mL-1) of 125I-anti CA19-9 antibody were added to 20 µL of crude cytosolic fraction (100, 75 and 75 µg protein) for benign (fibroadenoma) and malignant (premenopausal IDC and postmenopausal IDC) respectively, completed to a final volume 500 µL with 0.15 M PBS pH 7.2. After incubation for 3 hrs at 37 oC the bound CA19-9 was determined as mentioned in section (6.2.2).

Calculations

1. The (B/T) % values were determined as in section (6.2.2).

2. Values of (B/T) % were plotted versus the concentrations of 125I-anti CA19-9 included.

6.2.3.3. Effect of pH on the Binding

Reagents

All reagents prepared is described in section (2.1.6) and (2.2.3.1).

Procedure

Twenty microlites (100, 75 and 75 µg protein) of cytosolic fraction (fibroadenoma, premenopausal IDC and postmenopausal IDC respectively) were added to 25 µL (0.0565 mg.mL-1) of 125I-anti CA19-9 antibody respectively.The volume of the mixture was completed with PBS buffer of different pH (6.8, 7.0, 7.2, 7.4, 7.6, 7.8, 8 and 8.2) to a final volume 500 µL. After incubation for 3hrs at 37 oC, the bound CA19-9 was determined as mentioned in section (6.2.2).

Calculations

1. The (B/T) % values were determined as in section (6.2.2).

2. Values of (B/T) % were plotted versus the corresponding pH.

6.2.3.4. Effect of Temperature on the Binding

Reagents

All reagents prepared is described in section (2.1.6) and (2.2.3.1).

Procedure

Twenty microliters (100, 75 and 75 μg protein) of cytosolic fraction (Fibroadenoma , premenopausal IDC and postmenopausal IDC) were added to 25 μL (0.0565 mg.mL-1) of 125I-anti CA19-9 antibody respectively. The volume of mixture was completed to a final volume 500 μL with PBS buffer at pH 7.8 for fibroadenoma , pH 8.0 for premenopausal (IDC) and pH 7.0 for postmenopausal (IDC). The experiment was carried out at (5, 15, 25, 37 and 45oC) for 3hrs. After incubation the bound CA19-9 was determined as mentioned in section (6.2.2).

Calculations

1. The (B/T) % values were determined as in section (6.2.2).
2. Values of (B/T) % were plotted versus the temperature.

6.2.3.5 Effect of Incubation Time on the Binding

Reagents

All reagents prepared is described in section (2.1.6) and (2.2.3.1).

Procedure

Twenty microliters (100, 75 and 75 μg protein) of cytosolic fraction (fibroadenoma , premenopausal IDC and postmenopausal IDC) were added to 25 μL (0.0565 mg.mL-1) of 125I-anti CA19-9 antibody respectively. The reaction mixture was completed to a final volume 500 μL with PBS buffer pH (7.8 , 8.0 and 7.0) respectively. The experiment was carried out at 25 oC , 37 oC and 45 oC for fibroadenoma , premenopausal (IDC) and postmenopausal (IDC) respectively. The incubation was carried out at different time intervals (1, 2, 3, 4, 5 and 6 hrs). The bound CA19-9 was estimated as mentioned in section (6.2.2).

Calculations

1. The (B/T) % values were determined as in section (6.2.2).
2. Values of (B/T) % were plotted versus incubation time.

6.2.3.6. Effects of Different Halides on the Binding

Reagents

3. Phosphate buffer (PB) were prepared as described in section (2.1.6) without addition of NaCl .

4. Halid reagents were prepared in concentration of 0.01M PB at pH (7.8, 8.0 and 7.0) individually, by dissolving each of 0.021gm of NaF, 0.0292gm of NaCl, 0.0515gm of NaBr, and 0.075gm of NaI in a final volume 50mL of PB and the pH was adjusted.

5. The breast tumors homogenates (fibroadenoma , premenopausal IDC and postmenopausal IDC) were prepared as described in section (2.1.7), except using PB-buffer instead of PBS at the same pH and same concentration was carried out the homogenization.

Procedure

The experiment was carried out at optimum conditions as mentioned in section (6.2.3) using three groups of human breast homogenate (i.e., fibroadenoma, premenopausal IDC and postmenopausal IDC), by incubating 20 µL of the homogenate from each group containing (100, 75 and 75 µg protein) respectively with 25 µL (0.0565 mg.mL-1) of 125I-anti CA19-9 antibody. The reaction mixture was completed to a final volume 500 µL with PBS buffer pH (7.8, 8.0 and 7.0) containing 0.01 M of each of the following salts: NaF, NaCl, NaBr and NaI in each assay tube (A sample without the addition of any salt was used as a control). The assay tubes were incubated for (4,1 and 6 h) at 25 ,37 and 45oC for three group individually. The bound CA19-9 was estimated as mentioned in section (6.2.2).

Calculations

1. The (B/T) % values were determined as in section (6.2.2).

2. Values of (B/T) % were plotted versus 0.01 M of NaX.

6.2.3.7. Effects of Monovalent and Divalent Cations on the Binding

Reagents

1. Phosphate buffer (PB) were prepared as described in section (2.1.6) without addition of NaCl .

2. Monovalent and divalent cations (0.025 M) were prepared in PB buffer, and then the pH was adjusted to 7.8, 8.0 and 7.0 individually by dissolving each of 0.0931 gm of KCl, 0.0668 gm of NH4Cl, 0.2541 gm of MgCl2.6H2O, 0.1388 gm of CaCl2.2H2O, 0.2474gm of MnCl2.4H2O, 0.3150 gm of CuSO4.5H2O, 0.1703 gm of ZnCl2 , in a final volume 50 ml of PB and the pH was adjusted.

Procedure

The experiment was carried out at optimum conditions using three groups of human breast homogenate (i.e., fibroadenoma, premenopausal IDC and postmenopausal IDC) respectively.

The same steps mentioned on section (6.2.3.6) were followed to determine the effect of monovalent and divalent cations on the binding , except ; the buffer solution was PB (0.15 M) containing 0.025 M of the following salts: KCl , NH4Cl , MgCl2.6H2O, CaCl2.2H2O, MnCl2.4H2O, CuSO4.5H2O and ZnCl2.

Calculations

1.The (B/T) % values were determined as in section (6.2.2).

2.Values of (B/T) % were plotted versus the 0.025 M of monovalent and divalent cations.

6.2.3.8 Recovery of CA19-9

Reagents

All reagents prepared is described in section (2.1.6) and (2.2.3.1). Standared concentration of CA19-9 255 U.mL-1 was used.

Procedure

The experiment was carried out at optimum conditions. Known concentration of CA19-9 (255 U.mL-1) was added to the three group of benign (fibroadenoma) and malignant (premenopausal IDC and postmenopausal IDC) breast tissues homogenates. The experiment was carried ou at optimum conditions that was obtained in the experiment of section (6.2.3).

Calculations

1. The bound (c.p.m.) of the reaction mixture added to tissue homogenate with 125I-anti CA19-9 antibody, represent the measured value.

2. The bound (c.p.m.) of CA19-9 in tissue homogenate with 125I-anti CA19-9 antibody only , represent the expected value.

3. The recovery % (yield) calculated as follows:

$$Recovery\,\% = \frac{Measured\,values\,(c.p.m)}{Expected\,values\,(c.p.m)} x100$$

Results and Discussions

Determination of CA19-9 levels in Sera of Patients with Benign and Malignant Breast Tumors

Serum CA19-9 levels were measured by a solid-phase "sandwich" Immunoradiometric Assay (IRMA), which is specifically recognized by the anti CA19-9 monoclonal antibody.The monoclonal antibody is coated on the solid phase, or radiolabeled with the iodine 125 and used as a tracer. The radioactivity of the bound is directly proportional to the amount of CA19-9 presents at the beginning of the assay.

CA19-9 levels in sera of patients with benign breast tumors (group I) and (pre-and post-menopausal) malignant breast tumors (group II and group III) were measured by immunoradiometric assay. Three groups were matched with one group of control subjects. Table (6.2) shows the results obtained from this study. CA19-9 concentration of specimens and control were determined directly from standared curve in figure (6.1) .The level of serum CA19-9 in benign breast tumor patients was found to be 31.0 U.mL-1 ($p<0.05$), where that of (pre-and post-menopausal) malignant breast tumor patients were found to be 33.1 U.mL-1 ($p<0.05$) and 32.1 U.mL-1 ($p<0.0005$) respectively. While in control, the level was found to be 28.8 U.mL-1 .Matching case and control subject proved to be important for controlling undesired variability. The mean CA19-9 was significantly high in postmenopausal patients ($p<0.0005$) while in premenopausal and benign breast tumors the mean of CA19-9 was significantly low ($p < 0.05$ Student's t-test).

Table (6-2): Sera CA19-9 levels (U.mL-1) in patients with benign and malignant breast tumors. (All other details are explained in the text).

Group	Patients	No. of Cases	Age (year)	Serum CA19-9 U.mL-1 (mean ± SD)	P values
I	Benign breast tumors	10	18-35	31.0 ± 1.52	P<0.05
II	Premenopausal malignant breast tumors	10	35-43	33.1 ± 2.79	P<0.05
III	Postmenopausal malignant breast tumors	10	53-65	32.1 ± 0.13	P<0.0005
Control	Control	10	25-35	28.8 ± 0.631	

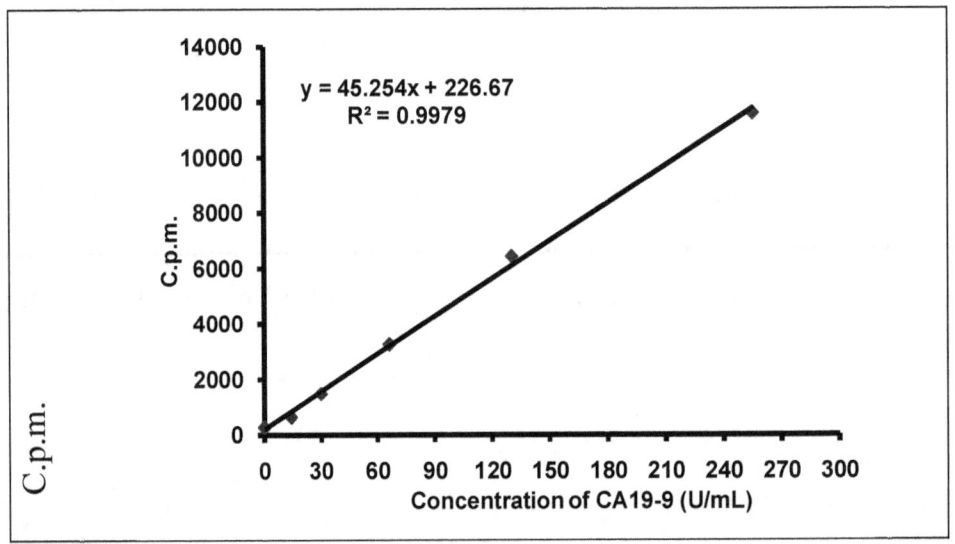

Figure (6.1): Standard curve of CA19-9. (All other details are explained in the text).

CA19-9 was at low concentration in sera of healthy individuals, these results are in agreement with several authors previously (120).

There were few studies to evaluate CA19-9 in breast tumors patients. Several investigators (222) detected CA19-9 in bone metastasis in breast cancer patients and in patients without documented metastases and reported that

CA19-9 level elevated in patients with metastases breast cancer.

When patients were analyzed with respect to the menopausal status, significant differences between the monastic and non monastic patients was detected (222).

Several studies proved the possibility of the role of carbohydrate antigen 19-9 as a tumor marker in colorectal cancer (132), pancreas (234), gastric (235), liver disease (236) and esophageal cancer (125). Recently, European group proved that CA19-9 monitored in patients with tumors of gastrointestinal tract and endometrial cancer could be used as a tumor marker and can be helpful (237) in monitoring patients with breast cancer. They observed significant increase of CA19-9 and CA15-3 levels (237) in all patients.

Preliminary Test of the Binding of CA19-9 with 125I-Anti CA19-9 Antibody

Supernatant and pellet obtained at speed (4000 r.p.m) were investigated in the three groups of human breast tumor homogenate (fibroadenoma, premenopausal IDC and postmenopausal IDC). In each fraction, CA19-9 was detected through the incubation of 125I-anti CA19-9 antibody with crude fraction supernatant and pellet individually for 3 h at 37oC in PBS buffer pH 7.2 as a medium to complete the reaction.

The separation of the bound antibody from unbound was carried out at 4000 r.p.m for 45 min. to precipitate the 125I-anti CA19-9 antibody/CA19-9 complex formed.

Table (6.3): Incidence of CA19-9 in supernatant and pellet fractions in three different breast homogenate. (All other details are explain in the text) .

Groups	Age(year)	B/T %	
		Supernatant fraction	Pellet fraction
Benign	34	5.32	1.43
Premenopausal (IDC)	43	5.48	2.03
Postmenopausal (IDC)	63	5.86	2.47

Table (6-3) shows the amount of binding B/T % values of pellet and supernatant fractions. The data revealed that CA19-9 in cytosolic fraction obtained from supernatant was higher in incidence than in pellet fraction, according to these results cytosolic fraction was collected. CA19-9 collected and the pellet was then discarded.

Factors Effecting of 125I-Anti CA19-9 Antibody Binding to CA19-9 in Breast Tumors Homogenates

Effect of Protein Concentration on the Binding

To obtain the optimum protein concentration of cytosolic fraction for the binding of CA19-9 with 125I-anti CA19-9 antibody, cytosolic fraction containing increasing amount of soluble CA19-9 in the presence of fixed amount of 125I-anti CA19-9 antibody was carried out as it was mentioned in section (6.2.3.1). Figure (6-2) represent the formation of (125I-anti CA19-9 antibody/CA19-9) complex in three cases (fibroadenoma, premenopausal IDC and postmenopausal IDC) and shows that (100, 75 and 75 µg protein) were the most appropriate concentration to give the maximum values of binding in crude fraction of three cases respectively. The decrease of the binding at high concentration of cytosolic fraction (in three cases) in the reaction mixture may be due to a conformational change in CA19-9 and 125I-anti CA19-9 antibody rather than the formation of reversible inactive

(125I-anti CA19-9 antibody/CA19-9) complex (238) and may be due to splitting antigen into large fragments with proteolytic enzymes (239).

In all subsequent experiments an amount of (100, 75 and 75 μg protein in three cases respectively), were used in the incubation mixture.

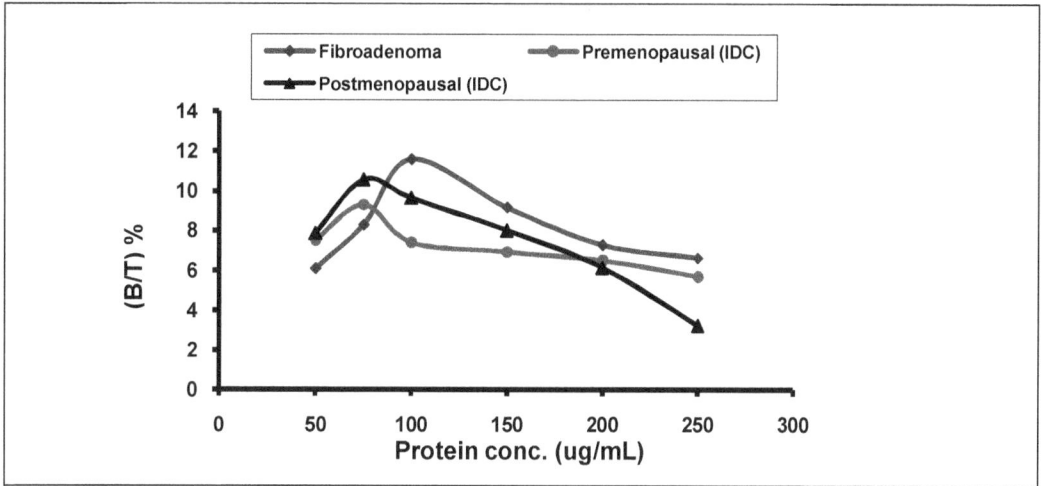

Figure (6.2): Influence of increasing protein concentrations on the binding of CA19-9 with 125I-anti CA19-9 antibody. (All other details are explained in the text).

Effect of 125I-Anti CA19-9 Antibody concentration on the Binding

One of the most important factors that effect binding is the concentration of 125I-anti CA19-9 antibody. To determine the suitable concentration of 125I-anti CA19-9 antibody, cytosolic sample (100, 75 and 75 μg protein) in the three cases (fibroadenoma, premenopausal IDC and postmenopausal IDC) respectively were incubated with increasing concentration of 125I-anti CA19-9 antibody, the incubation was carried out for 3 h at 37 oC. The results revealed that the optimum concentration of the 125I-anti CA19-9 antibody to give the maximum binding in all three cases was (0.0565 mg.mL-1). The results showed that an increase in the conc. of 125I-anti CA19-9 antibody caused a decrease in the binding %. This is because the soluble complexes, and the excess of antibody cover all antigentic sites, which leads to complex formation inhibition. Accordingly in all subsequent experiments, 0.0565 mg.mL-1 of 125I-anti CA19-9 was used as the optimum conc., which gives the highest binding %.

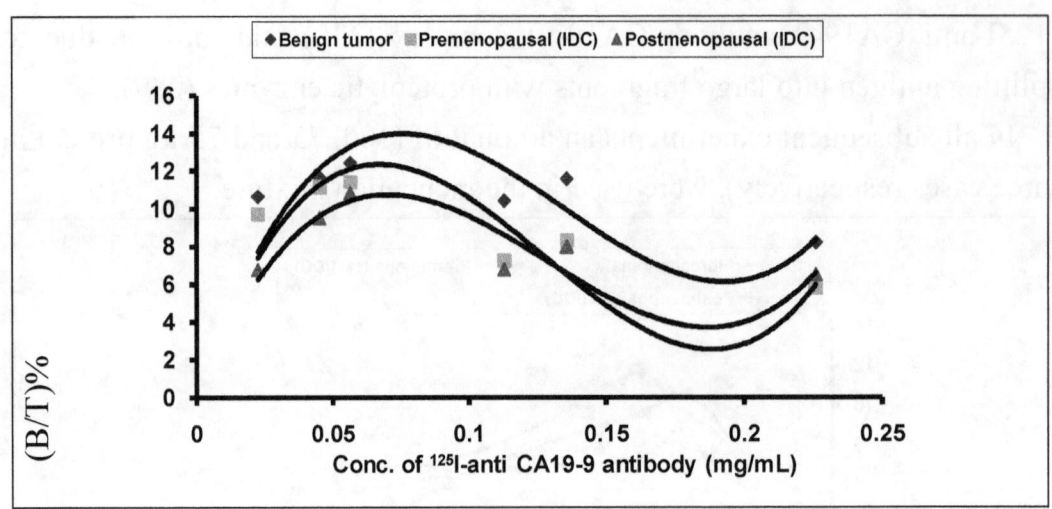

Figure (6.3): Effect of different concentration of 125I-anti CA19-9 antibody on the binding with CA19-9. (All other details are explained in the text).

Effect of pH on the Binding

The effect of pH on the binding of radioactivity CA19-9 to its antigen CA19-9 was investigated. Figure (6-4) shows that the maximum binding of 125I-anti CA19-9 antibody to its antigen CA19-9 was found to be (7.8, 8.0 and 7.0) in the three cases used (fibroadenoma , premenopausal IDC and postmenopausal IDC) respectively. The shift in pH of the environment may involve a protonation-deprotonation process occuring within the change of polar groups of the amino acids residues present in the binding domain (240) .According to these results, the pH of the buffer used in all subsequent experiments were (7.8, 8.0 and 7.0) for the three cases respectively.

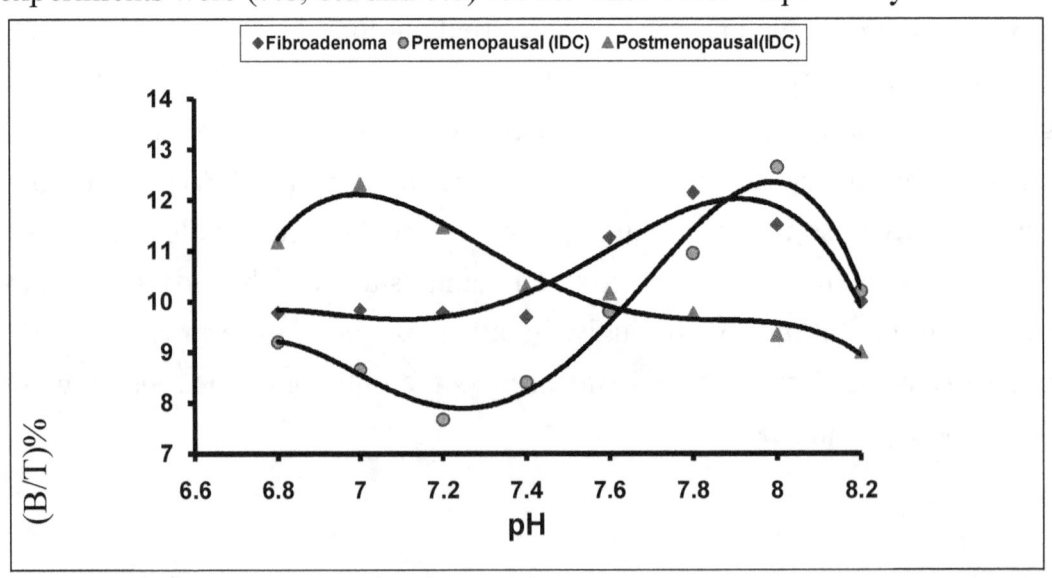

Figure (6-4): The effect of pH on the binding of CA19-9 with its antibody 125I-anti CA19-9 antibody with CA19-9. (All other details are explained in the text).

Effect of Temperature on the Binding

Temperature dependency of the association of 125I-anti CA19-9 antibody to its cytosolic fraction CA19-9 was investigated. Cytosol fraction of benign and malignant breast tumors was incubated for 3 hrs at different temperatures (5, 15, 25, 37 and 45 oC). Figure (6-5) reveals that the binding of 125I-anti CA19-9 antibody to its cytosol fraction CA19-9 was increased when the temperature was raised from 5 to 25 oC in fibroadenoma and the maximal binding was obtained at 25 oC and from 5 to 37 oC in premenopausal (IDC) and the maximal binding was obtained at 37 oC. Finally from 5 to 45 oC in postmenopausal (IDC) and the maximal binding was obtained at 45 oC. The decrease in the binding at temperature higher than the optimum temperature is probably due to denaturation of CA19-9 molecules (241) or due to proteolytic degradation of enzyme (150). According to these results (25 oC , 37 oC and 45 oC) respectively they will be used in all the subsequent experiments for the three cases used.

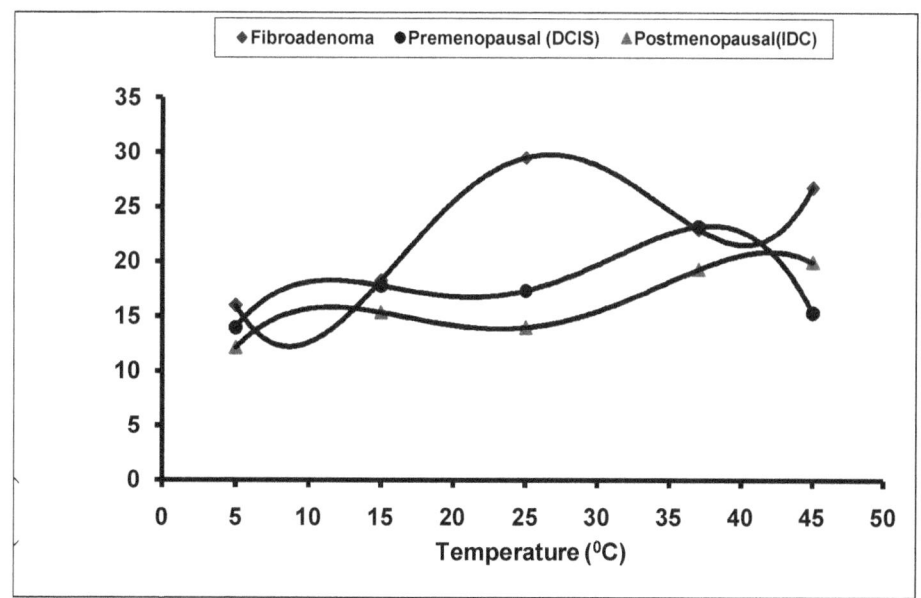

Figure (6-5): Effect of temperature on the binding of 125I-anti CA19-9 antibody with CA19-9. (All other details are explained in the text).

The Effect of Incubation Time on the Binding

To choose the most appropriate incubation time at (25, 37 and 45 oC) for the three cases used in this study (fibroadenoma, premenopausal IDC and postmenopausal IDC) respectively , the experiments were carried out at different time intervals. Figure (6-6) shows the results of this analysis. It seemed that the specific binding of 125I-anti CA19-9 antibody to cytosolic fraction homogenate for the three cases were maximal at (4,1 and 6 hrs) respectively. In view of these results, the incubation time used in all subsequent experiments were (4,1 and 6 hrs) respectively.

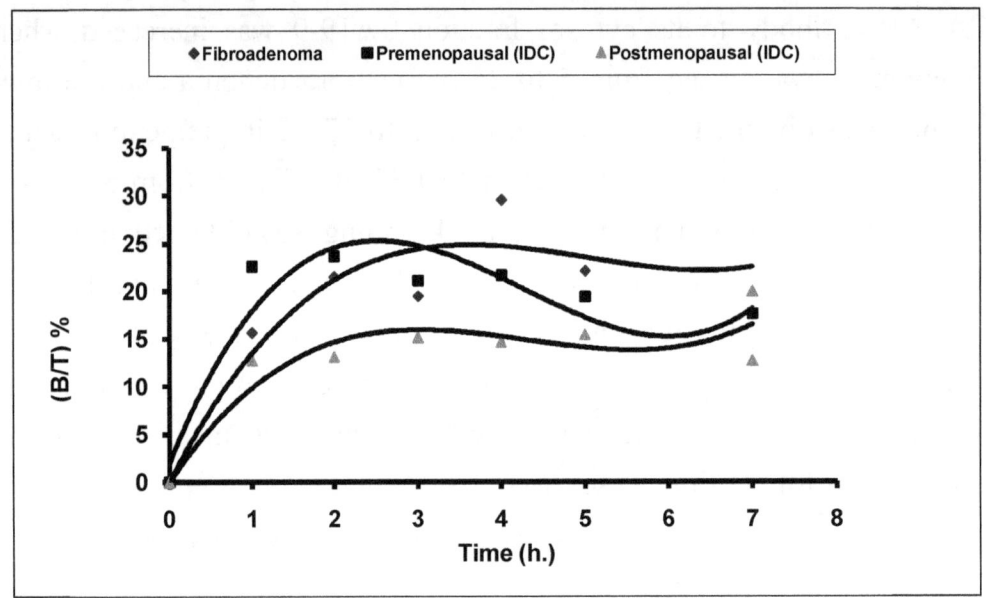

Figure (6.6): The effect of incubation time on the binding of 125I-anti CA19-9 antibody with CA19-9. (All other details are explained in the text).

Effect of Different Halides on the Binding

Figure (6.7) shows the effect of different halides salts (i.e., NaF, NaCl, NaBr and NaI) at 0.01 M concentration on the extent of 125I-anti CA19-9 antibody binding to their cytosol fraction homogenate in benign and malignant breast tumors. The sodium halides (ion radius) in the incubation mixture of benign and postmenopausal malignant breast tumors induced inhibition of the percent of binding according to the following sequence:

NaI>NaF>NaCl>NaBr

While the sodium halides in the incubation mixture of premenopausal malignant breast tumor (IDC) induced activation of the percent of the binding in the order:

NaF<NaCl<NaI<NaBr

164

Melander and Horvath (1977) reported that the effect of halide salt type on hydrophabic interactions is quantified by its molar surface tension increment (MSTI) which is a measure of the increasing in a surface tension by the salt (171) , also they found that parameter increases as the following sequence:

NaF>NaCl>NaI

The same researches found that halides with higher MSTI values will strengthen the hydrophabic interactions while halides with lower MSTI values reverse this effect. Thus the dependence of the extent of the binding in benign and malignant (pre-and post-menopausal) breast tumors on MSTI values of the corresponding halide further implicates the low involvement of hydrphobic forces in maintaining the stability of (125I-anti CA19-9 antibody /CA19-9) complex formed.

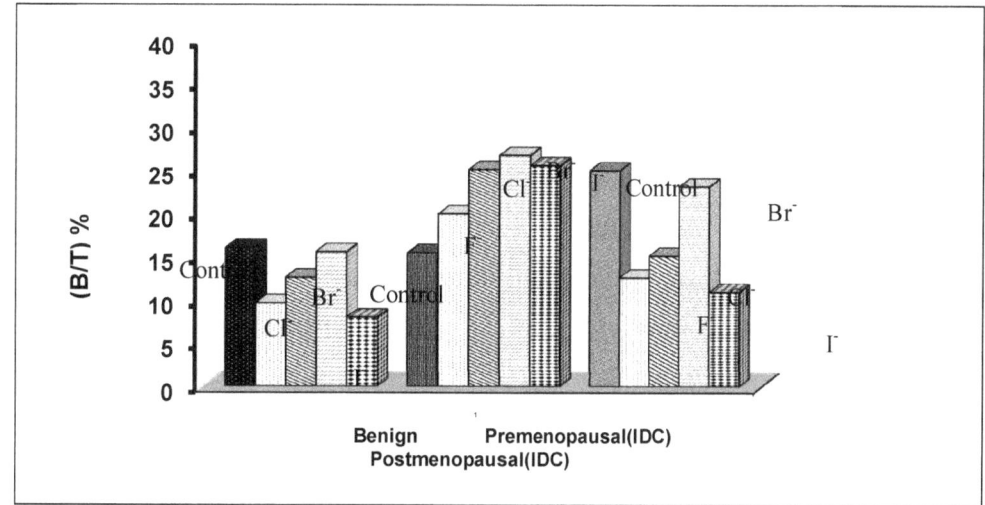

Figure (6-7): Effect of different halides on the binding of of 125I-anti CA19-9 antibody with CA19-9. (All other details are explained in the text).

Effect of Monovalent and Divalent Cations on the binding

Figure (6.8) and (6.9) show the effect of different divalent and monovalent cations respectively on the binding value in benign and malignant breast tumors. The results indicate that the binding process is sensitive to the presence of cation metal ions. CuSO4.5H2O at concentration 25 mM was showed to increase the binding two folds than the control as compared with other divalent cations.

CaCl2.2H2O induced activation in the binding in benign (fibroadenoma) and malignant (premenopausal IDC), while induced inhibition in the binding in malignant (postmenopausal IDC). ZnCl2 decreased the binding in two groups (fibroadenoma and premenopausal IDC) , while ZnCl2 increased the binding in malignant (postmenopausal IDC).

The frequency of the stimualtion of the binding of 125I-anti CA19-9 antibody to its cytosolic fraction CA19-9 homogenate of the three groups by divalent cations is according to the following:

Postmenopausal breast cancer tissue homogenate (IDC)

$Cu^{+2} > Zn^{+2} > Mn^{+2} > Mg^{+2} > Ca^{+2}$

Premenopausal breast cancer tissue homogenate (IDC)

$Cu^{+2} > Ca^{+2} > Mn^{+2} > Mg^{+2} > Zn^{+2}$

Benign breast tumor tissue homogenate (Fibroadenoma)

$Cu^{+2} > Ca^{+2} > Mg^{+2} > Mn^{+2} > Zn^{+2}$

The binding of metal ions to proteins is a function of pH among the different classes of groups, such as carboxyl, amino, imidozol and tyrosyl (the unshared electron pairs for nitrogen , oxygen and sulfur atoms) (242). The sites of binding of metal ions may range from elaborate chelate sites to simple complex formation which discrete single ligand groups in the protein. In short, chelation plays a dominant role in establishing the relative strengths of binding of a given metal ion by various sites in protein (243).

Figure (6-9) shows that monovalent cations inhibit the binding in benign and malignant premenopausal (IDC), while the monovalent cations induce activation of the binding in-group of malignant postmenopausal (IDC). The alternation of increased and decreased binding percent between these cations may be ascribed to the differences in tissues studied (244). The variation of results obtained between these divalent cations may be ascribed to the difference in tissue studied (245) .

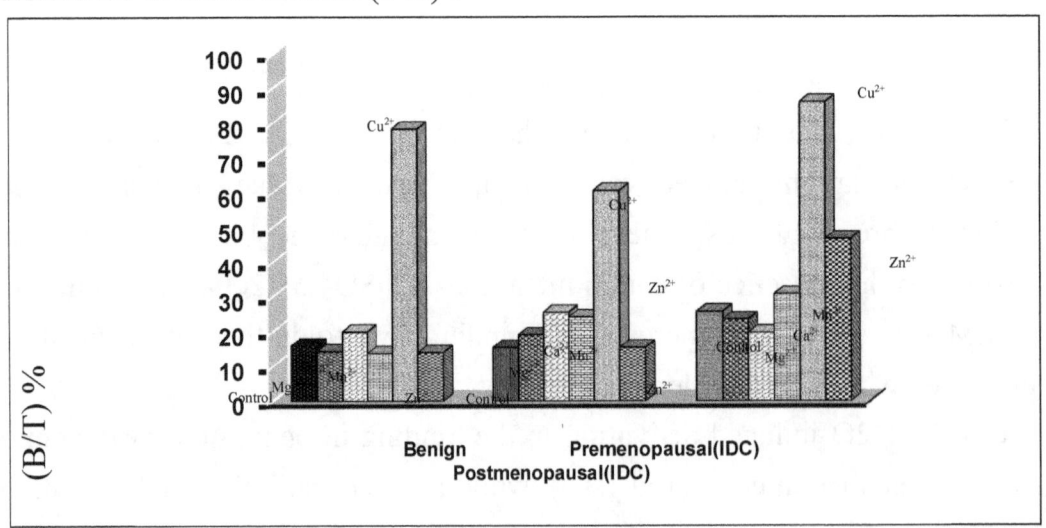

Figure (6.8): Effect of different cations on the binding of 125I-anti CA19-9 antibody with CA19-9 in different human breast tumor homogenate. (All other details are explained in the text).

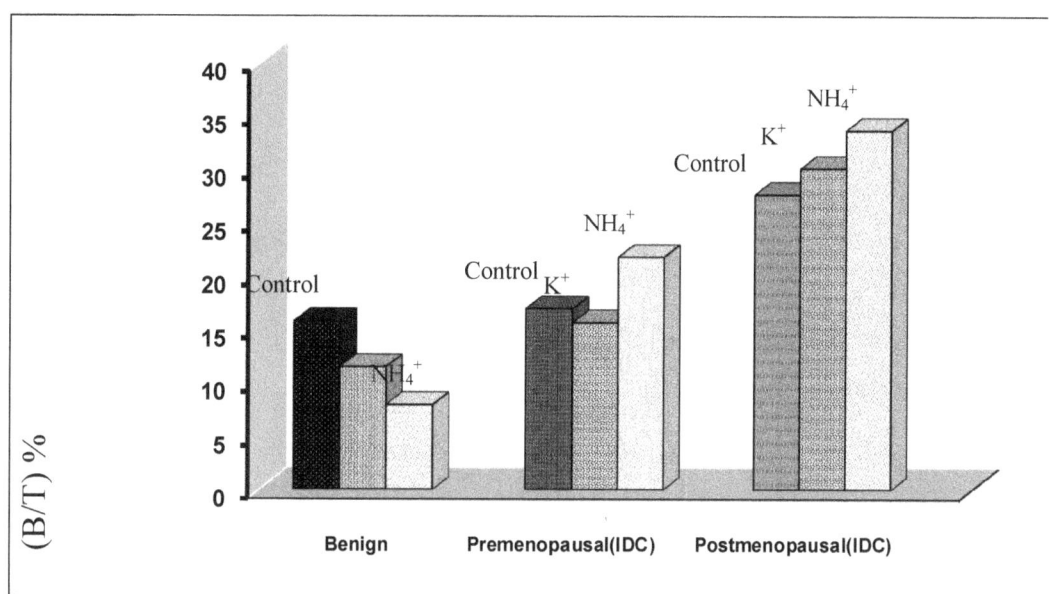

Figure (6.9): Effect of different monovaleat cations on the binding of 125I-anti CA19-9 antibody with CA19-9 in different human breast tumor homogenate. (All other details are explained in the text).

Recovery of CA19-9

The method used to estimate the percent recovery of cytosolic fractions of benign (Fibroadenoma) and malignant (pre-and post-menopausal IDC) breast tumors homogenates. The results summarized in table (6.4) indicate that CA19-9 extracted from malignant breast tumors homogenates was recovered less than CA19-9 extracted from benign breast tumor (fibroadenoma) and CA19-9 extracted from malignant premenopausal (IDC) malignant breast tumors homogenates was recovered more than CA19-9 extracted from postmenopausal (IDC) malignant breast tumors homogenates. Also the results indicate that total CA19-9 could determine through the developed method of immunoradiometric assay. The percent of recovery indicates the precision of the used method.

Table (6.4): Recovery of CA19-9 (AII details are explained in the text).

Type of CA19-9	Measured B/T %	Expected B/T %	Recovery % Measured/ Expected %
Benign	103	104	99
Premenopausal	192	195	98
Postmenopausal	166	175	95

Conclusion

1. CA15-3) complex in the two cases benign and malignant breast tissues have a characteristics spectrum.

2. The developed protocol for the assay of CA15-3 or CA19-9 is suitable for the assessment of CA15-3 or CA19-9 in tissues.

3. The magnitude of elevation in CA15-3 is more than CA19-9 in the same sera of patients with breast tumors. This indicates that CA15-3 as a tumor marker is more specific than CA19-9.

4. Partially purified CA15-3 from benign and malignant breast tumors homogenates shows high affinity to 125I-anti CA15-3 antibody than crude CA15-3.

5. Kinetic studies of 125I-anti CA15-3 antibody with partially purified CA15-3 in benign and malignant breast tissues homogenates show that the reaction is temperature and time dependent. Pseudo first order kinetics at (5, 15, 25, 37, and 45 oC) was observed, in all cases.

6. The results obtained from the thermodynamic studies on the association of 125I-anti CA15-3 antibody with partially purified CA15-3 in benign and malignant breast tissues homogenates indicate that the binding reaction occurs spontaneously $\Box G° < 0$, and entropically driven since $\Box S° > 0$.

References

1. Porth, C.M.; (1994). "Path physiology", 4th ed. Philadelphia. J.B. Lippincott.

2. Berkow, R.; Beers M.H; Fletcher A.J.; Bogin R.M.; (2001). "The Merck manual of medical information". Home ed. ht, Whitehouse Station. Merck & Co., Inc.; Chapter 238, Section 22.

3. World Health Organization: Hostological typing of breast tumors. In: International Histological Classification of Tumors; (1981). 2nd ed. Geneva, World Health Organization P19.

4. Cotran, R.S.; Kumar V.; Collins T.; (2000). "Robbins pathologic basic of disease", 6th ed. Philadelphia. W.B. Saunders Company. Chapter 25, pp 1102-1107.

5. Dixon, J. M. ; Mansel R.E.; (1995). "Congential proplems and aberrations of normal breast development and involution", In: Dixon J. M Jed) ABC of Breast Diseases. BMJ Publishing Group, London, Chapter1, pp6-11.

6. Flectcher, J.A; Pinkus G.S; Weidner N., Morton C.C.; (1991). Am. J. Pathaol., 138: 1199-1207.

7. Dupont, W.d., Page D.L., Parl F.F. (1994). N. Engl. J. Med., 311: 10-15.

8. Burnet, K.L.; (2001). "Holistic breast care", 1st ed. Harcourt Publishers Limited. Chapter 2, p23.

9. Belieu, R.M.; (1994). Obstet. Gynecol. Clin. North Am., 21:461.

10. Rosai, J.; (1996). "Ackerman's surgical pathology", 8th ed. Philadelphia. Mosby Yer Book, St. Louis, Vol 2, pp: 1565-1568.

11.Casciato, D.A; Lowitz B.B.; (2000). "Manual of clinical oncology" 4th ed. Philadelphia. Awolters Kluwer Company. Chapter 10, pp218-237.

12.Rubin, P.; Williams J.P; (2001). "Clinical oncology, a multidisciplinary approach for physical and students", 8th ed. Philadelphia. W.B.Saunders Company. Chapter 17, pp 267-299.

13.Iraqi cancer Registry (ICR) Results; (1993). Ministry of Health, (1989-1991). Iraqi Cancer Borad.

14.Iraqi cancer Registry (ICR) Results; (1995). Ministry of Health, (1992-1994). Iraqi Cancer Borad.

15. Iraqi cancer Registry (ICR) Results;(1998). Ministry of Health, (1995-1997). Iraqi Cancer Borad.

16. Iraqi cancer Registry (ICR) Results;(2000). Ministry of Health, Iraqi Cancer Borad.Unpublished.

17. Pazdur, R.; Coia L.R.; Hoskins W.J.; Wagman L.D.; (1996). "Cancer management: a multidisciplinary approach, medical, suergical & radiation oncology", 1st ed. Princeton. Bristol-Myers Squibb Company, Chapter2.p 22.

18. Osborne, C.K.; (1994). "Internal medicine", 4th ed. Philadelphia. Mosby, St. Louis, pp 924-926.

19. Miller, A.; Howe G.; Sherman G.; (1989). N. Eng. J. Med., 321: 328.

20. Lata C.; (2000) Cancer. http://www.vhihealthe.com

21. Kohlmeier, L. ; Mendez M.; (1997). Proc. Nutr. Soc., 56: 369.

22. Sala, E.; Warren R.; Duffy S.; (2000). Br. J. Cancer, 83(1): 121-6.

23. Fiedenreich, G.; Howe G.; Miller A.; (1993). Am. J. Epidemiol., 137:512.

24. Williams, E.L.; (1994). Am. J. Epidemiol., 140: 956-958.

25. Newcomb, P.; Storer B.; Longnecker M.; (1994). N. Eng. J. Med., 330:81.

26. Landis, A.; Murray D.; Bolden M.; (1999). Cancer Statistis, 49:1.

27. Frykery, E.R.; Masood S.; Copeland E.M.; (1993). Surg. Gynecol. Obstet, 177: 425-440.

28. Wood, W.C.; (1996). Semin. Oncol., 23:446-452.

29. Goldschmidt, R.A, Victor T.A.; (1996). Semin. Surg. Oncol., 12: 314-320.

30. Garne, J. P.; (1994). Cancer, 73:1438.

31. Di Costanzo,D.; Rosen P.P.; Gareen I.; (1990). Am. J. Surg. Pathol., 14:12-23.

32. Cavalli, F.; Hansen H.H.; Kaye S.B.; (1997). "Text book of medical oncology", 1st ed., London, Martin Dunitz Ltd, Chapter 4, p58.

33. Mann, C.V.; Russel R.C.G.; William N.S.; (1995). "Baily & Love's short practice of surgery", 22nd ed. London, Chapmann & Hall Medicine, Chapter 39, p543.

34. Giuliano, A.E.; (1995). Ann. Surg., 222: 394.

35. Tiernery, L.M.; Mc Phee S. J.; Papadakis M.A.; (1999). "A large medical book, current medical diagnosis & treatment", 38th ed. New Yourk. Appleton & Lange, Chapter 16, pp678-703.

36. Hellman, S.; (1994). J. Clin. Oncol., 12: 2229-2234.

37. Fisher, B.; Redmond C.; (1992). J. NCI., 11:7-13.

38. Veronesi, U et al; (1995), 222:612.

39. Devita, J.R.; Vincent T.; Sanmel H.; Hellman S.; Rosenberg S.A.; (1997). "Cancer principle and proactice of oncology", 5th ed. Philadelphia, J.B. Lippincott-Raven publishers, p1562.

40. Cooke, A.L.; (1995). Int. J. Radiat. Oncol. Bio. Phys., 31:777.

41. Guenther, J.M.; Tokita D.M.; Guiliano A.K.; (1994). Cancer, 73:2613

42. Haagensen, CD.; (1997).Int. Radiat Oncol. Biol. Phys., 2: 975-980.

43. Valero, V.; (1997). Oncology, 11:34.

44. Cameron, D.A.; Gabra H.; Leonard R.C.; (1994). Br. J. Cancer, 70: 120-124.

45. Christman, K., Muss H.B., Stanley V.; (1992), JAMA, 268: 57-62.

46. Muss, H.B., (1995). Semin. Oncol., 22:14-16.

47. Bondaontra, G., Zambetti M., Valagussa P., (1995), JAMA, 542-547.

48. Powels, T.J.; (1997), Semi Oncol; 24 (suppl): 48-54.

49. Sundeland, M.C., Osborne C.K.; (1991). J. Clin. Oncol.,9:1283-1297.

50. Russel, R.C.G.; Williams N.S; Bulstored C.T.K; (2000). "Bailey & Love's short practice of surgery" 23rd. arnold, a member of the holder healine Group. Chapter 46, PP 750-759.

51. Burtis, C.A.; Edward R.; Ashwood E.R.; (1999). "Tietz text book of clinical chemistry", 3rd ed. Philadelphia, W.B. Saunders Company. Chapter 23, pp 722-748.

52. Anderson, S.C.; Cockayne S.; (1993). "Clinicla chemistry, concepts and applications", An HBJ☐int.ed. Philadelphia, W.B. Saunders company. Chapter 16, pp 322-330.

53. Kalstan, M. B.; Onyekwere O.; Sidransky D.; Vogelstein B.; Craig R.W.; (1991). Cancer Res., 51: 6304.

54. Burtis, C.A.; Ashwood E.R.; (1994). "Tietz textbook of clinical chemistry", 2nd ed. Philadelphia, W.B. Saunders company. Chapter 21, pp 899-920.

55. Eurepean Group of Tumors Markers (EGTM). http://www.med.uni.muenchen.de/egtm/detail/4.htm.

56. Stearns, V.; Yamauchi H.; Hayes D.F.; (1998). Breast Cancer Res. Treat., 52: 239-259.

57. Schwartz, M. k.; (1994). Adv. Exp. Med. biol., 353: 47-53.

58. Hayes, D.F.; (1994). "Tumor markers for breast cancer: Current utilities and future prospects", In: Hayes DF (ed) Hematology/ oncology clinics of North Amirica: philadelphia, W.B. Sounders Company, pp 485506.

59. Safi F.; Kohler I.; Rottinger E.; Boger H.G.; (1990). Cancer, 68 (3): 574-582.

60. Ordonez, N.G.; (1989). Cancer Bull, 41: 142-151.

61. Sell,S.; (1990). Human Pathol., 21: 1003-19.

62. Soletormos, G.; Nielsen D., Schioler V.; (1996). Clin. Chem., 42: 564-575.

63. Bjorklond, B.; Bjorklund V.; (1957). Arch. Allergy, 10: 153-184.

64. Karaloglu, D.V.; Yasusever N., Dalay E.N; (1995). Eur. J. □□Gynaec. Oncol., 5:363-367.

65. Blijlevens, N.M.A.; Oosterhuis W.P.; Oosten H.R.; Mulder N.H.; (1995). Anticancer Res., 15: 2711-2716.

66. Schuurman, J.J.; Bong S.B.; Einarsson R.; (1996). Anticancer Res., 16: 2169-2172.

67. Locker, G. J.; Mader R.M.; Braun J.; Sieder N.E.; Marosi C.; Rainer H.; Jakesz R.; Steger G.G.; (1995). Oncology, 52:140-144.

68. Gion, M.; Mione R.; Leon A.E.; Dittadi R.; (1999). Clin. Chem., 45 (5): 630-637.

69. Correale, M.; Abbate I.; Gargano G.; (1992). In . J. Biol. Markers, 7: 43-46.

70. Jcobs, I.; Bast R.C.; (1989). Hum. Reprod., 4: 1-12.

71. Helzlsouer, K. D.; Bush T.L.; Albrg A. J.; Bass K.M.; Zacur H.; Comstock G.W.; (1993). J. Am. Med. Assn., 269: 1123-1126.

72. Kaneko, J.J.; harvey J.W.; Bruss M.J.; (1997). "Clinical biochemistry of domestic animals", 5th ed. New York, Academic Press, Chapter 26, p764.

73. Ward, B.G.; Mc Guckin M.A.; Rarnm L.E.; Coglan M.; (1993). Cancer, 71: 430-438.

74. Chatterjea, B.M.N.; Shinde R.; (1995). "Textbook of medical biochemistry" 2nd ed. India, Medical Publishers (P) LTD. Chapter 41, part III, pp1088-1098.

75.Yasasever, V.; karaloglu D.; Erturk N.; Dalay N.; (1994). Eur. J. Gynaec Oncol., 1: 33-36.

76.Hilkens, J., Kroezen V.; Bonfrer J.M.G.; Bakker M. D.; Bruning P.F.; (1986). Cancer Res., 46: 2582-2587.

77.Hilkens, J.; Buijs, F; (1988). J. Biol. Chem., 263(9): 4215-4222.

78.Galectine-4.   http:// www.albany.edu/chemistry/sarma/cwabstiactsmar98.htm.

79.The American Society of Clinical Oncology; (1996). J. Clin Oncol, 14(10): 2843-2877.

80.Tandon, A.K.; Clark G.M.; Chamness G.C.; (1990). N. Engl. J. Med., 322: 297-302.

81."http://www.asco.org/prof/pp/html/guide/tumor/m-tumor 15. htm"

82.Hilkens, J.; Buijs F.; Hilgers J.; Hayeman P.; Calofat J.; Sonnenberg A; (1984). Int .J. Cancer, 34: 197-206.

83.Thomas, C.M.G.; (1995). Ned. Tijdschr. Klin. Chem; 20 (6): 298-300.

84.SaFi, F.; Kohler I.; Rottinger E.; Beger H.G.; (1991). Cancer, 68(3): 574-582.

85.Sekine, H.; Ohno T.; Kufe D.W.; (1985). J. Immunol., 135: 3610-3615.

86.Shimizu, M.; Yamauchi K.; (1982). J. Biochem., 91: 515-524.

87.Gendler, S.; Papadimitrion J.T.; Duhig T.,; Rothbard J.; Burchell J.; (1988). J. Biol. Chem., 263: 1280-12823.

88.Cordell, J.; Richardson T.C.; Pulford K.A.; Ghosh A.K.; Gatter K.C.; Heyderman E.; Mason D.Y.; (1985). Br. J. Cancer, 52: 347-354.

89.Wesseling, J.; (1997). "The Carcinoma- Associated Mucin Episialin (MUC1) in Cell Adhesion and Tumor Progression" Ph. D. Thesis, Department of Tumor Biology, The Nether lands Cancer Institute.

90.Hou, M.F.; Huang T.J.; Hsieh J.S.; Huang Y.S.; Huang C.J.; Chan H.M.; Wang J.Y.; Chen Y.L.; Jong S.B.; Yang C.C.; (1995). Kaohsiung J. Med. Sci.; 11: 660-666.

91.Wojtacki, J.; Bokiniec A.D.; Kowalski D.M.; Zoltowska A.; Cresielski D.; Suszko M.; (1996). Neoplasma, 43 (4): 225-229.

92.Kufe, D.; Inghirami G.; Abe M.; Hayes D.; Justiwheeler H.; Scholm J.; (1984). Hybridoma, 3: 223-232.

93.Ligtenberg, M. J. L.; Vos H.L.; Genissen A.M.C.; Hilkens J.; (1990). J. Biol. Chem., 265: 5573-5578.

94. Swallow, D.M.; Gendler S.; Griffiths B.; Corney G.; Taylor P.J.; Bramwell M.E.; (1987). Nature, 328: 82-84.

95. Jentoft, N.; (1990). Trends. Biochem. Sci., 15:291-294.

96. Hilkens, J.; Wesseling J.; Vos H.L.; Storm J.; Van der Valk S.W.; Maas M.C.E.; (1995). Ned. Tijdschr. Klin. Chem., 20 (6): 293-298.

97. Hilkens, J.; Ligtenberg M.J.L; Vos H.L; Litvinov S.V.; (1992). Trends. Biochem. Sci., 17:359-363.

98. Tsubura, a.; Morii S.; Vdea S.; Sasaki M.; Zother S.; Watzing V.; Mooi W.; Hageman P.C.; Hilkens J.; Tweel J.V.; (1987). Arch. Dermatol. Res., 279:550-557.

99. Zoher, S.; Lossnitzer A.; Hageman P.C.; Delemarre J.G.; Hilkens J.; Hilgers J.; (1987). Lab. Invest., 57: 193-199.

100. Zotter, S.; Hageman P.C.; Lossnitzer A.; Mooi W.; Hilgers J.; (1988). Cancer Rev., 11-12: 55-101.

101. Zaretsky, J.Z.; Weiss M.; Tsarfaty I.; Hareuveni M.; Wreschner D.H.; Keydar I.; (1990). FEBS Lett; 265: 46-50.

102. Hilkens, J.; Buys F.; (1988). J. Biol. Chem., 263: 4215-4222.

103. Linsley, P.S.; Kallestadt J.C.; Horn D.; (1988). J.Biol.Chem., 263: 8390-8397.

104. Ligtenberg, M.J.L.; Buijs F.; Vos H.L.; Hilkens J.; (1992). Cancer Res., 52: 2318-2324.

105. Special, Article by the American Society of Clinical Oncology; (1996). J. Clin. Oncol., 14(10): 2843-2877.

106. Hilkens, J.; Ligtenberg M.J.L.; Vos H.L.; Litvinov S.V.; (1992). Trends. Biochem. Sci., 17: 359-363.

107. Braga, V.M.; Pew berton L.F.; Duhig T.; Gendler S.J; (1992). Development; 115: 427-437.

108. Vande Wiel-van Kemenade, E.; Ligtenberg M.J.L.; de Boer A.J; Buijs F.; Vos H.L.; Melief C.J.M; Hilkens J.; Figdor C.G.; (1993). J. Immunol., 151: 767-776.

109. Jerome, K.R.; Barnd D.L.; Bendt K.M.; Boyer C.M.; Taylor Papadimitriou J.; Mckenzie I.F.C; Bast R.C.; Finn O.J.; (1991). Cancer Res., 51: 2908-2916.

110. Jerome, K.R.; Domenech N.; Finn O.J.; (1993). J. Immunol., 151: 1654-1662.

111.   Ioannides, C.G.; Risk B.; Jerome K.R.; Irimura T.; Warton J.T.; Finn O.J.; (1993). J. Immunol., 151: 3693-3703.

112.   Takahashi, T; Makiguchi Y.; Hinoda Y.; Kakiuchi H.; Nakagawa N.; Imai K.; Yachi A.; (1994). J. Immunol., 153: 2102-2109.

113.   Hayes, K.F.; Tondini C.; Kufe D.W.; (1992). "Clinical applications of CA 15-3 In: Serological tumor markers". S.Sell. ed. Totowa, NJ, The humana Press, pp.281-307.

114.   Lamerz, R.; (1992). "CA 19-9 gastrointestinal cancer antigen In: serological cancer Markers" S.Sell. ed, Totowa, NJ, The Humana Press, pp 309-339.

115.   Takasaki, H.; Uchida E.; (1987). Pancreas, 2: 398-403.

116.   Furuya, N.; Kawa S.; Hasebe O.; Tokoo M.; Mukawa K.; Mukawa K.; Macjima S.; Oguchi H.; (1996). Br. J. Cancer, 73 (3): 372-376.

117.   Webb, A.; Scott M. P. and Bensted J.; (1996), Eur. J. Cancer; 23A(1): 63-68.

118.   Koprawski H.; Steplewski Z.; Mitchell K. and Heryn D.; (1979). Somatic Cell Cent.; 5: 957-972.

119.   Magnani, J.L.; Nilsson B.; Brockhaus M.; Zopf D.; Steplewsk L.; koprowski H.; Ginsburg V.; (1982). J. Biol. Chem; 257: 14365-14369.

120.   Del-Villano, B.C.; Brennan S.; Brock P.; Bucher C.; Liv V.; McClure M.; Rake B.; Space S.; Westrick B.; Schoemarker H.; Zurawski V.R.; (1983). Clin. Chem., 29 (3): 549-552.

121.   Ritts, R.I. ; Del-Villano B.G.; GoV L.W.; (1984). Int.J. Cancer,33:339-345.

122.   Beretta E. ; malesci A. ; Zerbi A.; (1987). Cancer, 60: 2428-2431.

123.   Glenn, J.;Steinberg W.M.; Kurtzman S.H.; (1988). J. Clin. Oncol., 6: 462- 468.

124.   Taniguchi, T. ; Kitamura M. ; lwasaki Y. ; Yamamoto Y. ; Igar: A. ; Toi M.; (1997). Br.J. Cancer, 75 (5): 673-677.

125.   Mcknight, A.; Mannell A.; Shperling I.; (1989). Br.J. Cancer, 60: 249-251.

126.   Maeta, M.; Yoshioka H.; Shimizu T.; Murakami A.; Hamazoe R.; koga S.; (1990). Oncology, 47(3): 229-233.

127.   Gupta, M.K.; Arciaga R. ; Bocci L.; (1985). Cancer, 56: 277-283.

128. Putzki, H.; Ledwoch J.; Student A.; (1988). J. Surg. Oncol., 37: 133-135.

129. Ohuchi, N.; Takahashi K.; Matoba N.; (1989). Jpn. J. Clin. Oncol., 19: 242-248.

130. Thomas, W.M.; Robertson J. F.R.; Price M.R.; (1991). Br. J. Cancer., 63: 975-976.

131. Iemura, k..; Moriya Y.; (1993). Eur. J. Surg. Oncol., 19: 439-422.

132. Kouri, M.; Pyrohonen S.; Kuusela P.; (1992). J. Surg. Oncol., 49: 78-85.

133. Burioka, N.; Suyama S.; Tatsukawa T.; Hori S.; Kometani Y.; Kawasaki Y.; Nakada N.; Sasaki T.; (1997). Yonago. Acta. Medica., 40: 147-151.

134. Steinberg, W.; (1990). Am J. Gastroenterol., 85: 350-355.

135. Sherif, M.S.; A.Razek A.A.H.; (1994). The New Egyp. J. Med., 10(4): 1821-1825.

136. Yasasever, V.; Dincer M.; Camlica H.; Karaloglu D.; Dalay N; (1997). Clin. Boichem., 30 (1): 53-56.

137. Wojacki, J.; Bokiniec A.D.; (1995). Libri. Oncol., 24: 147-152.

138. Van-Dalen, A.; (1999). Tumour Biol., 20 (3): 117-129.

139. Schuurman, J.J.; Bong S.B.; Einarsson Ri; (1996). Anti Cancer Res., 16: 2169-2172.

140. Locker, G.J.; Mader R.M.; Braun J.; Sieder A.E.; Marosi C.; Rainer H.; Jakesz R.; steger G. G.; (1995). Oncology.,l 52: 140-144.

141. CA 15-3 as a Marker for Breast Cancer" http://www.asco org/prof./pp/html/m-Tumor 10.htm.

142. Hayes, D.F.; Zurawski V.; kufe D.W.; (1986). J. Clin.Oncol.,4: 1542-1550.

143. Devitu, T.V.; Helllmen S.; Rosenberg A.S.; (1996). "Important advances in oncology", Philadelphia, Lippincotl-Raven.

144. Colomer, R.; Ruibal A.; Geuolia J.; (1989). Cancer, 64: 1674-1681.

145. Pal, S.; Sanyal V.; Chattopadhyay V.; (1995). Int. J. Cancer, 60: 759-765.

146. Kaplan, l.; and Pesce A.; (1989). "Clinical Chemistry", 2nd ed.; C.V.Mosby; p255.

147. Lawry, O.H.;Rosebrough N.J.; Farr A.L.; Randell R.J.; (1951). F. Biol. Chem., 193: 265-275.

148. Janson, J.C.; Ryden L.; (1998). "Protein purification", 2nd ed, New York, A John Wiley & Sons, Inc., Chapter 1 & 14.

149. Al –Khayat, T.H.; (1991). "Molecular characterization of prolactin receptors in human prostate" Ph.D. Thesis supervised by Al-Mudhaffar S.A., College of Science, and Baghdad University.

150. Scopes, R.K.; (1982). "Protein purification principles and practice", New York, Springer Verlag, pp 197, 162.

151. Geraghty, J. G.; Coveney E.C.; Sherry B., O' Higgins N.J.; Duffy M.J.; (1992). Cancer, 70: 2831-2838

152. Colomer, R.; Ruibal A.; Genoola J.; Salvador L.; (1989). Cancer; 64: 1674-1681.

153. Hayes, D.F.; Zurawski V.R.; Kufe D.W.; (1986). J. Clin. Oncol., 4: 1542-1550.

154. Nekulova, M,; Simickova M.; Pecen L.; Eben K.; Vermousek I.; Stratil P.; Cernoch M.; Lang B.; (1994). Neoplasma, 41: 113-118.

155. Vizcarra, E. ; Lluch A. ; Cibrian R. ; Jarque F. ; Alberola V. ; Belloch V.; (1996). Breast Cancer Res. Treat., 37: 209-216.

156. Duffy, M.J.; Sherry, F.; (1988). Ann. Clin. Biochem.; 25(Suppl.): 53s-4s.

157. Duffy, M.J.; (1999). Ann. Clin. Biochem., 36: 579-586.

158. Coveney, E.; Geraghty J.G.; Sherry F.; McDermott E.W.; O'Higgins NJ; Duffy M.J.; (1995). Int. J. Biol. Markers, 10: 35-41.

159. Robertson, J.F.R.; Pearson D.; Price M.R.; Selby C.; Badley R.A.; Pearson J.; (1990). Eur. J. Cancer, 26: 1127-1132.

160. Bon, G.S.; von Mensdorff. Povilly S.; kenemans P.; van kamp G.J.;Verstraeten R.A.; Hilgers J.; (1997). Clin. Chem., 43:585-593.

161. Hayes, D.F.; Zurawski V.R.; kufe D.; (1986). J. Clin. Oncol., 4: 1542-1550.

162. Ichihara, S.; Aoyamatt H.; (1994). Cancer, 73(8): 2181-2185.

163. Pons-Anicet, D.M.F.; Krebs B.P.; Mira R.; (1987). Br. J. Cancer, 55:567-570.

164. Crippa, F.; Bombardieri E. ; Seregni E.; (1992). J. Nucl. Biol. Med., 36:52-55.

165.   Bryant, N.J.; (1986). "Laboratory immunology and serology", 2nd ed., Philadelphia, W.B.Saunders company, Chapter 5, pp 49-52.

166.   Roitt, I.; Brostoff J.; Male D.; (1998). "Immunology". 5th ed, london, Mosby philadelphia st. Louis.

167.   Dad liker, W.B.; and Satussure V.A.; (1970). Immunochemistry; 7: 799.

168.   Steiner, A.L; Kipnis D.M.; Utiger R.; (1969). Proc. Nat. Acad. Sci. USA., 64: 367.

169.   Devlin, T.M.; (1986). "Text book of biochemistry with clinical correlation", 2nd ed. John Wiley and Sons, Inc, New York, pp125, 66.

170.   Scheraga, H.A.; (1961). "Protein structure", New York: Academic Press pp 365, 571.

171.   Melander,   W.;   Horvath   C.;   (1977). Arch..Biochem.Biophys.,183:200-215.

172.   William, E.P.; (1998). "Fundemental immunology", 4th ed. Philadelphia, Lippicott. Raven, chapter 4, pp 75-110.

173.   Mellor, Maley; (1947). Nature, 159: 370.

174.   Williams, R.J.P; (1959). "The Enzymes", 2nd ed., New York, Academic Press, vol I, pp391.

175.   Gendler, S.J.; Spicer A.P.; (1995). Ann. Rev. Physiol., 57: 607-634

176.   Webb, A.; Scott-Mackie P.; Cunningham D.; Norman A.; Andreyev J.; O'Brien M.; Bensted J.; (1996). Eur .J. Cancer, 32 A (1): 63-68.

177.   Gion, M.; Mione R.; Leon A.E.; Kittadi R.; (1999). Clin. Chem., 45 (5): 630-637.

178.   Stacker, S.A.; Tjandna J.J.; Xing P.X.; Walker I.D.; Thompson C.H.; Mckenzie I.F.C.; (1989). B. J. Cancer., 59: 544-553.

179.   Hilkens, J.; Kroezen V.; Bonfrer J.M.G.; Brunning P.F.; Hilgers J.; Eajkeren van M.; (1985). Protides of biological Fluids; 2: 651-653.

180.   Abe, M.; Dufe D.; (1987), J. Immunology, 139: 257.

181.   Bonfrer, J.M.; (1995). Ned. Tijdschr. Klin. Chem., 20: 301-304.

182.   Brostoff, J.; and Male D.; (1994). "Clinical immunology, an Illustrated outline", Philadelphia. Mosby, Section8, p112.

183.   Shiu,R.P.C.; Friesen H.G.; (1974). J. Biol. Chem.; 249:7902.

184.   Gallagher, T.S.; Voss. J r; (1969). Immuno. Chemistry, 6:573.

185.   Al-Atrakchi S.A.M.; (2002). "Protein engineering of carcinoembryonic antigen and their receptors located in malignant mammary tissues". Ph. D. Thesis, Supervised by Al-Mudhaffar S.A., College of Science, Baghdad University.

186.   Rosier, J.S.; Gokulrangan G.; Girault H.; Svojanovsky S.; Wilson G.S.; (2000). Langmuir, 16: 8489-8494.

187.   Seely, D.H.; Wang W.Y.; Salhanick H.A.; (1980). Boichem. Boiphy. Acta., 632: 535.

188.   Al- Mudhaffar, S.A.; (2000). Iraqi. J. Chem.; 26J1): 186-194.

189.   Al-Mudaffar, S.A.; (2000). Iraqi. J. Chem.; 26 (4): 892-905.

190.   Scatchard G.; (1949). Ann. N.Y. Acad. Sci, 51:660.

191.   Chamberlain J.; Jargarinece N.; Ofner P.; (1966). Boichem. J., 99:10.

192.   Adams A.; Karrott D.; (1985). Boichem. Biophy. Acta., 632:535.

193.   Weiland G.A.; Molinoff P.B.; (1981). Life Science, 29:313.

194.   Segel I.H.; (1979). "Biochemical calculation" 3rd ed., John willey & Sons, Inc. pp311.

195.   Williams, C.A.; Chase M.W.; (1971). "Methods in immunology and immuno chemistry". 5th ed.. New York: Academic Press, Vol. III, chapter 13.

196.   Nemeth, G.; Scheraga H.A.; (1962). J. Phys. Chem., 66: 1773.

197.   Waelbroeck, M.; Van-Obberghen E.; De-Meytes P.; (1979). J. Biol. chem., 254: 7736.

198.   Haro, L.S. ; Talamantes F. J.; (1985). Mol. Cell. Endocrinol, 43: 199.

199.   Blumenthal, D.K.; Stull J.T.; (1982). Biochemistry, 21:2386.

200.   Laport, D.C; Wierman E.M.; Storm D.I.; Biochemisrtry; 19:3814.

201.   Laskowski, M.; Leach S. ∂. ; Scheraga H.A.; (1960). J. Am. Chem. Soc., 5:71.

202.   Leach, S.J.; Scheraga H.A.; (1960). J. Boil. chem., 235: 2827-2829.

203.   Saif-Allah, P.H.; (2000). "Biochemical studies on prolactin and some tumor makers in breast tumors" Ph.D. thesis supervised by Al-Mudhaffar, S.A., College of Science, Baghdad University.

204.   Kiernan, J.A.; (1999). "Histological & histochemical methods theory & practice" 3rd ed., Reed Educational and Professional Publishing Ltd, Chapter 19, pp 391-398.

205.   Johustone, A.; Thorpe R.; (1996). "Immuno-Chemistry in practice", 3rd ed., Blackwell Science Ltd, pp292-311, 1-4.

206.   Williams, C.A.; Chanse M.W.; (1968). "Methods in immunology and immunochemistry", New York, Acadimic Press, Vol II, Chapter 10, pp163-174.

207.   Mathews, Ch.K.; Holde K.E.; (1990). "Biochemistry" California: The Benjamin/Cummings Publishing Company.

208.   Freifrlder, D.; (1982). "Physical biochemistry, application to biochemistry molecular biology", 2nd ed., San Francisco: W.H. Freeman & Company. Chapter 14, pp 494-591.

209.   Leach, S. J.; (1969). "Physical principles and techniques of protein chemistry", New York, Acadimic Press, Part A, Chapter 3, pp102-170.

210.   Axelsen, N.H.; (1983). "Hand book of immunoprecipitiation in gel thechniques", 3rd ed. London, W.A.Banjamin, Inc.

211.   Yang, J.T. ; Foster J.F; (1954). J. Am Chem Soc.; 76:1588.

212.   Tanforel, C.; Buzzell J.G.; Rands D.G.; (1955). J Am Soc; 77: 6421.

213.   Leach, S. J.; (1969). "Physical principles and techniques of protein chemistry", Part A. 5th ed. London, Academic Press; Chapter 3, p: 102.

214.   San, Y.; Bovey D.A; (1960). J. Am. Chem. Soc., 235: 2818.

215.   Brealy, G.J.; Kaska M.; (1950). J. Am. Chem. Soc., 77: 4462.

216.   Silvestien, R.M.; Bassler G.C.; Marril T.C.; (1981). "Spectrophotometric dentification of organic compounds", New York, John Wiley and Sons., p181.

217.   Herskowits, T.T.; Laskowski M.Jr.; (1962). J. Biol. Chem., 2481-2492.

218.   Donavan, J. W.; (1965). Boichemistry, 4:823.

219. Scherage, H.A.; (1961). "Protein structure" New York, Academic Press, pp: 175-287.

220. Koprowski, H.; herlyn M.; Steplewski Z.; Sears H.F.; (1981). Science,253.

221. Takasaki, H. ; Uchida E. ; Temero M.A; (1988). Cancer Res., 48: 1435-1438.

222. Aydiner, A.; Topuz E.; Disci R.; Yosasever V.; Dincer M.; Dincol K.; Bilge A.; (1994). Acta Oncologica, 33(2): 181-186.

223. Barbara, A.W.; Gerald k.; (1996). Newsletter, 4 (9): 1-7.

224. Pasquali, C.; (1994). I. J. Pancreat, 15: 171-177.

225. Szymedera, J.J; (1986). Tumour Boil., 7: 333-342.

226. Coreale, M.; Arnberg H.; Blockx P.; Bombardieri E.; Castelli M.; Encabo G.; Gion M.; Klapdor R.; Martin M.; Nilsson s.; (1994). Int. J. Biol Markers; 9(4): 231-238.

227. Kim, S.M.; Kim S.H.; Choi S.Y.; Kim Y.C.; (1992). J. Korean Med. Sci., 7(4): 297-303.

228. Reid, M.E; Lomas-Francis C.; (1997). "The Blood group antigen facts book", New York, Academic Press, pp 1-6.

229. Shimono, R.; Mori M.; Akazawa k.; (1994). Am .J. Castroenterol., 89: 101-105.

230. Strom, B.L.; Maislin G. and West S.L.; (1990). Int. J. Cancer; 8: 8-13.

231. Gion, M.; Ruggeri G.; Mione R.; (1993). Int .J. Biol. Markers, 8:8-13.

232. Filellax, Molina R. ;Pique J.M.; (1994). Tumor Biol, 15; 1-6.

233. Encabo, G.; Ruibal A.; (1986). Bull Cancer (Paris); 73:256-259.

234. Kovacs, I.; Toth P.; Arkosy P.; Hamori J.; Sapy P.; (1997). Acta. Chir. Hyng., 36(1-4): 172-173.

235. Harada, H. ; Tsukada Y. ; Karasawa Y.; (1994). Clin. Chim. Acta., 228(2): 101-112.

236. Zinser, J.W.; (1997). Rev. Gastroenterol. Mex., 62(3): 145-148.

237. Cwiertka, K.; Dapustova M.; machacek J.; Kohoutek M.; Minarik J.; (1998). Tumor Markers. Abstracts, 2(1); 010. quoted from internet.

238. Changeux, J.P.; (1966). Mol. Pharmacol.; 2: 369.

239. Roitt, I.M.; (1984). "Esential immunology", 5th ed., Oxford, Blackwell Scientific Publications. Chapter 1, p 4.

240. Haro, L.S.; Talaments F.G.; (1985). Molec. and Cellular Endoc.,43:199.

241. Daxembichler, G.; Grill H. J.; Wiesinger H.; Wittliff J.L.; (1997). "In: Multiple molecular forms of steroid hormone receptors", Agarwal M.L., editor. Elsevier, North-Holland Biomedical Press, p163.

242. Weiss, R.B.; (1989). Oncology, 3: 135-148.

243. Hvidt, A.; Nielsen S.O.; (1966). Advan. Protein Chem., 21:287.

244. Sjiu, R.P.C; Friesen H.G.; (1971). J.Biol. Chem., 294: 7902.

245. Melander, W.; hovarth C.; (1977). Arch. Biochem. Biophys., 183:200.

B

Tumor Markers in Molecular Diseases

Prof. Dr. Sami A.Almudhaffar
Dr.Bilal J. M. Al-Rawi

Part B

Alpha Fetoprotein in Human Colorectal Tumors

Prof. Dr. Sami Al Mudhaffar
Dr.Bilal Jasir Mohammed Al-Rawi

SUMMARY

1. The presence of alpha-fetoprotein was investigated in benign and malignant human colorectal tumors. The results obtained indicate a higher incidence for protein in the malignant tumors than in the benign tumors.

2. Immunoradiometric assay (IRMA) was developed and used to characterize the binding of ^{125}I-anti AFP antibody to AFP of colorectal tumor tissues. The data obtained revealed an increment of AFP in the supernatant fraction in comparison to the pellet part. Also a higher incidence of AFP in Malignant than those in benign colorectal tissues.

3. Alpha fetoprotein was isolated from colorectal tissue homogenate in benign and malignant cases using gel filtration technique, then the binding properties of partially purified AFP have been characterized.

4. The kinetic parameters k_{+1}, k_{-1}, k_a, and k_d, of the binding of ^{125}I-anti AFP antibody with AFP (crude and partially purified) in colorectal tissue homogenates were determined at different temperatures.

5. The thermodynamics of the ^{125}I-anti AFP antibody binding to AFP (crude and partially purified) were studied using Van't Hoff and Arrhenius equations. From these equations, the thermodynamic parameters of the standard state ($\Delta G°$, $\Delta H°$, $\Delta S°$) and the transition state (ΔG^*, ΔH^*, ΔS^*) and also E_a were determined.

6. Spectroscopic studies in the UV range (200-350) nm were carried out on (Human AFP), (^{125}I-anti AFP antibody), and (AFP/^{125}I-anti AFP antibody complex). The UV spectrum of each protein type was obtained and the effect of pH and the polarity on the spectra of these proteins were studied. Also the spectrophotometric titration and its effect on spectra was studied.

Chapter one

Introduction
1.1 The colon:

The colon is commonly considered to consist of five segments: the caecum, with the vermiform appendix at its base and the orifice of the ileocaecal valve above; then the ascending, transverse, descending, and sigmoid portions (Fig 1.1)[1]

The junction of the ascending and transverse colon - the hepatic flexure - is angulated and lies close to the undersurface of the liver. A second angulation between the transverse and descending colon -

the splenic flexure - lies close to the hilum of the spleen, and usually lies at a higher level than the hepatic flexure [2].

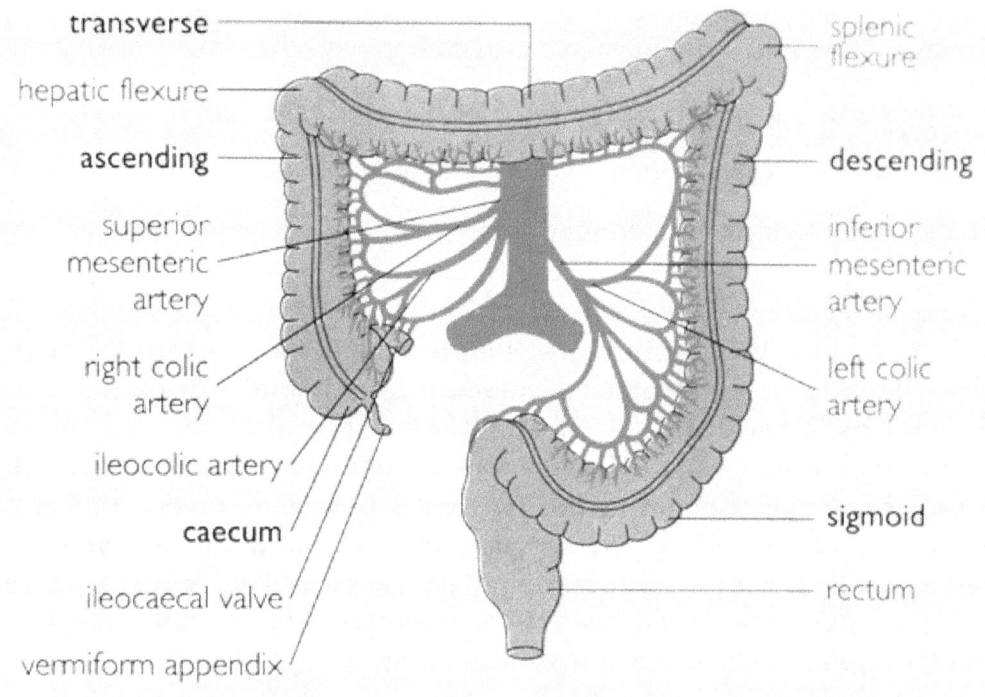

Fig. (1-1): Diagram showing the anatomy of the colon and its blood supply. *From Misiewicz R.L., "Clinical Gastroenterology" 2nd ed. Slide Atlas CD-ROM.

The proximal colon, extending to the distal transverse colon, is derived from the embryonal midgut and shares its blood supply (from the superior mesenteric artery) with the small intestine. This part of the colon is mainly concerned with absorption of water and electrolytes and to a lesser extent with reabsorption of bile acids [3].

The more distal colon arises from the embryonal hindgut and is supplied by the inferior mesenteric artery; it has less absorptive capacity and - with the rectum - functions mainly as a storage site for faeces prior to evacuation [3].

1.2 The rectum:

The rectum, the proximal third of which is intraperitoneal, is about 18cm in length and joins the sigmoid colon to the anal canal. The beginning of the rectum is ill-defined anatomically, but the rectosigmoid junction lies at the level of the sacral promontory and forms a moderately acute angle, which is usually obvious to the sigmoidoscopist [4].

The rectum is supplied by the superior rectal artery - the final branch of the inferior mesenteric artery - and venous drainage is via the superior rectal vein to the inferior mesenteric vein [5]. Lymphatic drainage is upwards unless obstructed, when lymph may drain to nodes on the pelvic sidewalls [6].

The colon and rectum are, together, a little over 1m in length, and the diameter diminishes from caecum to sigmoid, increasing again in the rectum. Most of the outer longitudinal muscle coat is gathered into 3 distinct bands - the taeniae coli - 6 to 10mm in width. These run from the top of the caecum to the rectum, where they merge to form a more continuous covering. The taeniae are shorter than the colon and therefore gather' it into sacculations, or haustra. The mucosa of the colon is thrown into folds, the plicae semilunares, which take on a triangular appearance in the transverse colon [7].

Unlike the small intestine, the colon is relatively fixed, particularly the ascending and descending segments. The more mobile transverse colon has a short mesentery, whilst the sigmoid, with a broader, longer mesentery is generally the most mobile. Covering the serosal surface of the colon are fatty structures arising within the mesentery known as the appendices epiploicae, calcification of which may occur with advancing years [7].

The histological architecture of the colon is similar to that of the more proximal bowel (Fig 1.2). The mucosa comprises parallel rows of epithelial tubules or crypts, surrounded by the connective tissue framework of the lamina propria [8].

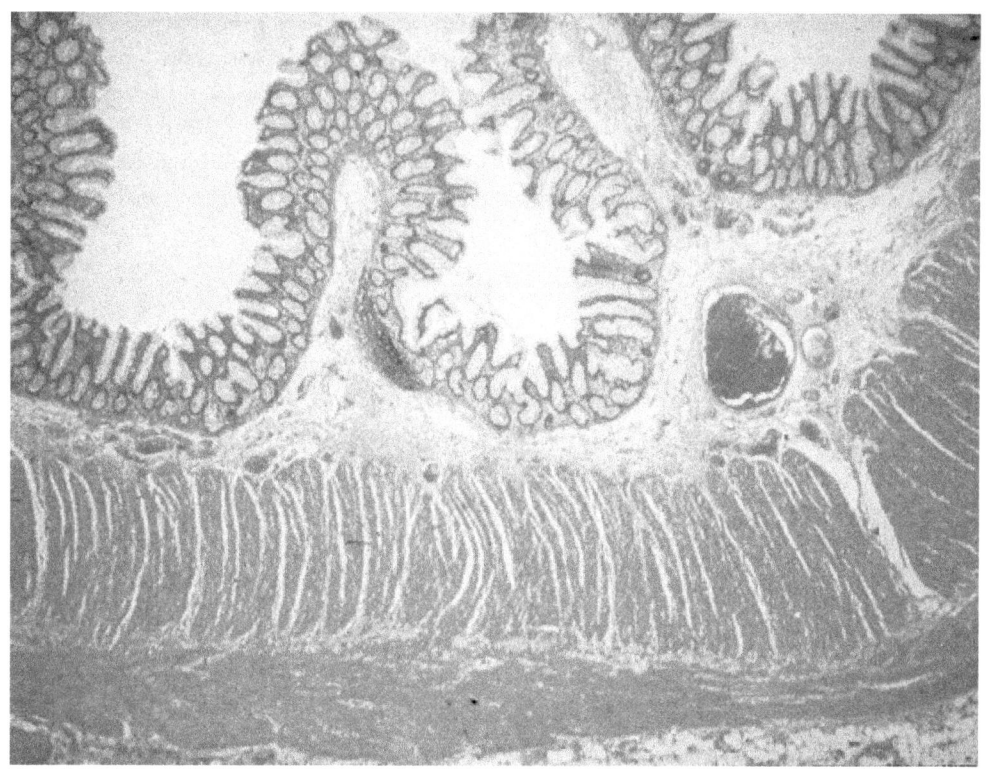

Fig. (1-2): Histological appearance of the large intestine showing the relationship of the mucosa to the muscle layers. *From Misiewicz R.L., "Clinical Gastroenterology" 2nd ed. Slide Atlas CD-ROM.

The crypts are lined by goblet cells, and have undifferentiated cells at the base; the surface epithelium between the crypts is mostly composed of columnar absorptive cells with occasional goblet cells. The goblet cells produce glycoproteins; in health, these are predominantly sulphomucins at the crypt bases with some sialomucins more superficially [9].

The colonic muscularis mucosae and the submucosa have fewer lymphatic channels than the small bowel - an important feature when considering the metastatic potential of neoplastic mucosal lesions [10].

1.3 Colorectal Tumors:
1.3.1 Benign Tumors (Polyps):

A polyp is growth extending into the lumen of the gastrointestinal tract, and is defined as a visible protrusion above the surface of the surrounding normal large bowel mucosa in the colon [11]. Although polyps occur in all sections of the gastrointestinal tract, but are most common in the colon. Polyps may be detected endosmotically by sigmoidscope or radiographically by the barium enema [12]. Colorectal polyps can be classified either according to their shapes or according to their histology [13]. The polyps can be classified hitologically as either neoplastic (adenomatous polyps) or non–neoplastic. Although all adenomatous polyps are non – neoplastic, and thus have no malignant potential. Lastly, sub mucosal polyps include lymphoid polyps, lipomas, and less common histological types [14].

Neoplastic Polyps (Adenomatous Polyps):

Adenomatous, or adenomas, are attached to the bowel wall by a stalk "pedunculated" or by a broad, flat base "sessile". The clinical significance of identifying adenomas is that it has malignant potential and may develop into a cancer [15]. It is generally accepted that all colorectal cancer originates from a precursor adenoma [16].

The national polyp study has demonstrated that colonscopic removal of adenomatous polyps significantly reduce the risk of developing colorectal cancer [16].

Adenomatous polyps occur more frequently in men than in women [17,18]. Older patients have an increased risk of having adenomas. Although adenomas vary greatly in size, most are small, measuring less than 1.0 cm in diameter. In the national polyp study of the 3.371 colonoscopically removed adenomas, 38% were only 0.5 cm or less, 36% were 0.6 to 0.1 cm, and 26% were larger than 1.0 cm [19]. In those found to have adenomatous polyps at colonscopy, about 60% have a single adenoma and 40% have multiple adenomas [20]. Increased age was associated with increased rate of multiple adenomas [18]. Adenomatous polyps exhibit the same predominantly left – sided colonic distribution that found with colorectal cancers. Several clinical studies have reported that more than 60% of colonoscopically removed adenomas are located distal to the splenic of flexure [19, 21].

Other benign colon diseases [22]:

➢ Leiomyoma
➢ Pnuo omatosis cytoides intestinalis
➢ Colitis cystica profunda
➢ Polyposis Syndrome
➢ Cowden's disease
➢ Muri-Torre Syndrome
➢ Cronkhite-Canada Syndrome

1.3.2 Malignant Tumors:

Colorectal cancer encompasses carcinoma of both colon and rectum [23]. Colorectal cancer is the fourth most common incident cancer (after breast, lung and prostate) and the second most common cause of cancer death after lung [24] in the United State, with 111,000 new cases per year and 51,000 death. Rates vary approximately 20 – folded around the world .The highest rates are seen largely in the developed world, North America, Western Europe, and Australia, with age adjusted (world standard) incidence rates of 25 to 35 per 100,000 in the late 1980s [25].

In the 1960s and 1970s, colorectal cancer was essentially the only cancer that occurred approximately equal frequency in men and women [26]. In North America and Australia (high rates), Japan and Italy (rapidly rising rates) in particular, the male age adjusted rates now appear to exceed those of females, sometimes by more than 20% [27] .The difference is less marked in England and Wales .The male and female rates in New Zealand non – Maoris (around 30 per 100,000) are equal [23]. There is a tendency for the rates to be similar between the sexes or to show a male excess after that age. The risk of cancer varies by sub site within the colon and rectum. The sub site risk varies between the sexes and further by age, women have higher rates of right-sided neoplasm's than men and tend to develop cancer at an earlier age [28].

The population of colorectal cancer in Iraq [29,30], like other Asian and Middle East countries were lower than that in the North America and Western countries, however it was witnessed an increasing in an alarming way in the percentage at the last ten years. It was also showed that colorectal cancer when combined colon and rectal cancer was among the commonest ten cancers in Iraq for the last ten years.

1.4 Etiology for Colorectal Cancer:

Carcinogenesis in the colon and rectum has been described in terms of an initiation – promotion model based on observation in laboratory animals. The first step involves initiating factors that directly interact with cellular DNA to induce mutations in genome. After words the process is driven by promotional factors, which are not mutagenic by themselves but enhance cellular proliferation of previously mutated cells. The human diet contains a lot of naturally occurring mutagens or substances that can be metabolized into mutagens [31,32].

In contrast to pancreatic cancer, smoking does not play an etiologic role. Dietary factors however may play important causative and protective role. Fat intake has been the most consistently inverse association and fiber intake the most consistently inverse association. Comparing the incidence in different countries, the rates of colon cancer are strongly associated with the intake of animal fat and meat [33].

During the past several years more and more is known about the molecular events leading to the development of colorectal cancer. It is believed to result from a series of genetic alternations leading to the progressive disorder of the normal mechanisms controlling growth. An important genetic alternation that has been demonstrated is a mutation within codon 12 of the K – ras – proto oncogen. About half of colorectal carcinoma and similar percentage of adenomas larger than 1 cm in diameter have been found to have the ras – gene – mutation compared to less than 10 % of patients with adenomas smaller than 1 cm [31,34].

Harden [35], implicated in all cancers and their importance would vary depending on environmental effect strength. The current etiologies of colorectal cancer are:

❖ Genetics [36,37].

❖ Environmental factors (Diet including fat, fibers, and meat).

❖ Biological factors [33].

❖ Pathological disorders [33].

❖ Family History [33].

❖ Viruses [38,39].

1.5 Staging System of Colorectal Cancer [40]:

1.5.1 TNM Classification:

The TNM classification for colorectal cancer table (1–1) of the American Joint Committee on cancer and the Union International Contra le cancer (AJCC/UICC) is compatible with the Dukes classification is that each of the three subsets of tumoral, nodal and metastatic involvement makes no assumption about the status in another part of the system that is, it allows for the exceptions to the general role of orderly progression of tumor spread.

The different Dukes classifications do not encompass all three prognostic factors in one staging system.

1.5.2 Dukes Classification:

The first practical staging was the classification outlined initially in 1930 by Cuthbert Dukes, based on an earlier clinical categorization of rectal cancer by the surgeon, Lockhart – Mummery. Dukes classified rectal tumors from A to C:

* Stage A– penetration into but not through the bowel wall.

* Stage B– penetration through the bowel wall.

* Stage C involvement of lymph nodes, regardless of the extent of the bowel wall penetration. Subsequently, many modification, were made by Dukes, as well as by others [40].

A fourth stage D was added to indicate disease beyond the limit of surgical resection. Duke's classification is still widely used in staging colorectal cancer because of its simplicity and ease of use particularly by the surgeon and oncologist. However, the TNM system is increasingly being applied internationally.

Table (1–1): TNM clinical classification system for staging colorectal cancer

Primary tumor (T)				
T_x	Primary tumor cannot be assessed.			
T_o	No evidence of primary tumor.			
T_{is}	Basement membrane (intraepithelial) or lamina propria (intramucosal).			
T_1	Tumor invades the sub mucosa.			
T_2	Tumor invades the muscular is propria.			
T_3	Tumor invades through the muscular is propria into the serosa or into the non–peritonealized pericolic or perirectal tissue.			
T_4	Tumor directly invades other organs or structures and/or pre forates the visceral peritoneum tissues.			
Regional Lymph nodes (N)				
NX	Regional lymph nodes cannot be assessed.			
No	No regional lymph nodes metastasis.			
N1	Metastasis in 1–3 pericolic or perirectal lymph nodes.			
N2	Metastasis in >4 pericolic or perirectal lymph nodes.			
Distant Metastasis (M)				
M1x	**Presence of distant metastasis cannot be assessed.**			
Mo	No distant metastasis.			
M1	Distant metastasis.			

Stage Grouping				
	AJCC/VICC			Dukes[a]
Stage O	T is	N0	M0	–
Stage I		N0	M0	A
		N0	M0	–
Stage II	T_1	N0	M0	B
		N0	M0	–
Stage III	T2	N1	M0	C
	T3	N2	M0	–
Stage IV	T4	Any N	M1	–
	Any T			
	Any T			
	Any T			

Dukes^a B is composite of better (T$_3$, N$_0$, M$_0$) and worse (T$_1$, N$_0$, M$_0$) prognostic groups as in Dukes C (any T, N, M$_0$ and any T, N$_2$, M$_0$).

1.6 Clinical of Colorectal Cancer:

Symptoms depend to some extent on the site of the primary tumor. Cancers of the proximal colon usually grow longer before they produce symptoms than those of the left colon and rectum. Constitutional symptoms "fatigue, shortness of breath, angina" secondary to microcytic hypochromic anemia may be the principle manner of presentation of right colon tumors. Less often, blood from right colon cancers is admixed with stool and appears as "mahogany feces" .As a tumor grows, it produces vague abdominal discomfort or presents as a palpable mass. Obstruction is uncommon because of the large diameters of the caecum and ascending colon, although rectal cancer may block the ileocecal valve and cause distance small bowel obstruction [4].

Adenocarcinoma of the colon and rectum may be present as long as five years before symptoms appear; however, persons with a symptomatic disease often have occult blood loss from their tumors, and the bleeding rate increase with tumors size and degree of ulceration [41].

The left colon has a narrower lumen than the proximal colon and cancers of the descending and sigmoid colon often involve the bowel circumferentially and cause obstructive symptoms. Patients may represent with colicky abdominal pain, particularly after meals, and changes in bowel habits. Constipation may alternate with increased frequency of detection, as small amounts of retained stool moved beyond the abstracting lesion. Hematochezia is present more often with distal lesions than the proximal ones, and bright red blood passed per rectum or coating the surface of the stool is common with cancers of the left colon and rectum. Rectal cancers also cause obstruction and changes in bowel habits, including constipation, diarrhea, and tenesmus .Rectal cancer may invade locally to involve the bladder, vaginal wall, or surrounding nerves, resulting in perinea or sacral pain, but this is a late occurrence [42].

Colorectal carcinoma should be considered when a patient, especially older than 40 years presents with hypochromic microcytic anemia or frank hematochezia and rectal bleeding. Too often, anemia in the elderly is ascribed to "chronic disease", only to be diagnosed later as a sign of colorectal cancer [42].

194

Symptomatic patients with colorectal cancer are often misdiagnosed. Symptoms are described to begin conditions such as diverticular disease (abdominal pain, bleeding, change in stool caliber), irritable bowel syndrome (abdominal pain, change in bowel habits), or hemorrhoids (rectal bleeding).

Abdominal pain – in any form–and bleeding also merit evaluation from cancer in this age group. Large bowel cancer affects younger patients. Particularly those are with inflammatory bowel disease or strong family history for colorectal and other cancers [43].

1.7 Treatment of Colorectal Cancer:

Surgery:

The goal of surgery is wide resection of the involved segment of bowel together with removal of its lymphatic drainage vessels. The resection should include a segment of colon at least 5 cm on either side of tumor.

Surgical resection is the treatment of choice for most colorectal cancer. Preoperative colons copy should be performed, if possible, to rule out synchronous lesions and serum carcino embryonic antigen (CEA) should be measured to inform staging and post follow – up.

In-patient with colorectal cancer the primary tumor should be resected, even in the presence of distant metastasis to prevent obstruction or bleeding. In patients with advanced disease and multiple medical problems, repeated palliative fulguration or rectal tumors may be preferable to surgery. Newer modalities, such as laser photo ablations are being tested as alternative means of palliation in these patients [44,45].

Endoscopic Therapy:

Endoscopic Therapy using the neodymium– yttrium–aluminum garnet (Nd:YAG) laser has been used to recanalize the rectum as palliative therapy in patients with abstracting rectal cancers who are poor surgical risks or who have advanced stages of malignant disease [45].

Radiation Therapy:

Radiation therapy is used preoperatively or post operatively to decrease local recurrence in those with high–risk rectal and recto sigmoid cancers (Dukes' B2 and C lesions), or in combined pre–operative and post operative "sandwich approach". Given the rectal demonstration of decreased recurrence and increased survival in patients with rectal cancer receiving combined post operative radiation and chemo therapy, this would appear to be the treatment of choice for high risk patients with transmural tumor extension or lymph node metastases [46].

Chemotherapy:

The mainstay of chemotherapeutic palliation is 5–fluorouracil (5–FU). 5–FU is a pyrimidine antimetabolite action in the "S" phase of the cell cycle [47]. The cytotoxic effects of 5–FU are mediated by active metabolites, which inhibit the synthesis of thymidine, DNA and proteins [48]. The clinical effectiveness of 5–FU is schedule dependent. Bolus regimens give consistently poor results with response rates of 5–25% [49-53]. Response rates improve markedly when 5–FU is administered by i.v. Infusion over 5 days or by protracted i.v. infusion over a period of many months. Several [54,55] studies have demonstrated that in vitro 5–FU cytotoxic activity may be enhanced significantly by many drugs such as interferon, folates, and other antineoplastic drugs. In particular it has been shown that folonic acid (FA) strengths the binding of the 5–FU active metabolite, 5–FdUMP, to its target enzyme, forming a stable ternary complex that dissociates very slowly, increasing the fluoropyrimidine cytotoxic activity [56]. The ability of folonic acid (FA) to enhance 5–FU anti neoplastic activity has been confirmed in some prospective randomized trials in colorectal carcinoma patients [57,58]. Other drugs for chemotherapy are: Methortrexate (MTX) [59], PALA [N–phosphonoacety], L–aspartic acid [60], Hydroxy urea [61], Cisplatin [62], Dipyridanol [63], Tomudex, Oxalipaltin [63], Doxifluridine [63], Taxotere [61].

Photodynamic Therapy (PDT):

PDT has also been used to treat patients who are poor surgical risks. Patients are sensitized with a hematoporphyrin derivative, which is taken up by the tumor. Photography is then performed using a tunable dye laser and a flexible optical fiber, which can be inserted into the tumor [44].

1.8 Tumor Marker:

A tumor marker is a substance produced by a tumor or produced by the host in response to a tumor that can be used to differentiate a tumor from normal tissue or to determine the presence of tumor based on measurement in the blood or secretions[64].

Tumor markers are found in cells, tissues, and body fluids such as cerebrospinal fluid, serum, plasma, and milk. The ideal markers would be useful in diagnosis, staging and prognosis of cancer, provide an estimation of tumor burden, and serve for monitoring effects of therapy, detecting recurrence, localization of tumor, and screening in general population [64].

The first tumor marker reported was the Bence – Jones protein, since its discovery in 1847 by precipitation of a protein in acidified boiled urine. The measurement of Bence – Jones protein had been considered as a diagnostic test for multiple myeloma (a tumor of plasma cells). More than one hundred years after its discovery, Edelman and Poulik identified the Bence–Jones protein as the monoclonal light chain of immunoglobulin secreted by tumor plasma cells. Monoclonal Para proteins appear as sharp bands in the globulin area in electrophoretic patterns of serum. Diagnosis of multiple myeloma is often based on this finding or on the presence of an elevated level of monoclonal immunoglobulin in the serum[65].

1.8.1 General Aspects of Tumor Marker:

The tumor marker can be measured qualitatively or quantitatively by chemical, Immunological, or molecular biological methods to identify the presence of a cancer [64].

Most (if not all) tumor markers do not fit the ideal profile. The reason for this can be the relative lack of sensitivity and specificity of the available tests. It should be noted that vitually any protein or chemical has the potential to be tumor marker. As tumor cell grow and multiply, some of their substances increase in tumor tissues and/or leak into the bloodstream or other fluids. Depending upon the tumor marker, it can be measured in blood, urine, stool or tissue [65].

Morphologically, cancer tissue has been recognized by pathologists as resembling fetal tissue more than normal adult differentiated tissue. Tumors are graded according to their degree of differentiation as being (1) well differentiated (2) poorly differentiated, or (3) an plastic (without) forms. Tumor markers are the biochemical or immunological counterparts of the differentiation state of the tumor. In general, tumor markers represent re–expression of substances produced normally by embryogenically closely related tissues[65].

Few markers are specific for a single individual tumor (tumor – specific markers); most are found with different tumors of the same tissue type (tumor – associated markers). They are present in higher quantities in cancer tissue or in blood from cancer patients than in benign tumors or in the blood of normal subjects[66].

1.8.2 Features of Ideal Tumor Marker:
The ideal tumor markers have the following properties[67]: –

* High clinical sensitivity.

* High clinical specificity.

* Tumor markers levels proportional to tumor volume.

* Reflect tumor heterogeneity.

* Low levels in healthy population.

* High levels in benign diseases.

* Discriminatory to identify tumor and metastasis from benign to healthy states.

* Provide adequate lead times for early diagnosis and early treatment.

* Assay sensitivity to detect stage cancer.

1.8.3 Application of Tumor Markers:
Many applications exist for tumor markers in clinical oncology. Clinically important utilization of markers includes[68,69]:

Early detection of tumor.

-

Differentiating benign from malignant conditions.

-

Evaluating the extent of the disease.

-

Monitoring the response of the tumor to therapy.

-

Prediction or detecting the recurrence of the tumor.

-

1.8.4 Clinical Application of Tumor Markers:
The clinical usefulness of tumor markers is summarized in table (1.2).

Table (1-2): Clinical Usefulness of tumor markers[70,71].

Tumor Marker	Biochemical properties	Molecular weight	Primary clinical applications
Alpha-fetoprotein (AFP)	Glycoprotein, 4% carbohydrate; considerable homology with albumin	~70 kDa	Diagnosis and monitoring of primary hepatocellular carcinoma and germ cell tumors. Prognosis of germ cell tumors.
Cancer antigen 125 (CA 125)	Mucin identified by monoclonal antibodies	~200 kDa	Monitoring ovarian carcinoma. Prognosis after chemotherapy
Cancer antigen 15-3 (CA 15.3, BR 27.29)	Mucin identified by monoclonal antibodies	>250 kDa	Monitoring breast cancer
Cancer antigen 72.4 (CA 72.4)	Glycoprotein identified by monoclonal antibodies	~48 kDa	Monitoring gastric carcinoma
Cancer antigen 19-9 (CA 19-9)	Glycolipid carring the Lewisa blood group determinate	~1,000 kDa	Monitoring pancreatic carcinoma
Carcinoembriyonic antigen (CEA)	Family of glycoproteins, 45%-60% carbohydrate	~180 kDa	Monitoring gastrointestinal and other adenocarcinomas
CYFRA 21-1	Fragments of cytokeratin	~30 kDa	Monitoring bladder and lung carcinoma
Estrogen receptor	Nuclear transcription	65 kDa	Predicting response to endocrine therapy in breast cancer
Human chorionic gonadotrophin (hCG)	Glycoprotein hormone consisting of tow non-covalently bound subunits (α and β)	~36 kDa	Diagnosis and monitoring non-seminomatous germ cell tumors, choriocarcinomas, hydtidiform moles, seminomas. Prognosis of germ cell tumors.
Neuron specific enolase (NSE)	Dimer of the enzyme enolase	~87 kDa	Monitoring small cell lung carcinoma, neuroblastoma, apudoma.
Placental alkaline phosphatase (PLAP)	Heat-stable isoenzyme of alkaline phosphatase	~86 kDa	Monitoring of germ cell tumors (seminomas)
Progesterone receptor	Nuclear transcription factor	A from: 94 kDa B from: 120 kDa	Predicting response to endocrine therapy in breast cancer .

Prostate specific antigen (PSA)	Glycoprotein serine protease	~36 kDa	Diagnosis , screening and monitoring prostatic carcinoma
Squamous cell carcinoma antigen (SCC)	Glycoprotein sub-fradion of tumor antigen T4	48 kDa	Monitoring squamous cell carcinomas
Tissue polypeptide antigen (TPA)	Fragments of cytokeratin 8,18 and 19	~22 kDa	Monitoring bladder and lung carcinoma
Tissue polypeptide specific antigen (TPS)	Fragment of cytokeratins 18	~22 kDa	Monitoring metastatic breast carcinoma

1.8.5 Tumor Markers in Colorectal Cancer:

There are several tumor markers correlated with the incidence of colorectal cancer, but the most important markers are:

* CA 19 – 9:

It is a monoclonal antigen raised against colon carcinoma cells [72]. The carbohydrate antigen is a glycolipid [63], it is shown to react against a monosialoganglioside antigen [72] .The antibody did not react against a panel of other ganglioside, but it did not react to elements in human meconium [73].

In colon cancer patients with the elevated value had a four folds increase in death compared to patients with lower values (p<0.001) [74].

* CA – 50:

It is a monoclonal antigen developed against the human adenocarcinoma cell line. CA – 50 was a marker for colorectal cancer and pancreatic .The clinical use of CA–50 is in the prognosis and monitoring therapy [65].

* CA – 242:

Studies of cancer Antigen CA – 242 have been described [75] .It has a little advantage over CEA alone. But when it used in panel of multiple markers may provide enhanced sensitive and specificity to serve as a monitoring tool [75].

* CEA

It is an oncofetal Antigen, CEA is currently in wide spread use as a marker for colorectal cancer [76].

Elevation can be observed in a variety of benign gastrointestinal disease [76], in smokers [77], and other malignancies including breast [78], lung [79], gastric [80], pancreatic [81], and gynecologic cancers [82].

Because of the lack of specificity for colon cancer and resulting false–positive results, CEA cannot be recommended for use in cancer screening.

The major clinical role for CEA is in monitoring response to surgery and detecting recurrence. The most effective indicator of recurrent disease is a progressive increase in serial CEA levels [83]. CEA may provide additional prognostic information at the time of staging [84].

* CA 72 – 4:

It is a monoclonal Antibody purified from human colon carcinoma xemograft. CA72–4 and CEA value may be complementary. The clinical use of CA72–4 is in prognosis and monitoring therapy [63].

* Cathepsin B:

It is a lysosomal cystein protease that can degrade matrix components, and result in a higher metastatic potential and a worse prognosis [85,86]. It was found that Cathepsin B in colon cancer correlated with the stage, that it had less elevated in stage I/II than in stage III/IV.

In addition there are several tumor markers for colorectal cancer are under study. Walach et al.[87] found that leukocyte alkaline phosphatas was more sensitive for detecting metastases than CEA. Verazin et al.[88] compared total sialic acid (TSA)/total proteins ratio to CEA in 146 consecutive patients undergoing colorectal reactions. TSA/TP was more frequently elevated than CEA levels in cancers in earlier stages (A, B2) [88]. In a study comparing TPA and CA19 – 9 to CEA, the combination of markers was more sensitive for detecting recurrence than CEA alone [89].

*AFP:

Some studies refer to use AFP as a tumor marker in monitoring colorectal cancer [90].

This tumor marker, that is, AFP is also elevated with other tumors such as, tumor of colon [90] and rectum [91].

Alpha Fetoprotein has been correlated with increased risk for recurrence and poor survival, particularly in later stage, and in colorectal cancer [92].

1.9 Alpha Fetoprotein (AFP) :

Alpha Fetoprotein is an oncofetal protein (embryonal serum protein) that appear in the fetal stage, or in cases of malignant neoplasms [93]. It is first useful tumor marker in gastrointestinal (GI) cancer, that was recognized [94]. Thus, the presences of circulating AFP in serum levels render it a useful marker for fetal distress, tumor growth, and hepatic dysfunction [95]. In addition to serum, AFP has been identified immunologically in the ascitic fluid, cerebrospinal fluid (CSF), bile, and urine of fetuses, neonates, and patients with hepatocellular carcinoma [96,97].

During the development in mammals, products termed "oncofetal proteins or feto specific serum proteins" are produced at various stages of embryonic, fetal life [95,98].These proteins are present in high concentrations in sera of fetuses and decrease to low levels or disappear after birth. In cancer cases, these protein appear to originate within tumor cell and enter the circulation as a result of secretion by the tumor or as breakdown products of tumor cell [99-101].

1.9.1 General Description of (AFP):

A. Composition:

Alpha Fetoprotein is classified as a member of an albuminoid gene family "superfamily" which consists of four members: Albumin (Alb), vitamin D-binding (Gc) protein (Gc DBP), AFP, and alpha-Alb (αAlb), termed a famine in humans [102-104]. The family members display structural similarities homologous amino acid sequence stretches, and similar cysteine disulfide bridge clusters [105]. All albuminiod gene family is structurally characterized by cysteine residues that are folded into layers that form loops dictated by disulfide bridging, resulting in a triplet domain, U-shaped molecular structure. The three domains of these gene family members have been confirmed by x-ray crystallography [106,107]. In human, the four albumioid genes lie in tandem on chromosome 4, encompassing 15 exons and 14 introns [108].

Alpha Fetoprotein is an alpha-1-glycoprotein with a molecular weight of (67-72) KD. It consists of a single polypeptide chain containing (3-5)% carbohydrates [109-112] represented by one oligosaccharide residue [113-116], which consists of hexose, hexosamine and sialic acid [117,118]. Carbohydrate compositions of AFP are: mannose, galactose, N-acetylglucosamine, and sialic acid [119]. The microheterogeneity of AFP results from the oligosaccharide structure and mainly determined by differences in the glycosylation process in different tissues [120]. The amino acid sequence deduced from the nucleotide sequence revealed that there are 590 amino acid residues in mature human AFP [121,122].

It is a protein member of albuminoid gene family consisting also of albumin and vitamin-D-binding group component protein that present as globular proteins comprised of three major domains [111,123-126]. AFP shares many similarities with albumin [127], including sequence homology [128-130], immunological cross-reactivity, and physiological properties [131-133].

B. Structure:

Alpha Fetoprotein molecule has 30 cysteine residues that are characteristically spaced and that form 15 disulfide bridges. AFP molecule is "U-shaped" possessing three domains, figure (1.3), in which two compact rigid domains (N-terminal domain I and C-terminal domain III) are connected by relatively labile domain II [134]. The structure of domain II could be approximated by a "molten globule" state. The secondary structural conformation of h-AFP shows compositions for α-helix of 49%, β-form of 17% and random coil of 34% [113]. Morinaga et al.[135], have shown a notable absence of disulfide flexibility between domains II and III. The N-terminus region (domain I) containing residues 1-79 is also characterized by flexibility of the polypeptide backbone [107]. Domain III exhibits greater amino acid sequence identity among the various mammals than do domains 1, and 2 [105]. Domain III is also known to contain a major hydrophobic-binding site and a proposed dimerization motif [136,137]. Domain I, and II contain binding sites for both fatty acids and bilirubin; however, domain II also display amino acid sequences related to cellular and extracellular matrix adherence regions and bears sites for the carbohydrate attachment via asparagines [138].

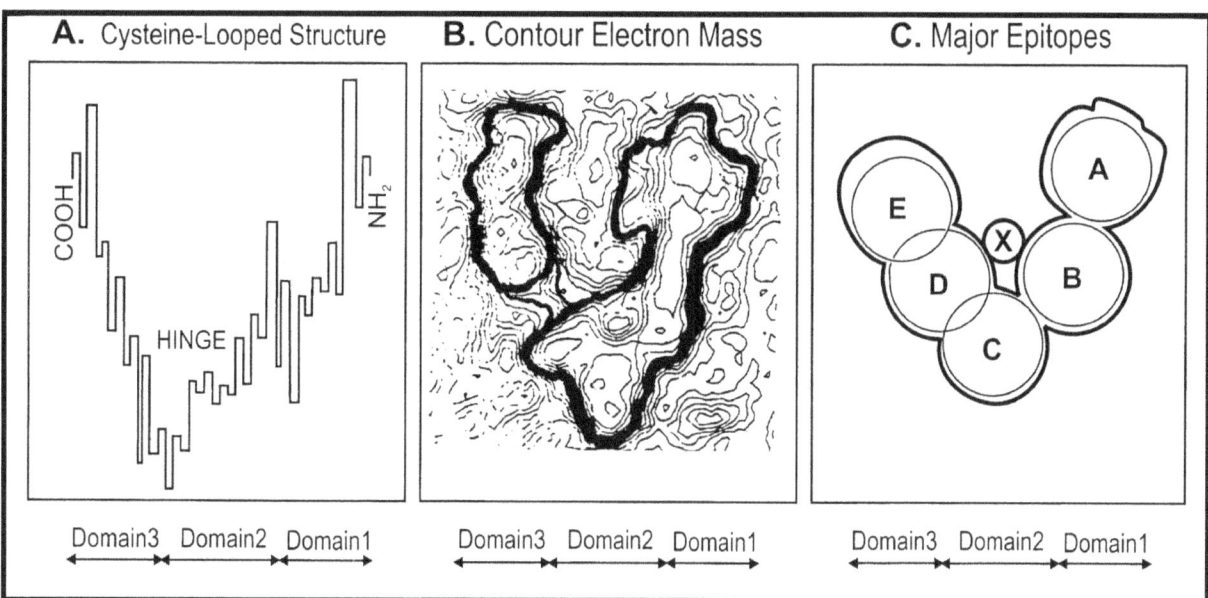

Figure (1.3): A three-panel diagrammatic mapping of the tri-domain structure of h-AFP.

(A) represents the cysteine-loop configuration, while (B) shows the electron dot contour massmap of h-AFP. In comparison, (C) predicts an epitopic map of h-AFP derived from 30 or more MoAbs [139,140]. The precise epitope domain locations have been detected, but have yet to be localized. The three-panel diagram represents a composite redrawn from [135,140,139-141].

1.9.2 Molecular Forms of (AFP):

AFP has long been reported to exist in bound and non-bound (free) molecular forms in both mammlian sera and tissues cytosols [143-145]. Circulating "free" AFP has been characterized as an immunoreactive with a molecular form of 70 KD [146,147]. In contrast, bound AFP is described as a non-immunoreactive form with a total molecular weight reportedly approaching 200 KD [148,149]. This larger molecular weight is attributed to a protein moity which binds to AFP and masks its antigenic epitopes, thus eluding immunological detection. AFP has also shown to exist in cells both as secretory and non- secretory forms. Several models in human have further display a molecular form of AFP, which is synthesized but not secreted by the cell [143,150].

1.9.3 Biological Roles of (Alpha Fetoprotein):

Although the physiological and structural properties of the oncofetal protein AFP has been largely documented, only *in vitro* functional roles of this protein have been ascertained to date [151-153]. Because AFP is synthesized during the cell cycle G1 and S-phase, it has been hypothesized that it affects cell growth [154,155] and it is known as a growth-promoting agent and modulating growth factor mediated cell proliferation [156]. Also possesses a growth inhibitory motif recently identified as occult epitopic segment of the molecule [157]. It has emerged as a dual regulator of cellular growth, capable of both enhancement and suppression including liver [154], Placental [158], lymphatic cells [159], ovarian [160], uterine [161,162], epidermal [160], fetal fibro-blasts [163], and hepatic phagocytic, bone marrow [151] in addition to various neoplastic cells [163].

The observation that AFP is able to bind estrogen led to the suggestion that AFP plays a role in sexual differentiation of the brain by protecting the fetus from the effects of circulating estrogen [164]. In addition to binding estrogen, AFP is able to bind other steroids as well as endogenous and exogenous substances such as unsaturated fatty acids [165], bilirubin [166], and various ligands including arachidonic acid, drugs, dyes, retinoic acid, and heavy metals [136,152,167], suggesting that AFP play a general transportation role [168]. AFP could also interact or bind cytoplasmic chaperone proteins that normally escort nuclear factors or transcription cofactors through the cytoplasm toward organelle interface [163,169].

1.9.4 Clinical Application of (AFP):

Since AFP is not strictly specific for a certain types of carcinoma, its determination is primarily used in medicine for monitoring the effects of therapy and surgery on the course of malignant condition, which initially showed increased levels of serum AFP [170].

It is used as a marker for both the diagnosis and monitoring of patients suffering from hepatocellular carcinoma [171]. Measurement of AFP is used to assess the completeness of surgical resection and response to therapy or recurrences [172]. It is also a diagnostic and therapeutic marker of endodermal sinus tumor, and showed good correlation to the tumor growth [173].

AFP also serves as a useful adjunct in follow-up of some patients being treated for germinal cell carcinomas of testis or ovary [174]. It was found that AFP with h-CG, and free h-CG are the most marker for the diagnosis, prognosis, and monitoring of patient with testicular germ-cell tumors. AFP is good indicator of patients likely to be refractory to treatment [175,176].

1.9.5 Detections Methods of (AFP):

Several methods have been used to detect or measure AFP levels, which include: radioimmunoassay (RIA) [154,177,178], immunoradiometric assay (IRMA)[179,180], double electroimmunodiffuiosn in agar gel (Ouchterlony's method)[181,182], immuno-electrodiffusion (IED)[183], immunoelectrophoresis[181], enzyme-linked immunosorbent assay (ELISA)[184-186], immunohistochemical [187,188], immunofluorescence [189], immuno-cyto chemical method [190,191], chemiluminescent immuonoassay (CLIA)[192,193] and radio-immunoscintigraphy [194].

1.9.6 AFP and Colorectal Cancer:

The relationship between AFP and colorectal cancer have been studied previously through the following studies:

a) In an adult who is not pregnant, the AFP level is elevated with certain cancers include: certain cancers of liver, colorectal cancer, ovarian cancer, pancreatic cancer, and testicular cancer.[195]

b) It is interesting that AFP is expressed by tumors such as primitive gastrointestinal and colorectal cancer. [196-198]

c) AFP has been correlated with increased risk for recurrence and poor survival, particularly in later stage, and in colorectal cancer [92].

d) This tumor marker, that is, AFP is also elevated with other tumors such as, tumor of colon [90] and rectum [91].

e) AFP has been reported to display regulatory activity on different types of tumor, as well as, normal cell [199].

Aim of The Work:

The aims of this work include:

1- Determination of AFP levels in sera of patients with malignant and benign colorectal tumors by immunoradiometric assay (IRMA).

2- Application of (IRMA) method for determination AFP from tissue homogenates in malignant and benign colorectal tumors.

3- Determination of the optimum conditions for the binding of ^{125}I-anti AFP antibody with cytosolic AFP in malignant and benign colorectal tumors.

4- Characterization of the binding of ^{125}I-anti AFP antibody with partially purified AFP in benign and malignant colorectal tumors.

5- Determination of kinetic and thermodynamic parameters for the binding reaction of crude and partially purified AFP with its specific antibody.

6- Spectroscopic studies on standard h-AFP, anti-AFP, and (AFP/Anti AFP) complex.

Chapter Two

Abstract:

Alphafeto protein (AFP) levels were measured in sera and tissue of two groups of patients with colorectal tumors by

immunoradiometric assay (IRMA). Group I consists of 13 patients with benign colorectal tumor and group two (II) consists of 21 patients suffering from colorectal cancer. In addition, a group of control subjects consists of 18 patients was included. The obtained results revealed that the level of Alphafeto protein (AFP) in group (II) was increased significantly (P<0.001), as compared with healthy subjects sera.

The optimum conditions of the binding of AFP in nuclear and cytosolic fractions were studied in order to develop the IRMA procedure to examine the presence of AFP in tissues of patients with colorectal tumors. The results showed a higher incidence of AFP in the group of malignant tumor and cytosolic fraction more than nuclear fraction. The optimum conditions found for estimation of AFP in the groups were as follows: protein amounts (300μg) for benign and (200μg) for malignant, tracer antibody (2.38 mg/ml) for benign and (1.47mg/ml) for malignant group. The optimum pH was 7.0 for benign and 7.2 for malignant. The optimum time and temperature were (180 min and 25oC) for benign and (180 min and 4oC) for malignant groups.

The uses of different halides were shown to cause increasing of the binding of [125]I –anti AFP Antibody to AFP in all studed groups. Also, studies show that increase the binding of AFP with its antibody by using divalent cation in both groups.

Introduction:

Different studies confirm that the level of AFP is elevated in tumors of Gastrointestinal Tract (GIT) systems [200]. Alpha fetoprotein is found in everyone's blood. Higher levels are found in the blood of pregnant women, fetuses, and young children. AFP levels can also elevate in the blood of people with certain diseases and conditions. In adults who are not pregnant, the AFP levels are elevated with certain cancer [201].

It is clear from the above information that there is a relationship between AFP and colorectal cancer, therefore, this part of the study was carried out in order to find the optimum conditions of AFP binding with its antibody in case of colorectal tumors, as well as the effect of different parameters that affect on the binding reaction.

2.1 Chemicals, Instruments and Samples:

All chemicals and reagents used in this study were of analar grade

Table (2-1): Chemicals used and companies provided with

Chemical.	*Company.*
1 Immunoradiometric assay for AFP was purchased from Immunotech (a Beckman company), France.	
2–Bovine Serum Albumin (BSA) ,Tris (hydroxy methyl amino methane) hydrochloride ,$MgCl_2$, $CaCl_2$,EDTA , and Sucrose	Fluka (Switzerland)
3– $CuSO_4.5H2O$, Na,K–tartrate ,NaOH , HCl , Na_2CO_3 ,NaF ,NaCl ,NaBr, $MnCl_2$, and Folin – ciocalteaue	BDH ,Limited pool (UK)
4– Blue dextran (2000) , Sephadex G 200	*Pharmacia Fine Chemicals (Sweden)*

2.2 Instruments:

Table (2–2): Instruments used and companies provided with:

Instruments	company
1 – Gamma counter Type 1270–rack GammII	LKB
2 – Cintra 5 UV/visible Spectrophotometer, SM–Shaker	England
3 – PH M62 Standard pH meter	Denmark
4 – Sartorius analytical balance BL 210 S	Germany
5 – Cooling centrifuge type 202–; with maximum Speed 13500 r.p.m.	Sigma

2.3 Patients:

A total of 34 colorectal patients involved in this study with benign and malignant tumors subjected to curative surgery. Their mean age was 49 years ranges (16–88 years). Two groups of colorectal tumor patients were involved in this study; one group with benign colorectal tumors and the other was with malignant. These groups were matched with a group control subjects (group III).

According to the histopathological examination of the resected pieces, the patients were grouped into the following:

Group (I): Consisted of 13 patients with benign colorectal tumors.

Group (II): Consisted of 21 patients with colon and rectum cancer.

Blood samples were also collected from all patients involved in this study, and also from normal donors with range between (18–62)

Years. The patients were admitted for treatment and diagnosis to the following hospitals in Baghdad:

*Iraqi College of Medicine, Teaching Hospital.

*Baghdad Teaching Hospital (Medical City).

*AL–Kindy Teaching Hospital.

*AL–Yarmook Teaching Hospital.

Patients with diseases that may interfere with this study were excluded. All surgical operations of malignant and benign tumors were done under the supervision of surgeons:

Dr. Zuhair AL–Bahraini, Dr.Saaeb Sedeq, Dr.Faleh AL–Aubaidy, Dr.Abd AL–Salam AL–Taai', and Dr.Maa'd Medhat.

2.4 Blood Sampling:

Blood samples (3–5) ml were obtained from patients of groups mentioned above, by vein puncture before surgical operation .The whole blood was left for (10–20) min. at room temperature. After coagulation, the serum was separated by centrifugation at 3000 r.p.m. for 10 min . Serum specimens were then frozen at -20°C until been used for different experiments assayed.

Table (2–3): The host information of patients and healthy which are used in this study:

Patients	Number	Type of Tumor	Age Range Year
Group I	13	Benign tumor	21–38
Group II	21	Colon and rectum cancer	41–66
Group III	18	Control	18-62

The weight of resected tissue samples range between (1.6-18) gm.

2.5 Specimens Collection:

The specimens were surgically removed from patients of colon and rectum (CR). They were immediately rinsed with ice–cold saline solution, and immersed in the same solution. They were collected and stored at -20°C until homogenization.

2.6 Preparation of Tissue Homogenate:

The frozen tissue were washed with ice-cold normal saline and then weighed. The samples were minced, pulverized, with a scalpel scissors in the Petri dish on ice bath, and then homogenized at 4°C in tris buffer (0.05M , pH 7.4) with ratio of 1 : 4 (weight : volume) using hand homogenizer .

The homogenates were filtered through a nylon mesh sieve in order to eliminate fiber connective tissue, and then centrifuged at 4°C. The supernatants and pellets were considered cytosolic and nuclear fractions respectively. The pellet (sediment) was discard, and the cytosolic (supernatant) was used in experiments involved cytosolic cancer antigen AFP source.

Solutions:

TES Buffer solution (0.05M, pH 7.4) was prepared as follows: (3.0285gm) of tris (hydroxy methyl amino methane), 0.93060 of Ethylene diamine tetra acetate disodium salt (EDTA) and (42.7875gm) of sucrose were dissolved in 400 mL of deionized distilled water, then the pH was adjusted with HCl (1M) at 7.4 and the solution was completed to 500 ml with deionized distilled water.

2.7 Protein Determinations:

The method of Lowry et al [151] was used to determine total proteins in tissue and sera, using bovine serum albumin (BSA) as standard protein.

Solutions:

1–Standard bovine serum albumin (BSA) (1mg/ml).

2– Solution A, Alkaline Sodium carbonate solution (2% Na_2CO_3 in

0.1N NaOH).

3–Solution B (Copper Sulphate –Sodium Potassium tartrate solution (0.5 % $CuSO_4$ .5H_2O in 1% Na–K tartrate.

4– Solution C, Alkaline copper solution, 50 mL of solution A was mixed with 1mL of solution B discard after one day.

5– Solution D, Folin ciocalteau solution, prepared by the dilution of the commercial solution with an equal volume of distilled water on the day of use.

Procedure:

1- One milliliter of standard bovine serum albumin (BSA) containing (0, 25, 50, 75, 100, 150, 175, 200) µg /ml protein was pipetted in a set of duplicate tubes.

2- A set of duplicate tubes containing 150 µL of cytosolic fraction of tissue specimens, and the volume were made to one mL with distilled water.

3- Five milliliter of solution C was added to all tubes. Then the contents were mixed by vortexing, and allowed to stand for 10 min. at room temperature.

4- Half milliliter of solution D was added drop by drop with mixing. The mixture was left to stand for 30 min. at room temperature.

5- The absorbance of the developing color was read at 600 nm against the blank.

6- The standard curve was obtained by plotting the absorbance against the corresponding concentrations of standard protein and used to determine the unknown protein concentration of tissue homogenate specimens and serum as shown in Fig. (2–1).

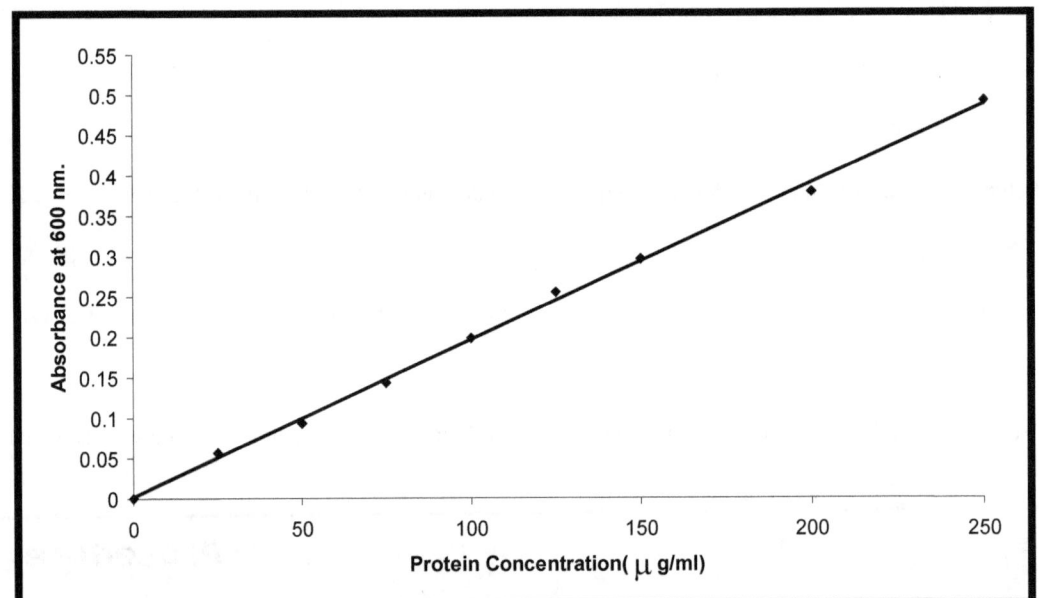

Figure: (2–1) Standard curve of protein determination (All other details are explained in the text).

2.8 Determination of Cancer Antigen AFP Level In Sera of Patients With Benign and Malignant Colorectal Tumors:

Serum AFP levels were measured by ImmunoRadiometric Assay (IRMA). The assay is a two site "Sandwich" assay in which two mouse monoclonal antibodies directed against two different epitopes of the molecule were used.

Samples or standards were incubated in tubes, coated with the first monoclonal antibody, in the presence of the second, ^{125}Iodine – labeled monoclonal antibody. Following incubation the liquid contents of the tubes are aspirated and the excess unbound, labeled antibody was removed by washing.

Solutions:

The solutions used in this assay were provided with the kit described as follows:

1- Monoclonal ^{125}I– labeled anti AFP tracer Antibody, one vial contains 22 µl (320 kBq) of ^{125}I-immunoglobulin in buffer with proteins, and sodium azide (<0.1%).

2- Anti– AFP monoclonal antibody coated tubes.

3- AFP standards, six vials (0.5 ml each). The vials contain AFP concentrations of (0, 3.20, 10.6, 42.4, 159, and 424) IU/ml. The standard have been calibrated using International Standard of alpha fetoprotein. The standard are in serum and preserved by sodium azid (<0.1%).

4- Control serum, two vials contain lyophilized in the presence of preservative.

5- Phosphate buffer: one vial contains 30 ml preserved by sodium azide (<0.1%).

6- Wash solution: one vial contains (50 ml) concentrated solution has to be diluted with 950 ml of distilled water before use.

Procedure:

The assay details in the following procedure described in the leaflet provided from a Beckman Company France Company is:

1- Number a duplicate series of ^{125}I – anti AFP Antibody coated tubes.

2- Add sequentially either 100 µL of standard control or samples to the bottom of tube, followed by 300µL of tracer, vortex gently.

3- To two additional tubes add µL of tracer in order to obtain total c.p.m (T).

4- Incubate all tubes for four hours at room temperature (18–25°C) with moderate horizontal shaking (4000r.p.m)

5- Aspirate contents of all tubes carefully, except of those for total c.p.m.

6- Add 2 mL of wash solution to each tube (except tubes for total c.p.m) and immediately aspirate contents of the tubes. Repeat this operation twice. No trace of dye should remain.

7- Measure radioactivity of tubes for counts bound (B) and total (c.p.m) (T).

Calculations:

1– The mean net count c.p.m for each group after the subtracting the background of each pair of duplicate tubes were counted in gamma counter for one minute.

2– The (B/T%) ratio was computed for each standard and unknown sample as follows:

$$(B/T\%) = \frac{\text{Standard or sample mean count}}{\text{Total activity mean count}} \times 100$$

3– The standard curve was drawn by plotting the percent value for each standard (vertical axis) versus AFP concentration (horizontal axis) on as shown in fig. (2–2).

4–AFP concentration of unknown were calculated from the standard curve using their duplicate counts.

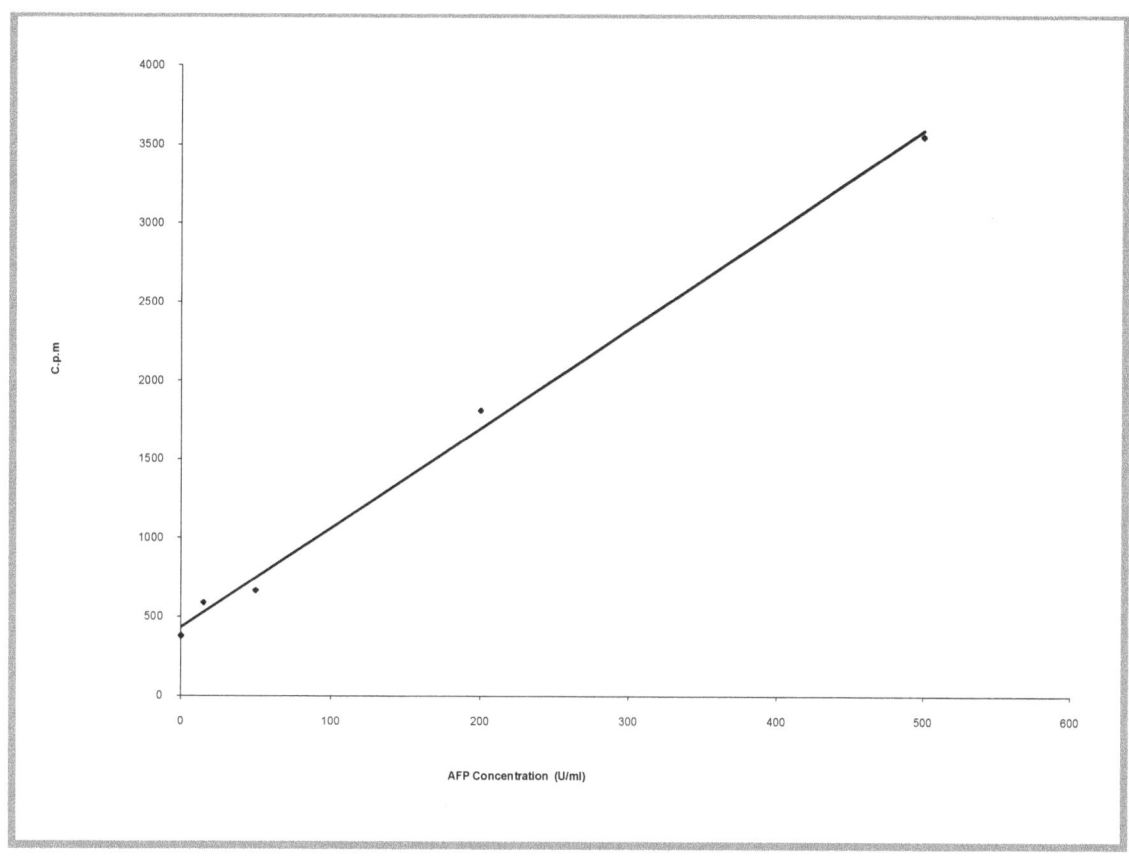

Figure (2–2): Standard curve of AFP determination in human sera
By IRMA method (All other details are explained in

the text).

2.9 Binding Studies of Cancer Antigen AFP In Colorectal Tumors Homogenate With ^{125}I–Anti AFP Antibody:

The detection of cytosolic and nuclear AFP were carried out using an ordinary tubes (not coated with Mab.)

1- 100µl of crude colorectal cytosolic homogenate (i.e., benign and malignant Colorectal tumor) having (1100 and 1000) µg protein was incubated with 50µl of ^{125}I–anti AFP Antibody, the volume of mixture was completed to 250µl with tris buffer (0.05M, pH7.4).

2- All tubes were incubated at 25 ºC for four hours.

3- After incubation, the tubes were centrifuged at 4000 r.p.m for one hour at 4ºC using cooling centrifuge.

4- The supernatant was discarded and the complex formed was counted in gamma counter for one minute.

215

5- Nuclear AFP was estimated by dissolving the sediment of nuclei in tris buffer (pH7.4) with a ratio 1:10 (weight: volume) with shaking.

6- In order to detect the nuclear AFP, 100µl of colorectal homogenate i.e. (benign and malignant colorectal tumor having (183,167mg protein) were incubated with 50ml of ^{125}I–anti AFP antibody and the volume were completed to 250µl with tris buffer (0.05M, pH7.4).

7- Steps 2,3,4 above were repeated.

Calculations:

1-The count radioactivity in each tube (expressed in c.p.m) represents the bound fraction B (i.e. ^{125}I – anti AFP Antibody/ AFP) complex.

2- The count ratio activities in the tubes containing ^{125}I–anti AFP antibody alone represent the total activity.

3- The ratio of (B/T)% for each tube was calculated as follows:

$$(B/T) = \frac{\text{Sample mean count (B)}}{\text{Total activity mean count (T)}} \times 100$$

Solution:

Tris buffer (0.05M, pH7.4) was prepared by dissolving (0.6057gm of tris (hydroxy methyl amino methane) in 50 ml of deionized distilled water and the pH was adjusted with HCl (1M) at pH7.4, the volume was completed to 100ml with deionized distilled water.

2.10 Factors effecting ^{125}I–Anti AFP Antibody Binding To AFP In Colorectal Tissue Homogenate:

2.10.1 The Effect of Different Protein Concentration of The Homogenate on The Binding:

1- Fifty micro liter of ^{125}I–anti AFP Antibody were added to 100µl of crude cytosolic homogenate (benign and malignant

216

Colorectal tumor) containing increasing amounts of protein concentration (50,100,200,300,400,500) μg completed to a final volume of 250ml with tris buffer (0.05M,pH7.4).

2- The tubes were incubated for four hours at 25 ºC.

3- Two additional tubes containing 50μl of ^{125}I–anti AFP antibody only, for total activity computation were set a side until counting.

4- After incubation, the tubes were centrifuged at 4000r.p.m for one hour at 4ºC using cooling centrifuge.

5- The supernatant was decanted and the radioactivity of the complex formed was counted.

Solutions:

Tris buffer (0.05,pH7.4) prepared as described in the experiment of (binding studies of Cancer antigen in colorectal tumors homogenate with ^{125}I – anti AFP Antibody (2-9).

Calculations:

1 – The B/T percent were determined according to the experiment (2-9)

2 – The percent of binding values B/T were plotted versus increasing amount of protein of the colorectal tissue homogenate.

2.10.2 The Effect of 125I–Anti AFP Antibody Concentration on The Binding:

1– Increasing amounts of ^{125}I–anti AFP Antibody (10,20,30,40,50)μl were added to 100μl colorectal tissue homogenate (300)μg protein for benign and 200μg protein for malignant colorectal tumor.

2– The volume was made up to 250μl with tris buffer pH7.4.

3 – Steps 2,3,4,5 in the experiment (2-10-1) were repeated.

Calculations:

1-Values (B/T)% were calculated as described in the experiment of (binding studies of Cancer antigen AFP in colorectal tumors homogenate with ^{125}I – anti AFP Antibody).

2-Values of (B/T)% were plotted versus concentration of labeled antibody (^{125}I–anti AFP Antibody).

2.10.3 The Effect of pH on Binding:

1- One hundred micro liter of human colorectal homogenate (benign and malignant) containing (300 and 200 μg protein) were added to (40 and 30μl) respectively i.e. (2.38 and 1.47 mg.ml^{-1} of ^{125}I–anti AFP Antibody).

2- Each tube in step1 was completed to 250μl with tris buffer (0.05M, pH7.4) with different pH (6.8 – 8).

3- Steps 2,3,4,5 in experiment of (2-10-1) were repeated.

Calculations:

1-Values (B/T)% were calculated as described in the experiment of (2-9).

2- Values of (B/T)% were plotted versus different pH.

Solutions:

Tris Buffer used as described in the experiment of (binding studies of Cancer Antigen AFP in colorectal tumors homogenate with ^{125}I – anti AFP Antibody).

1-One hundred micro liter of colorectal tissue homogenate (benign and malignant) containing (300 and 200 µg) protein were added to (2.38 and 1.47 mg.ml⁻¹) of ^{125}I – anti AFP Antibody.

2-The volume was completed to 250 µl with tris buffer (0.05 M, pH 7.0 and 7.2) respectively.

3- All tubes were incubated at 25°C at different time intervals (1, 2 , 3 , 4 , 5 , 6 , 7 , and 8) hours .

4-To determine the time course of AFP binding to ^{125}I–Antibody at different temperatures. Steps 1,2 in the same experiment were repeated at different temperature (4, 37 and 45) °C.

5- Steps 3,4,5 in experiment (2-10-1) were repeated.

Calculations:

1 – (B/T)% was calculated as described in the experiment of (2-9). At each time and temperature.

2 – The values of (B/T)% were plotted versus the time at different temperature.

Solutions:

Tris buffer was used as described in the experiment of (2-9).

2.10.5 The Effect of Different Halides on Binding:

1- At the optimum conditions of binding of AFP in colorectal tissue homogenate in both groups 100µl of each group of homogenate (300 and 200µg protein) and (2.38,and 1.47 mg.ml⁻¹) of ^{125}I–anti AFP antibody were incubated.

2- Fifty microliter of the following halides (0.01M) (NaI , NaBr , NaCl, and NaF) were added to the mixture.

3- The volumes of the mixture were completed to 250µl with tris buffer (pH 7.0 and 7.2) for each group of homogenate respectively.

4- All the tubes were incubated at the following temperature and time for the groups (3hr.at 25 °C for benign and 3 hr. at 4 °C for malignant). A sample of each group of tissue homogenate, as a control, was left without any addition of any halide and incubated at the optimum conditions for each group.

5- Steps 3, 4, 5 mentioned in experiment (2-10-1) were repeated.

Calculations:

1 –The value of (B/T)% was calculated as mentioned in the experiment of (2-9).

2 –The value of binding percent was plotted versus halides concentration.

Solutions:

1- Tris buffer was prepared as described in the experiment of (2-9) and was adjusted to the corresponding pH for each group of tissue homogenate.

2- Halides solutions were prepared in concentration (0.01M) in tris buffer solution, and the pH was adjusted to the corresponding pH for each group of tissue homogenate.

3- Four types of halides were prepared by dissolving 0.021gm of NaF, 0.0292gm of NaCl, 0.0515gm of NaBr, and 0.075 gm of NaI. Each type of halides were dissolved in tris buffer and the pH were adjusted to the corresponding pH of each group of tissue homogenate then the volume were completed to 50 ml with tris buffer.

1- **Fifty microliter of (0.025M) of the following divalent cations (MgCl₂, MnCl₂, CuCl₂, and CaCl₂.2H₂O) were added to each group of tissue homogenate at the optimum conditions for each group.**

2- **The incubation was made up for each group at the optimum time and temperature.**

3- **A sample of each group of tissue homogenate at the optimum conditions was left a side without addition of any salt was used as a control.**

4- **Steps (3, 4, 5, 6) in experiment of (The effect of different protein concentration of homogenate on the binding) were repeated.**

Calculations:

The value (B/T)% was calculated as mentioned in the experiment of (2-9) The value of binding percent was plotted versus each divalent cations.

Solutions:

1- **Tris buffer prepared as described in the experiment of (2-9) and adjusted to the corresponding pH for each group of tissue homogenate.**

2- **Divalent were prepared by dissolving (0.2541gm of MgCl₂.6H₂O, 0.1388gm of CaCl₂.2H₂O, 0.2474gm of MnCl₂.4H₂O, 0.3150gm of CuSO₄.5H₂O) in tris buffer, the pH was adjusted to the corresponding pH for each group of tissue homogenate at the optimum pH for each group, and the volume was completed to 50ml with tris buffer.**

Results and Discussion

Two groups of colorectal tumor were included in this study. These groups were classified according to the type of tumor (benign and malignant) and each type confirmed by histopathological examination. The homogenates were centrifuged at 9000 r.p.m and 4°C to obtain the pellet and supernatant. The pellet represents the nuclear source, while the supernatants were used to obtain the cytosolic source [202]. The Homogenization and centrifugation were carried out at 4C° in order to avoid protein denaturation [203].

The filtration of the tissue homogenate through several layers of nylon gauze was used to remove any suspended pieces of unhomogenized fragments and blood vessels, while the centrifugation of homogenate at 9000 r.p.m removed the unruptured cells and intact nuclei of the ruptured cells [204].

Determination of AFP Level In Sera of Patients With Colorectal Tumors:

AFP levels in sera were measured with an immuno radiometric assay (IRMA) in two groups of colorectal tumors matched with one group of control subject. Group I consisted of thirteen patients with benign colorectal tumors, group II included twenty-one patients with colon and rectum cancer as summarized in table (2–3). The groups were matched with a group of control subjects (group III).

Table (2–4) shows that the AFP levels in malignant group were highly significant elevated ($P < 0.001$) and low significant elevation ($P < 0.05$) for benign colorectal tumors as compared with the control according to the student's T–test analysis [205].

The mean serum AFP levels of the control were found to be (15.4 ng±1.5) which is equal to 11.396 ±0.47 as shown in table (2 – 4). For calibration, the equation was used; 1ng AFP = 0.74 IU AFP.

The elevation of AFP in patients with colon and rectum cancer was in agreement with other studies found in this field [200,206].

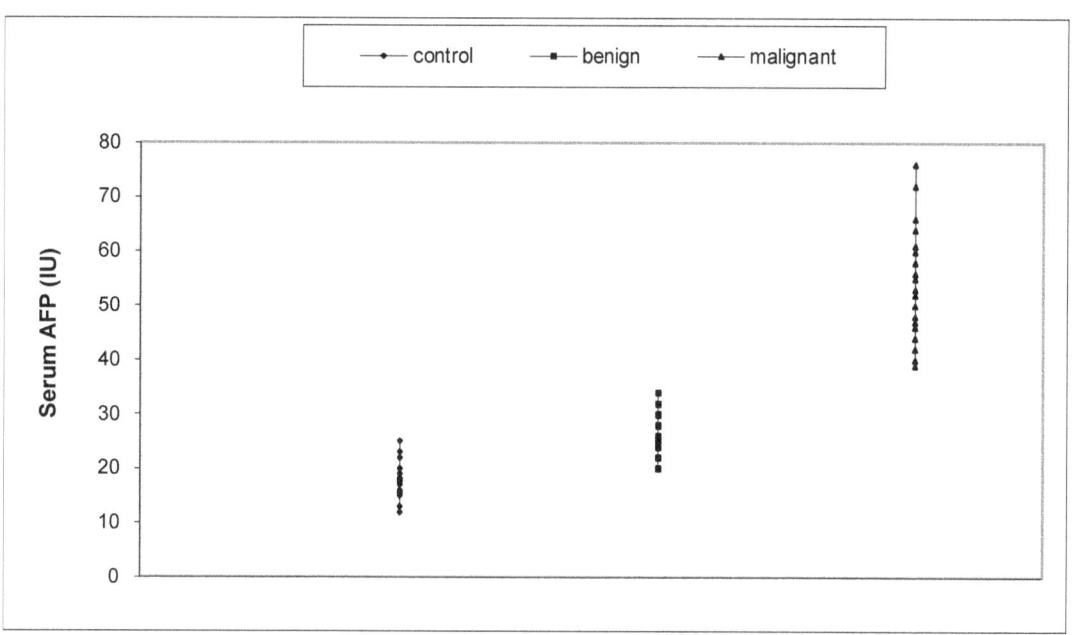

Figure (2 – 3): Distribution of AFP level (IU) in human sera by IRMA. (All other details are explained in the text).

Table (2–4): Sera AFP levels (IU) in patients with benign and malignant colorectal tumors.

Group	Patients	No. of cases	Age range year	Sera AFP IU±SD	P values
I	Bengin tumor	13	21–38	25.86±1.38	$P < 0.05$
II	Colon and rectum cancer	21	41–66	52.26±1.21	$P < 0.001$
III	Control	18	18-62	11.396± 0.47	

Binding Studies on AFP In Colorectal Tissue with 125I–Antibody

Preliminary Test of The Binding of AFP With 125I–Anti AFP

Cytosolic and nuclear AFP were investigated in groups of colorectal tumors homogenate (benign and malignant). In each group, AFP was detected by the incubation of ^{125}I–anti AFP Antibody with both cytosolic and nuclear fractions for 4 hr. at 25C° and the reactions solutions were completed with tris buffer. The separation of the bound antibody from the unbound was carried out at 4000 r.p.m for 1 hr. to precipitate the (^{125}I–Antibody / AFP) complex formed.

Table (2–5): Preliminary conditions of AFP in cytosol and nuclear fraction in different colorectal homogenate.

Group	B/T %	
	Cytosolic fraction	*Nuclear fraction*
Benign	4.5	0.8
Malignant	5.6	1.3

Table (2–5) shows the amount of binding B/T % values in both cytosolic and nuclear fractions. The data revealed that AFP was higher in cytosolic fractions than in nuclear, and in general AFP concentration in malignant tumor homogenate was more than that in benign tumors tissue. This observation is in good agreement with the results of several authors [207]. From these results, it can be said that the developed method was useful for determination AFP in colorectal tumor homogenates using ^{125}I-anti AFP antibody.

Optimum Conditions For The Binding of Cancer Antigen (AFP) In Colorectal Tissue Homogenate With 125I–Anti AFP:

Optimum Protein Concentration:

The increasing amount of homogenates (i.e. increasing amount of AFP) is good way to determine the best protein concentration. To estimate the suitable concentration of homogenates, 100 μL of ^{125}I–anti AFP antibody were incubated with increasing concentration of cytosolic homogenate, according to the details in experiment of (The effect of different protein concentration of homogenate on the binding).

Figure (2 – 4) represents the formation of (^{125}I–anti AFP Antibody / AFP) complex in the groups of colorectal tumors (benign and malignant).

The results revealed that the binding of AFP to ^{125}I–anti AFP Antibody increases with increasing protein concentration. Figure (2–4) shows that 200μg, 300μg protein were the most appropriate concentrations to give the maximum values of binding in benign and malignant respectively. The decrease in the binding after reaching the maximum binding may be due to the solubilization of the complex formed by the excess of AFP added [208], or may be due to the conformational changes in AFP and ^{125}I–anti AFP Antibody rather than the formation of reversible inactive (^{125}I–anti AFP Antibody / AFP) complex [209], another auther reported that when the precipitation of the complex out of the solution, due to the multivalent nature of both molecules [210]. The radioactive antibody has two binding sites, it can cross link antigenic sites of two different AFP molecules and can form maximum amount of the complex, therefore maximum precipitate will occur [210].

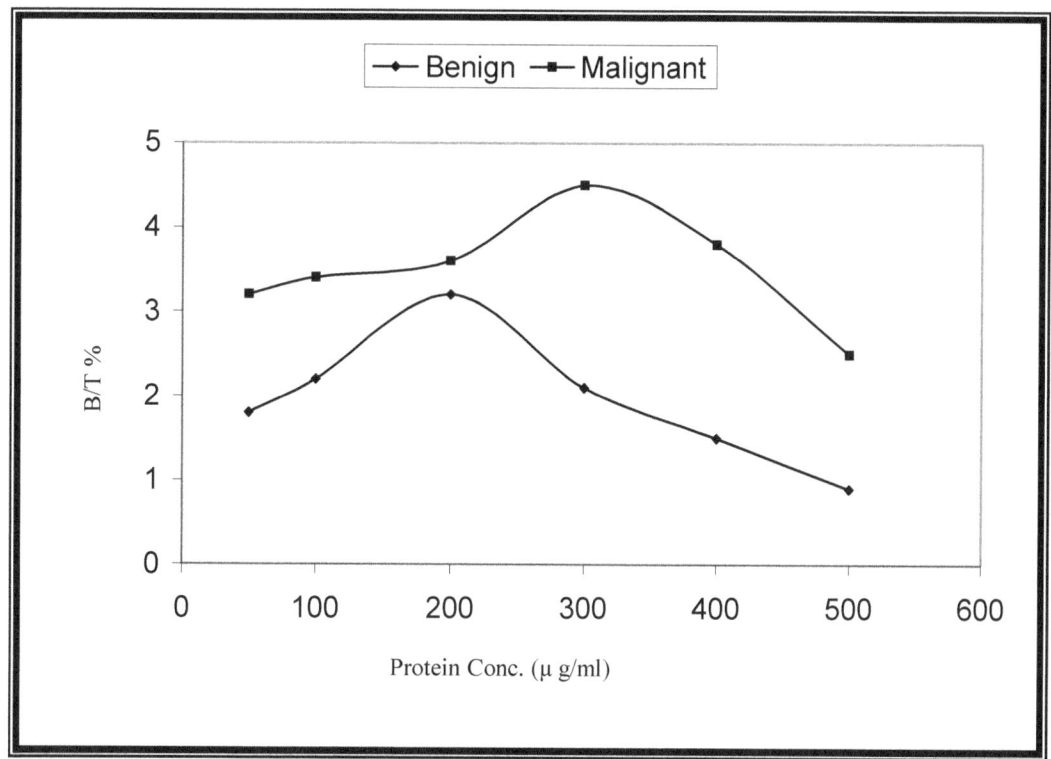

Figure (2–4):Influence of protein concentration on the binding with [125]I– anti AFP Antibody (All other details are explained in the text)

The Effect of 125I–Anti AFP Antibody Concentration on The Binding:

This experiment was carried out in the presence of fixed amount of protein concentration of the colorectal tumors homogenate and increasing concentration of ^{125}I–anti AFP Antibody. The results obtained are shown in figure (2–5), It is clear that the amount of (^{125}I–anti AFP Antibody / AFP) complex rises gradually and until the colorectal tumors homogenates was saturated with ^{125}I–anti AFP Antibody.

The maximum binding occurred when the concentration of the antibody were (2.38 and 1.47 mg.ml^{-1}) for the groups of colorectal tumors (i.e. benign and malignant) respectively. Then the binding percent decreased as the amount of ^{125}I–anti AFP antibody increased. The reason is due to the all-antigenic sites covered with antibody and the complex formation is inhibited [211]. These results indicate that the binding is principally dependent on the amount of the antibody in the reaction mixture [212], because one of the factors affecting the binding percent of Antibody – Antigen reaction is the concentration of the Antibody. According to the results of this experiment, the above concentrations of ^{125}I-anti AFP antibody were used in the subsequent experiments.

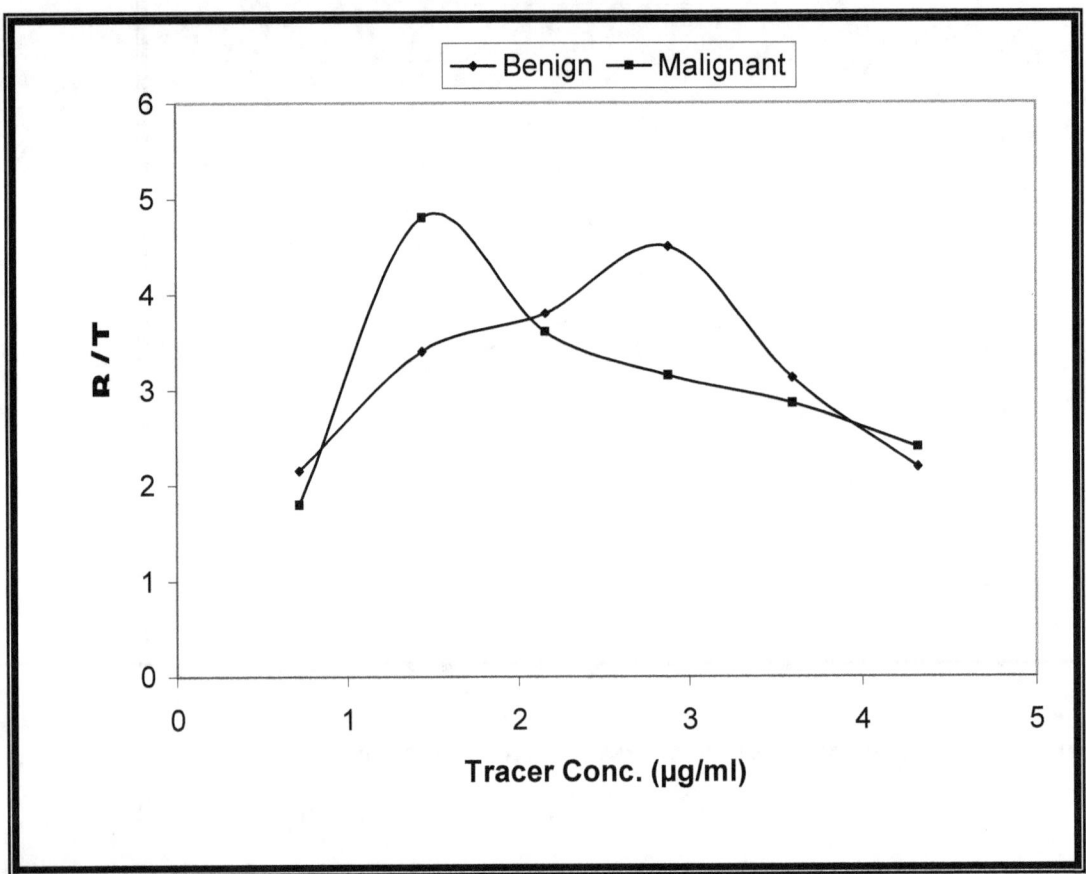

Figure (2–5): The effect of different concentrations of ^{125}I– anti AFP Antibody (All other details are explained in the text).

The Effect of pH on The Binding:

The effect of pH on the binding of AFP with ^{125}I–anti AFP Antibody was illustrated in figure (2 – 6). It is obvious that the binding occurs when the pH were (7.0 and 7.2) for the groups of colorectal tumors (benign and malignant) respectively.

Immunoprecipitates is usually performed at pH between 7.0 and 9.0; the immunoprecipitates were soluble below pH 4.5 ,and above pH 10.0. However these results indicate that the binding was pH dependentand the shift in the pH of environment might affect the properties of the macromolecules that involved in the binding. This effect may include the protenation-deprotenation process occurring within the possible ionization groups of the amino acids present in the binding domain of the molecules [213]. The protein AFP binding site which is composed of ionizable groups should be in the proper ionic form in order to maintain a proper conformation of the binding site to bind ^{125}I–anti AFP Antibody [214]. In addition, ^{125}I–anti AFP antibody itself may have ionizable groups and only at a certain pH the Antibody will have ionic form where it can bind to AFP [215].

Figure (2–6): Effect of pH on the binding of ¹²⁵I–anti AFP Antibody with AFP (All other details are explained in the text).

Time and temperature of The Binding of 125I–Anti AFP Antibody to AFP In Colorectal Tumors:

Figure (2–7) show the results obtained from the time course pattern at different temperatures (4, 25, 37 and 45°C). The maximum binding occurred at 4°C after incubation for 3 hours in crude fractions of benign and at 25°C after incubation for 3 hours in malignant tumor homogenates. These results indicate that ^{125}I–anti AFP Antibody binding to crude fraction of AFP are temperature and time dependent process.

The decrease of the binding may be due to either the degradation of AFP or irreversible dissociation of the (^{125}I–anti AFP Antibody / AFP) complex. At higher temperatures, denaturation and destruction tertiary structure may occur leading to loss of activity and conformational changes. At lower temperature, heat is not enough to overcome the energy barrier, even for the catalyzed reaction [216]. Heating more than 45C° disrupt the folded structure of the protein by increasing the vibrational motions of atoms [217].

The optimum conditions concluded out were used in all subsequent experiments.

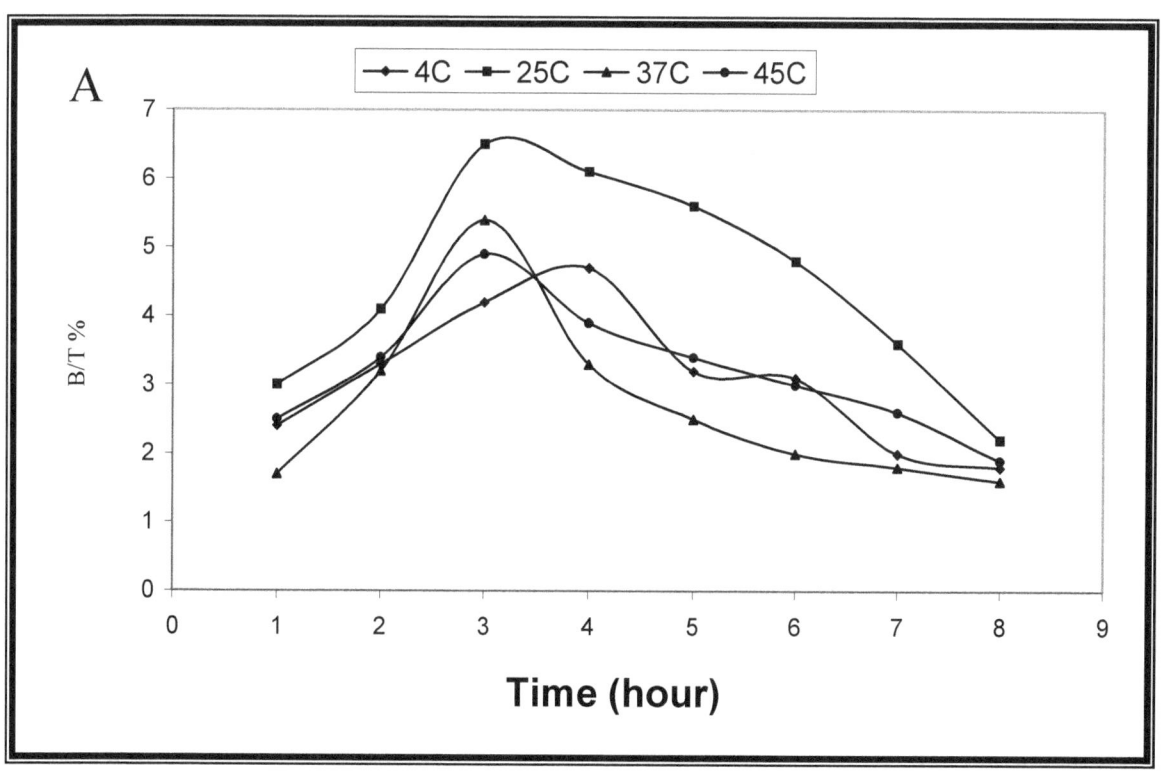

**Figure (2–7): Time course of the of ^{125}I–anti AFP Antibody binding to AFP in
(A) Benign Colorectal tumor. (B) Malignant colorectal tumor.**

(All other details are explained in the text).

The Effect of Different Halides on The Binding:

Figure (2–8) shows the effect of different halides on binding of ^{125}I-anti AFP antibody with AFP in colorectal tumor tissues. To study the effect of these halides on the binding of antigen-Antibody, (0.01 M) concentration of (NaI, NaCl, NaF) were used. The results indicated that the sodium halides increase the Antibody–Antigen (Ab – Ag) binding according to the following order for the (benign and malignant colorectal tumor): –

$$NaF < NaCl < NaI$$

Melander and Horvath [218] reported that the effect of halide salt type on hydrophobic interaction is quantified by its molal surface tension increment (MSTI), which is measure of the increase in surface tension, by the salt; also they found that this parameter increases as in the following sequence:

NaF> NaCl> NaI

The same researchers found that halides with higher MSTI values will strengthen the hydrophobic interactions while halides with lower MSTI values reverse this effect In contrast Edigintgton [219] found that chaotropic ions (I,F,Cl) distort,particularly the three- dimensional structure leading to the disruption of antibody-antigen interaction surface. This may be done in presence of high concentrations.

Figure (2–8): Effect of different halide on the ^{125}I–anti AFP Antibody binding to AFP in:

(A) Benign colorectal tumor
(B) Malignant colorectal tumor
(All other details are explained in the text).

The Effect of Divalent Cations on The Binding

The importance of ionic environment for the binding of ^{125}I–anti AFP Antibody to AFP in colorectal tissue homogenate is shown in figure (2 – 9, A and B). The presence of divalent cations (i.e., $CaCl_2.2H_2O$, $CuSO_4.5H_2O$, $MnCl_2.4H_2O$ and $MgCl_2.6H_2O$) at 25mM concentration increased the binding. The reason may be due to that the salt that may alter the nature of hydrophobic forces controlling stabilization of the complex formed and these vary depending on the nature of the interacting groups [220]. From the results illustrated in figure (2–9, A and B), it is suggested that these salts may provide some conformational changes in the AFP and the charged groups of the binding domain of the Antibody and Antigen molecule [221, 222], hinder maximal binding are shielded. If the interaction is dominated by ionic strength, high salt concentration lowers the affinity.

On the other hand, the presence of $MgCl_2$ at 25mM concentration inhibited the binding. This may be due to increased (Ab–Ag) complex solubility in the presence of these cations. The interaction of these ions with ionic groups of the (Ab – Ag) complex diminishes the Ab, Ag interactions, and therefore, increasing solubility of complex [223].

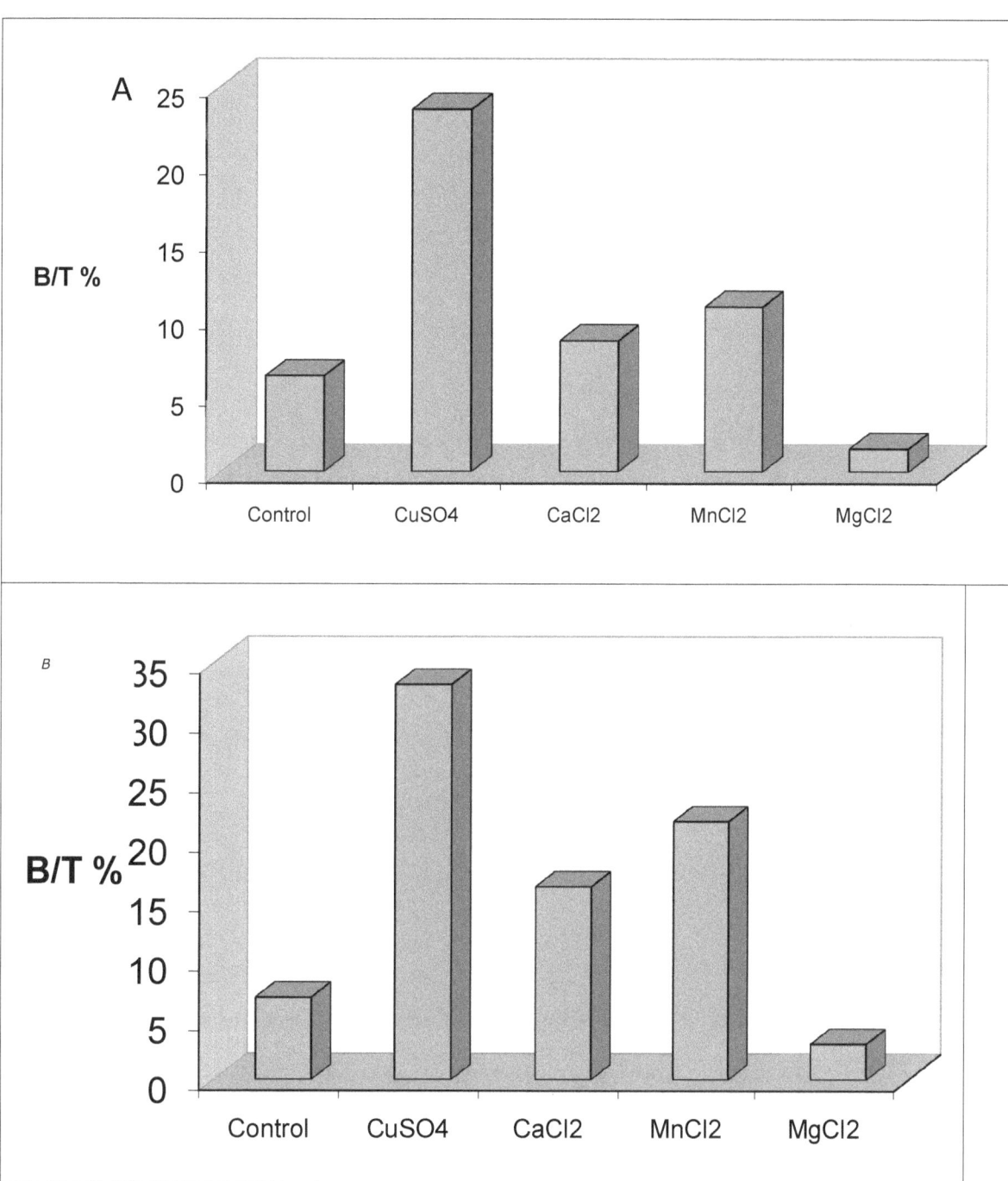

Figure (2-9): Effect of different divalent cation on the binding of 125I–anti AFP Antibody with AFP in:

(A) Benign colorectal tumor

(B) Malignant colorectal tumor

(All other details are explained in the text).

Chapter Three

Abstract:

Gel filtration was used to partially purify AFP from homogenate of colon tumors. It is useful to use gel filtration to isolate the ^{125}I- anti AFP/AFP bound (complex) from free (unbound) ^{125}I- anti AFP antibody.

The results revealed that the elution profile gave two peaks by using Sephadex G 200, the first peak with high molecular weight presenting the complex and the second peak represents a free antibody.

The optimum conditions of binding of the partially purified AFP with ^{125}I-anti AFP antibody were carried out.

Many papers study the methods of separation and isolation of Alphafeto protein (AFP) from the bound complex after complete the binding reaction between antigen and antibody [224, 225].

The literatures [224, 225] refer that the molecular weight of AFP is approximately 68 Kda [226], therefore it is useful to use gel filtration to carry out the separation, especially Sephadex G 200 that is suitable for the molecular weight.

The study of optimum conditions for binding of partially purified AFP with its labeled antibody is necessary step to characterize the protein molecules and note the differences in case of minimizing the competitors by purification.

Chromatography encompasses a diver and important group of separation methods that permit the scientist to separate, isolate and identify related components of complex mixtures.

In this chapter, gel filtration was used to separate ^{125}I-anti AFP antibody bound to AFP in cytosolic fraction of colorectal tumor homogenate from unbound (free) ^{125}I-AFP antibody. The factors that affect the binding of partial purified AFP to its antibody were also studied.

Materials and Methods:

3.1 Materials:

Chemicals:

All chemicals and reagents mentioned in chapter two were used in the experiments of the chapter.

3.2 Instruments:

All instruments mentioned in chapter two were used in the experiments of this chapter.

3.3 Patients:

The same patients of colon cancer and benign tumor mentioned in chapter two were used in the experiments of this chapter.

3.4 Partial Purification of AFP by Sephadex G 200 Column:

3.4.1 Preparation of The Column:

The dimensions of the column were chosen according to the following equation:

$$\text{Diameter} = \sqrt[3]{m / 10}$$

Where:

m: amount of protein in mg.

L= 30x diameter.

Where

L: length of column.

3.4.2 Preparation of The Buffer:

Tris buffer (0.05M) was prepared by dissolving 3.0285 gm of tris (hydroxy methyl amino methane, 0.9306 gm of EDTA and 0.1 gm of sodium azide in 400 ml, the volume was completed to 500 ml with deionized distilled water, the pH was adjusted to 7.2.

3.4.3 Preparation of Gel:

The gel was prepared by allowing the preswollen gel to swell again in tris buffer pH7.2, then left to settle and the excess of buffer was decanted. The step was repeated several times. Suction was then used to degas the gel then the slurry was left for 24 hrs. to equilibrate with buffer.

The swollen gel was suspended and carefully poured into vertical glass column (0.9 x 27) down the wall using a glass rod.

After the gel had settled the column was equilibrated with this buffer for 72 hrs.

3.5 Determination of The Void Volume:

The void volume of the column was determined using blue dextran 2000 at concentration of 2mg.ml–1 dissolving in tris buffer pH 7.2, the elution was carried out with the same buffer at a flow rate of 10ml, hr–1

Fractions of 1ml were collected and their absorbance was measured at 600nm. Figure (3–1) shows the elution profile of blue dextran 2000.The volume of the buffer required to elut the blue dextran, which represents the void volume, was (10 ml).

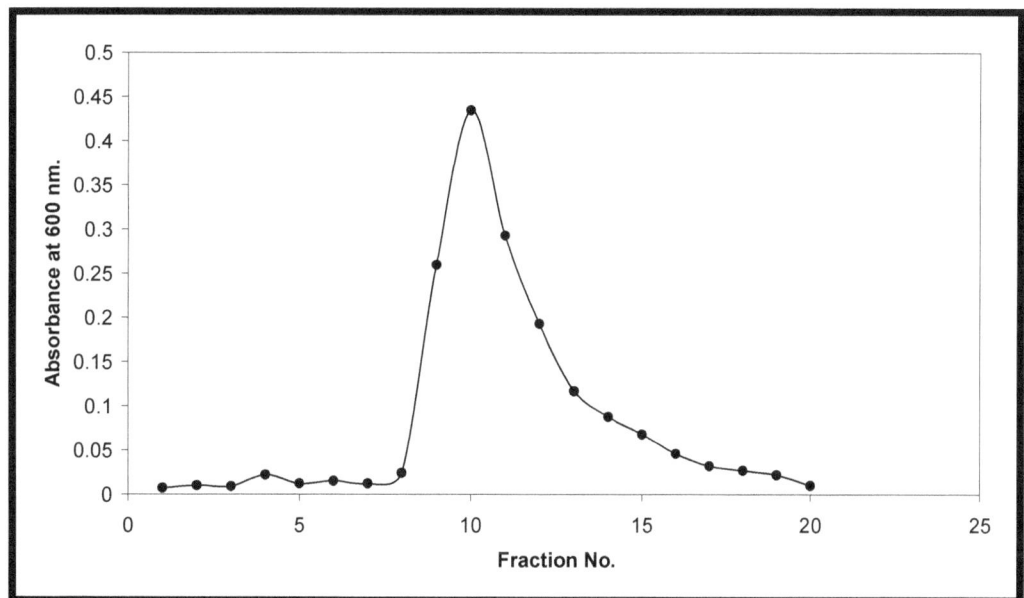

Figure (3–1): The elution profile of blue dextran 2000. (All other details are explained in the text).

3.6 Purification Procedure:

Reagents:

Tris buffer pH (7.2) contained 0.02% sodium azide was prepared as described previously in experiment (3-4-2).

Procedure:

The sample of tissue homogenate (720 µl) of colon and benign tumor containing approximately (8 mg) proteins was applied to the surface of the gel. The elution was carried out using tris buffer (pH7.2) with a flow rate of 10 ml.hr^{-1}, and fraction of one (ml) was collected, the elution was made at room temperature.

Calculations:

7- In each fraction, the protein concentration was determined according to Lowry et.al[151] and the total binding was estimated at the optimum condition of colon cancer and benign tumor, as described in chapter two (Factors effecting ^{125}I – anti AFP Antibody binding to AFP in colorectal tissue homogenate). The binding of each fraction was calculated and plotted against the elution volume. The specific binding activity percent was estimated from the following equation:

$$\text{Specific binding activity} = \frac{\text{Total binding B/T}}{\text{mg of protein}} \times 100$$

8- The isolation fold for AFP was determined using the following equation:

$$\text{Isolation fold of AFP} = \frac{\text{Specific binding of purified AFP}}{\text{Specific binding of crude AFP}} \times 100$$

3.7 Dialysis for Concentration

After preparing a dialysis tube, the fraction that contained high levels of the binding activity were collected, pooled, and concentrated by dialyzing against sucrose at 4°C for 2 hours, to obtain the required concentration to be used in the next experiments.

3.8 The Choice of Optimum Conditions For The Binding of The Partial Purified AFP To ^{125}I–Anti AFP Antibody:

3.8.1 The Choice of The Optimum Protein Concentration:

1- A volume of 40 μl of ^{125}I–anti AFP Antibody (2.38 mg.ml^{-1}) was added to increasing amounts (5, 10, 15, 20, 25 and 30 mg.ml^{-1}) of a pooled fractions under the first peak of the partial purified colon cancer and (9, 18, 27, 36, 45 and 54 mg.ml^{-1}) of a pooled fractions under the first peak of benign colorectal tumor, the volume was completed to 250 μl with 0.05 M of tris buffer pH 7.8.

2- All tubes were incubated for 4 hours at 25°C.

3- Two additional tubes containing 40 μl (2.38 mg.ml^{-1}) of ^{125}I–anti AFP Antibody only for total activity were set–aside until counting.

4- Steps 4,5 mentioned in chapter two the experiment of (The effect of different protein concentration of homogenate on the binding) were repeated.

Calculations:

The (B/T)% was calculated as described in chapter two , the experiment of (binding studies of AFP in colorectal tumors homogenate with ^{125}I – anti AFP Antibody). Then was plotted against protein concentration.

Solutions:

Tris buffer (0.05 M) was prepared by dissolving 0.6057 gm of tris (hydroxy methyl amino methan in 50 ml of distilled water and the pH was adjusted with HCl (1 ml) at pH (7.8). The volume was completed to 100 ml with distilled water.

3.8.2 The Effect of ^{125}I–anti AFP Antibody:

1- Increasing volume of ^{125} I–anti AFP Antibody (10, 20, 30, 40, 50, and 60 μl) containing (0.720, 1.44, 2.16, 2.88, 3.60 and 4.32 mg.ml^{-1}) was incubated with 25 and 18 μg.mL^{-1} of partially purified AFP for the cancer case and benign respectively, the volume was completed to 250 μl with 0.05M tris buffer pH 7.8.

2- All tubes were incubated for 180 min at 25°C.

3- Steps 4,5 mentioned in chapter two in the experiment (2-10-1) were repeated.

Calculations:

The (B/T)% was calculated according to chapter two the experiment of (binding studies of AFP in colorectal tumors homogenate with ^{125}I – anti AFP Antibody) and was plotted versus ^{125}I–anti AFP Antibody concentration.

3.8.3 The Choice of Optimum pH:

1- To choice the optimum pH for the partially purified AFP from colon colorectal tumor homogenate (25μg.ml^{-1}) and (75 mg.ml^{-1}) of the malignant and benign of partially purified AFP were incubated with (30, 20 μl) (2.16, 1.44 mg.ml^{-1}) respectively of^{125} I–anti AFP Antibody. The volume of all tubes was completed to 250 μl with tris buffer (0.05M) of different pH (6.8–8.0).

2- All tubes were incubated for 3 hours at 25°C.

3- Steps 4,5 mentioned in chapter two, the experiment of (binding studies of AFP in colorectal tumors homogenate with ^{125}I – anti AFP Antibody) were repeated.

Calculations:

The (B/T)% was calculated as described in chapter two the, experiment of (binding studies of AFP in colorectal tumors homogenate with ^{125}I – anti AFP Antibody) and was plotted against the corresponding pH.

Solutions:

Tris buffer (0.05M) was prepared as described the experiment of (The choice of the optimum protein concentration) and the pH was adjusted from (6.8 – 8).

3.8.4 The Time Course of Partially Purified AFP:

1- To determine the time course of the partially purified AFP from colorectal tumor homogenate, 25 µg.ml^{-1} protein of the malignant case of partially purified AFP was incubated with 30 µl of ^{125}I–anti AFP Antibody, while 18 µg.ml^{-1} protein of the benign case of the partially purified AFP was incubated with 20 µl of ^{125}I–anti AFP Antibody, the volume was completed to 250 µl with tris buffer (0.05M, pH 7.4).

2- All tubes were incubated at 25°C at different time intervals (1, 2, 3, 4, 5, 6, 7, 8, 9, and 10) hours.

3- To determine the time course of the two groups of partially purified AFP at different temperature, steps 1,2 in the same experiment were repeated at different temperature (4, 37and 45°C).

4- Steps 4,5 mentioned in chapter two the experiment of (The effect of different protein concentration of homogenate on the binding) were repeated.

Calculations:

1- The (B/T)% was calculated as described in chapter two, the experiment of (binding studies of AFP in colorectal tumors homogenate with ^{125}I – anti AFP Antibody) at different time and temperature.

2- The values (B/T) % was plotted against the time at different temperature.

3- The concentration of (^{125}I–anti AFP / AFP) complex formed after time t was calculated from the following equation:

$$^{125}\text{I - anti AFP / AFP} = \frac{\text{count(c.p.m) of }^{125}\text{I - anti AFP specifically bound after time(t)}}{\text{Total counts(c.p.m) of }^{125}\text{I - anti AFP used in the incubation}} \times \text{Concentration of }^{125}\text{I - anti AFP in the incubation (mg.ml}^{-1}\text{)}$$

in mg.ml^{-1} after time(t)

Solutions

Tris buffer (0.05M) was prepared as described in the experiment (The choice of the optimum protein concentration) and the pH was adjusted to 7.4.

Results and Discussion:

Partial Purification of AFP:

Isolation of cytosol AFP antigens was performed by gel exclusion chromatography technique. Colorectal tumor homogenate was applied to Sephadex G 200 (0.9x27 cm). The void volume (Vo) of this column was (10 ml) as calculated from the elution profile of the blue dextran.

Figure (3 – 2) shows the elution profile of AFP from malignant colorectal tumor and benign. The resultant fraction of the homogenate was collected, pooled and detected for the binding with ^{125}I–anti AFP Antibody.

All trials of gel filtration revealed two peaks profile. The first peak represents ([125]I- anti AFP antibody / AFP) complex, while the second peak represents unbound (free) ^{125}I-anti AFP antibody.

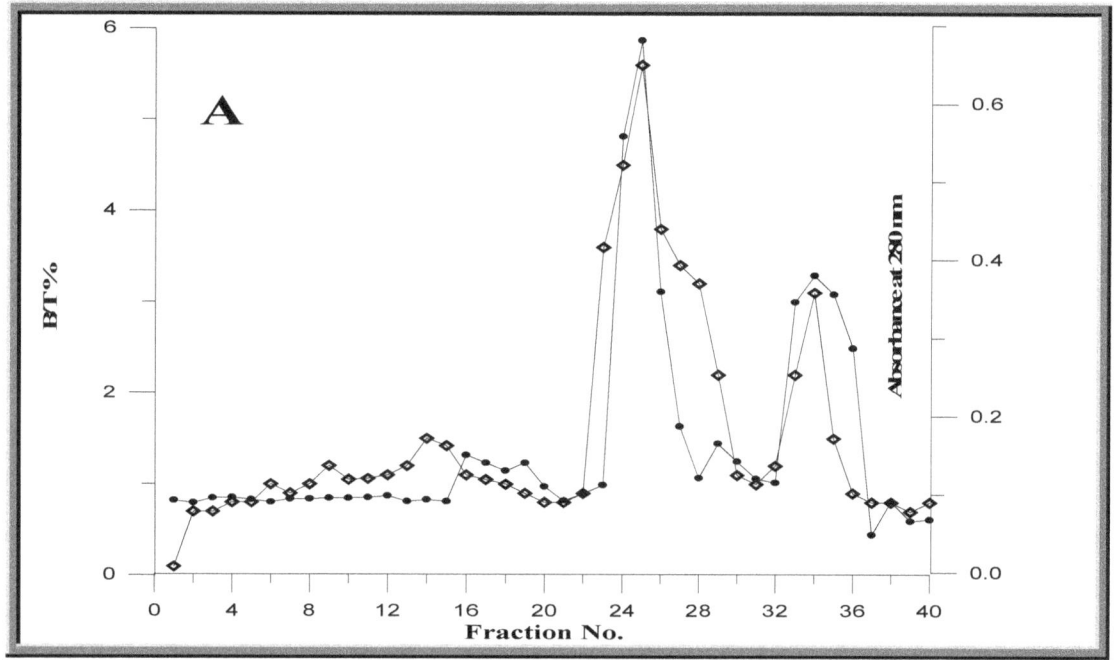

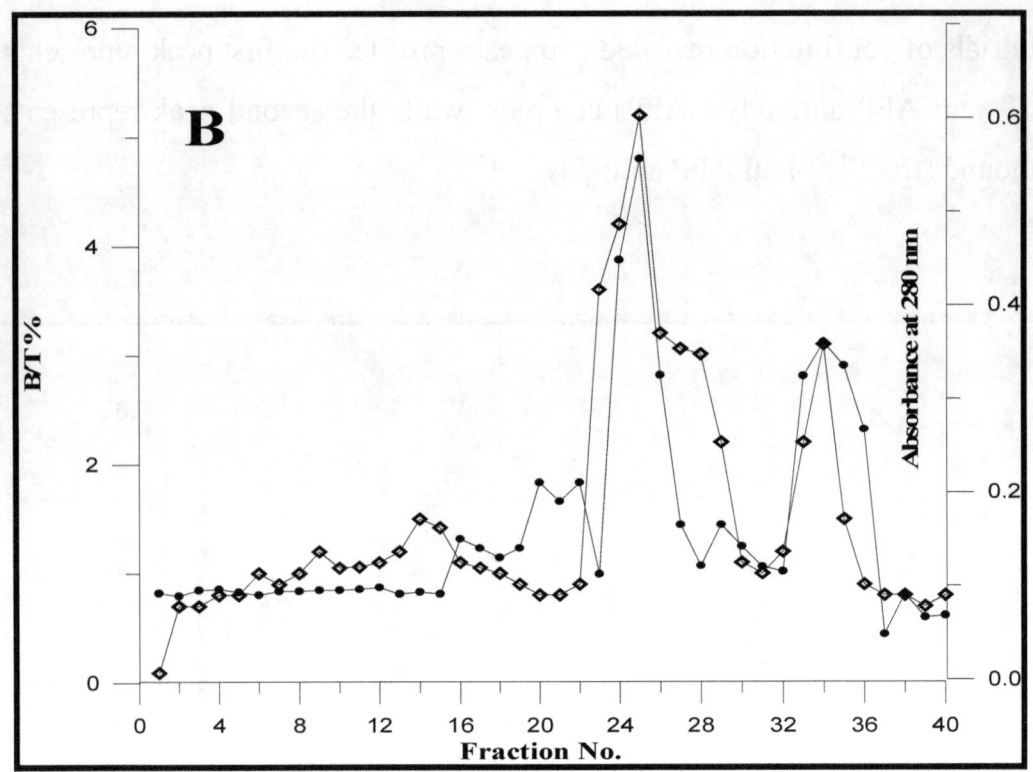

Figure (3–2): The elution profile of Human AFP '(●) means absorbance and

(◇) means B/T%'.

    A)   Benign colorectal tumor

    B)   Malignant colorectal tumor.

(All other details are explained in the text).

The Choice of the Optimum Conditions For The Binding of Partially Purified AFP with ^{125}I–anti AFP Antibody:

The Choice of the Optimum Protein Concentration:

    Figure (3–3) shows the optimum protein concentration for the isolated AFP of the malignant and benign colorectal tumor homogenate. This experiment was carried out by adding increasing amounts of the isolated forms to fixed amounts of ^{125}I–Antibody to produce (^{125}I–anti AFP Antibody/ AFP) complex. The maximum binding occurred at 25 μg.mL^{-1} for (benign) isolated form, while 18μg.mL^{-1} was the optimum protein concentration for the binding of (malignant) isolated form.

Further additions of AFP gave rise to solubilization of complex formed [227]. The excess of added antigen, which is represented by excess of protein, gives small chance to make a lattice between antibody and antigen [228].

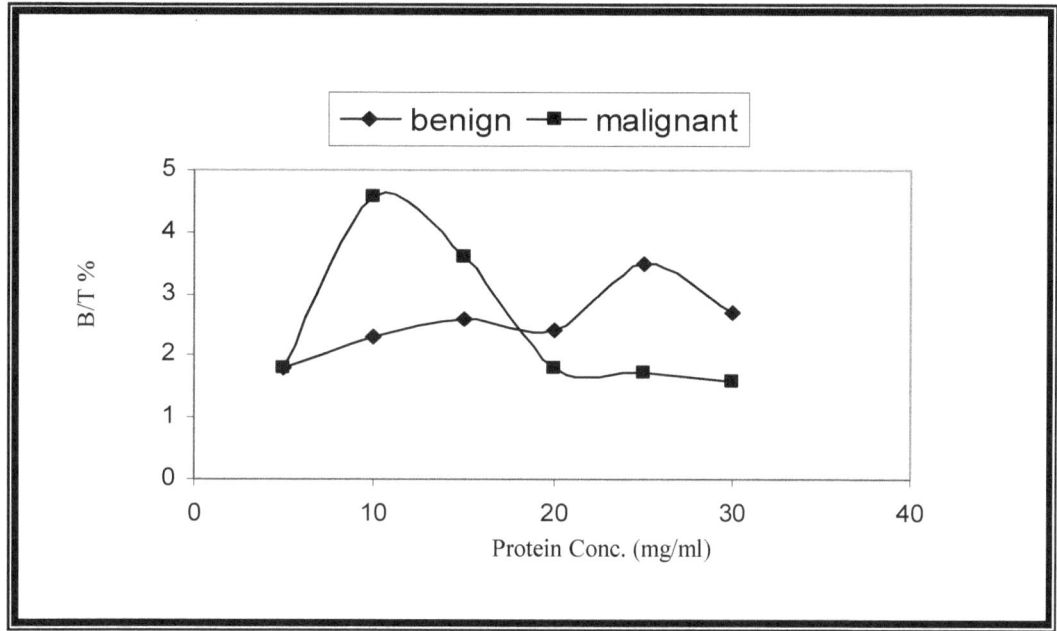

Figure (3–3) :Influence of Protein Concentration on the binding of ^{125}I–anti AFP Antibody with partially purified AFP from Colorectal tumor homogenate. (All other details are explained in the text).

The Effect of 125I–anti AFP Antibody:

Figure (3 – 4) shows the effect of ^{125}I – anti AFP Antibody concentration on the binding with isolated forms of the benign and malignant colorectal homogenate.

The maximum binding obtained at 2.16 mg.ml^{-1} for (malignant) and 1.44 mg.ml^{-1} for (benign). It was found that the amount of ^{125}I – anti AFP Antibody required to bind with their isolated Antigen forms is less than in crude homogenate. This may be due to the increment of the epitop (the part of an antigen molecule that binds to any single antigen combining site)[229].

245

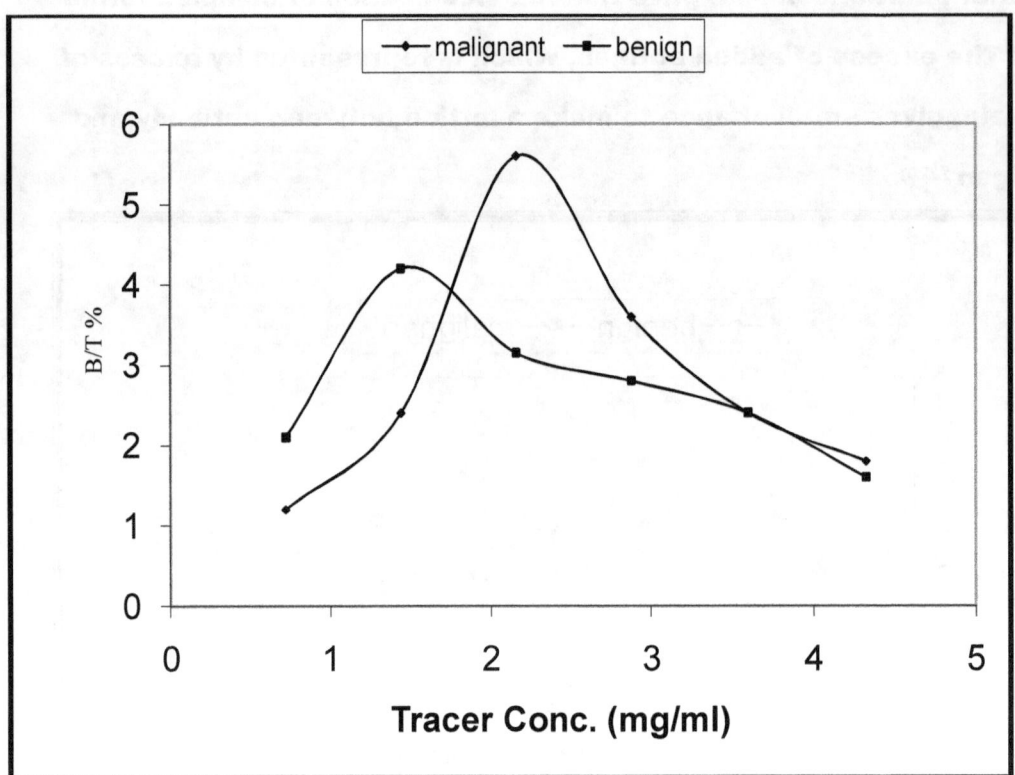

Figure (3–4): Effect of ^{125}I–anti AFP Antibody concentration on the binding with partially purified AFP from Colorectal tumor homogenates. (All other details are explained in the text).

The Choice of The Optimum pH:

In order to choice the optimum pH 27 μg.ml^{-1} and 19 μg.ml^{-1} for (malignant, benign) respectively of the two isolated forms of the colorectal tumor homogenate were incubated with 2.16,1.44 mg.ml^{-1} of ^{125}I–anti AFP antibody respectively. Figure (3 – 5) shows the optimum pH of the two isolated antigens forms. The results revealed that the optimum pH for both isolated antigens binding to its antibody was 7.4.

The similarity in pH (7.4) suggests that the AFP isolated forms possess the same epitopes in both cases. That means the induction of protonation – deprotonation process occurs with the same changed polar groups on the amino acid residues present in the binding domain [230].

malignant benign

Figure (3–5): pH effect on the binding of ^{125}I–anti AFP Antibody with partially purified AFP from Colorectal tumor homogenates. (All other details are explained in the text).

The Time Course of Partial Purification AFP:

Figures (3 – 6) and (3 – 7) illustrate the time course of the binding of isolated antigen from malignant and benign colorectal tumor homogenate to their antibody.

The malignant antigen binds to its antibody in highest state after 3 hours at 25 °C, while benign Antigen binds after 3 hours at 37°C.

In comparison with crude homogenate the optimum binding, so the binding of ^{125}I – anti AFP Antibody to its Antigen is a time and temperature dependent process. [231]

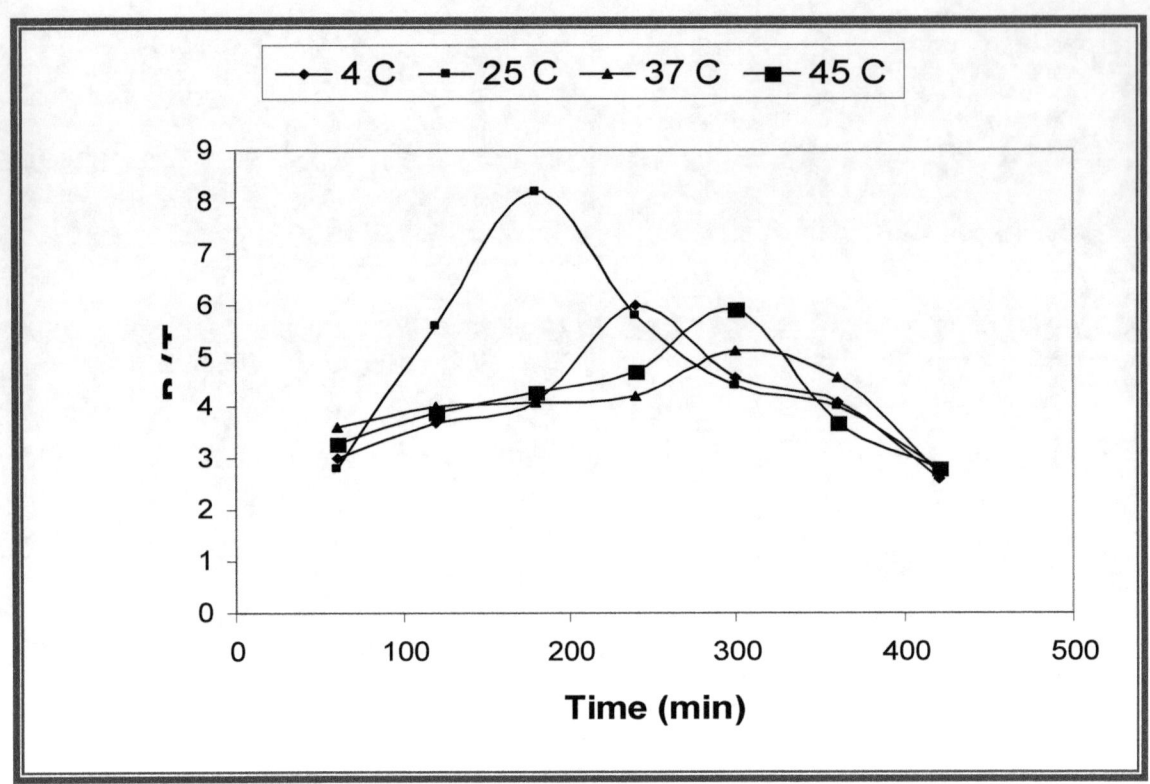

Figure (3–6): Time – Course of ^{125}I–anti AFP binding to partially purified (malignant) Antigen from Colon Cancer. (All other details are explained in the text).

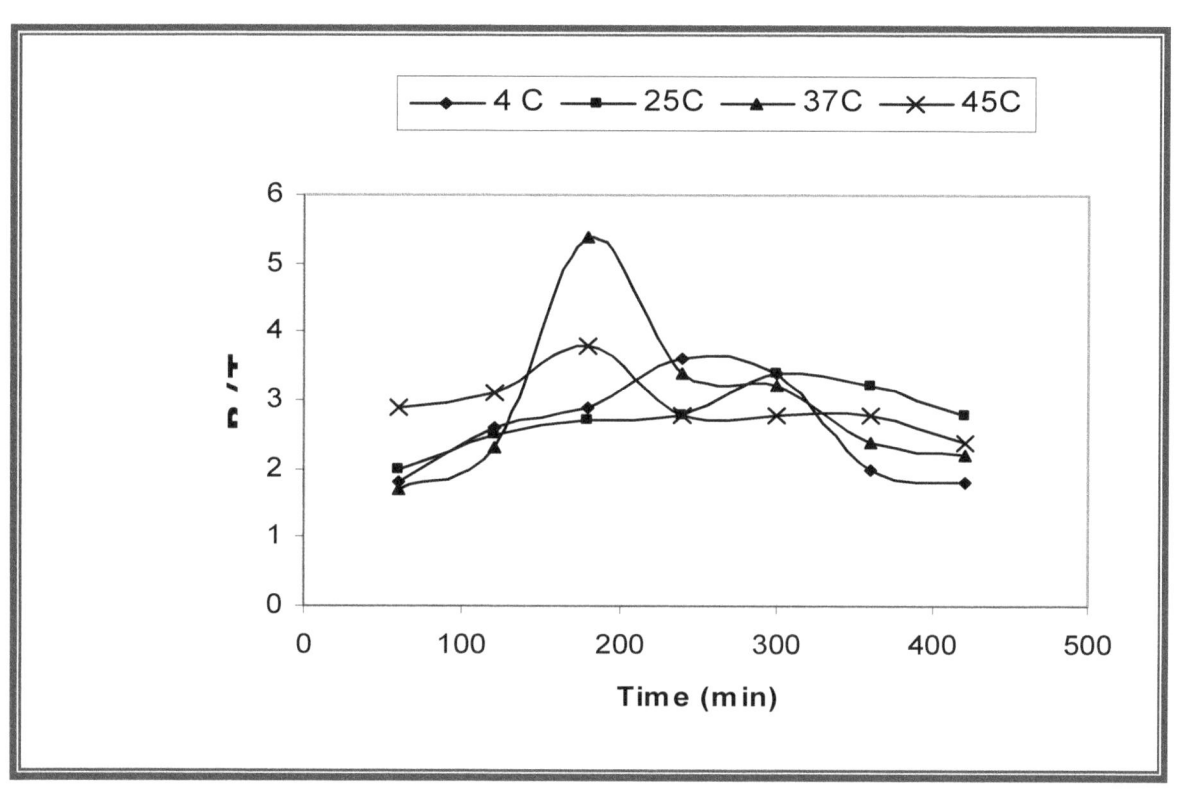

Figure (3–7): Time – Course of ^{125}I–anti AFP binding to partially Purified (benign) Antigen from Colorectal tumor. (All other details are explained in the text).

249

Chapter Four

Abstract:

Kinetic and thermodynamic parameters associated with the binding of ^{125}I- anti AFP Antibody to AFP in both crude colorectal homogenate and partially- purified fractions were investigated .It was shown that the reaction in all studied groups follow pseudo- first order reaction kinetics.

The value of kinetic parameters K_a, K_d, K_{obs}, K_{+1}, K_{-1}, $(t_{1/2})_{ass.}$, $(t_{1/2})_{diss.}$ and maximal binding capacity (B_{max}) at 25^oC for the binding of anti ^{125}I-AFP antibody with its cytosolic antigen in benign colorectal tumor were found to be : 0.0543×10^{10} M^{-1}, 15.064×10^{-10}M, 0.0158 min^{-1}, 4.086×10^6 M^{-1}.min^{-1}, 64.76×10^{-4} min^{-1}, 43 min, 104.28 min and 29.14 Pmol/mg protein respectively, while 0.0538×10^{10} M^{-1}, 15.142×10^{-10} M, 0.0163 min^{-1}, 4.301×10^6 M^{-1}.min^{-1}, 67.63×10^{-4} min^{-1}, 40 min, 79.66 min and 46.25 Pmol/mg respectively for the binding of anti 125I-AFP antibody with its cytosolic antigen in tissues of malignant colorectal tumor and at 4^oC.

The maximum binding of partially purified AFP occurred at 25^oC. The Ka and K_{+1} values increased with increasing temperature.

The Van't Hoff plot demonstrated linear relationship between ln Ka and 1/T, using crude colorectal homogenate partially purified AFP as AFP source. Plotting between ln K_{+1} and 1/T gave linear relationship called Arrheniuns relationship. The thermodynamic parameters ΔH^o, ΔG^o and ΔS^o for the formation of (^{125}I- anti AFP Antibody / AFP) complex at the standard state and Ea, ΔH^*, ΔG^* and ΔS^* which representing the transition state were determined.

Introduction

AFP displays both cellular uptake and transmembrane passage via specific cell surface receptors [232], and cytoplasmic binding protein accumulating in either receptome or lysosomal pathways. Many reports have identified and characterized various binding proteins associated with AFP in different cellular compartments [233, 143]. There is a large body of literary evidence on identification and characterization of various AFP binding proteins present in the free form in physiological fluids and in the membrane bound form on the surface of certain types of human cells [234,235,236].

Macromolecular interactions involve cooperative, independent and contiguous binding regions. The complexity caused by this, increased by the dynamic structural changes. The interaction and the equilibrium condition are influenced by the algebraic sum of the energies involved in reversible interacting energy components, namely electrostatic and hydrophobias, is best understood by study of the thermodynamic parameters of the interaction .The method most widely used for such investigations is isothermal titration calorimetry [237-239], which directly determines the heat changes during the binding process derives the thermodynamic parameters, enthalpy (ΔH), free energy (ΔG) and entropy (ΔS) . In the last two decades the method has been used to unravel the intricacies of the interaction between several ligand – ligand pairs, like protease – protease inhibitor [240], receptor – ligand [241], and Antigen – Antibody [242].

The idea of this chapter is to describe the basic mathematical analysis that could be used to explain the mechanism of binding of AFP to its Antibody to form (^{125}I – anti AFP Antibody / AFP) complex in human colorectal tissues, using benign and malignant colorectal tissue homogenate and also partially purified colon tumors fractions, as AFP source.

Materials and Methods:

4.1 Chemicals:

All chemicals and reagents mentioned in chapter two were used in the experiments of this chapter.

4.2 Instruments:

All instruments that described in chapter two were used in experiments of this chapter.

4.3 Kinetic Studies:

Scatchard Analysis of Determination Affinity Constant Ka and the Maximal Binding Capacity (Bmax) of:

(A) AFP in Colorectal Tissue Homogenate associated with 125I- anti AFP Antibody:

8- One hundred µL of colorectal homogenate (benign and malignant colorectal tumor) containing (300 and 200 µg .ml $^{-1}$) protein respectively were pipetted in each type of homogenate.

9- Increased volumes of ^{125}I – anti AFP Antibody for each group of colorectal homogenate i.e., (4, 8, 12, 16, and 20) µL, in the case of benign colorectal tumors and (8, 16, 24, 32, and 40) µL for Colorectal cancer were added to each assay tubes for each case.

10- The volumes of all tubes were completed to the final volume of 250 µl with tris – buffer (pH 7.0 and 7.2) respectively to the groups of tissue homogenate.

11- The time of the incubation required to reach the equilibrium state are reported in table (4 – 1)

12- After incubations of each group at each time and temperature required, all tubes were centrifuged at 4000 r. p. m. for one hour at 4°C by using cooling centrifuge .

13- The supernatant was discarded and the complex formed was counted in gamma counter for one minute.

Table (4- 1): The time of incubation for benign and malignant colorectal tumor homogenate at different temperatures.

Temp. °C	Time (hour)	
	Benign	Malignant
4	4	3
25	3	6
37	3	4
45	3	5

1- The B/T ratio was computed for each tube, where:

B: is the bound radioactivity (mean counts c.p.m), which represents the ^{125}I – anti AFP / AFP) complex.

F: is the free radioactivity (mean counts c.p.m), which represents (unbound or unreacted, ^{125}I – anti AFP).

T: is the total activity (mean counts)

F= T (total counts) – B (bound radioactivity).

2- The concentration of (^{125}I – anti AFP / AFP) complex in mg.ml^{-1} that formed after time (t) was calculated from the following equation :

$$B(mg.ml^{-1}) = \frac{B\,(c.p.m)}{T\,(c.p.m)} \times \text{concentration of } ^{125}I\text{ - antibody in the incubation medium in mg.ml}^{-1}$$

The affinity constant and maximal binding capacity were determined according to Scatchard equation [65,243].

$$\frac{B}{F} = \frac{1}{K_d}(B_{max} - B)$$

$$Ka = \frac{1}{K_d} = \frac{K_{+1}}{K_{-1}}$$

Where:

K_a = affinity Constant.

Kd = dissociation Constant.

B_{max} = maximal binding capacity

The value of the affinity constant of the binding Ka at each temperature can be calculated from the slop of the straight line , While

the value of the total concentration of AFP (Bmax) in colorectal tissue for each group was calculated from the intercept on X – axis .

(B) Partially Purified AFP in Colorectal Homogenate Binding with 125I – anti AFP Antibody:

1- **Increasing volume of ^{125}I – anti AFP Antibody (10 , 15 , 20 , 25 and 30) µL were incubated with 35 µg . ml $^{-1}$ of the of partially purified AFP from benign tumor tissue , while 18 µg . ml $^{-1}$ of partially purified AFP from malignant tumor tissue were incubated with increasing volume (4 , 8 , 12 , 16 and 20) µL of ^{125}I- anti AFP Antibody .**

2- **All tubes were completed to 250µL with tris buffer (pH 7.0 and 7.2) for benign and malignant tumor tissue respectively.**

3- **The time of incubation required to reach the equilibrium state were listed in table (4 – 2).**

4- **Steps 4, 5 in the experiment of (Kinetic studies of AFP in colorectal tissue homogenate associated with ^{125}I - anti AFP Antibody) were repeated.**

Table (4- 2): The time of incubation for the partially purified AFP from colorectal tumor tissues.

Temp. ºC	Time (hour)	
	Benign	Malignant
4	4	4
25	5	3
37	3	5
45	3	5

1- **The steps of calculations outlined in experiment of (Kinetic studies of AFP in colorectal tissue homogenate associated with ^{125}I - anti AFP Antibody), was followed exactly to obtain the values of Ka and Bmax at each temperature.**

4.4 The Thermodynamic Studies:

The thermodynamic of 125I- anti AFP Antibody binding to its Antigen in Colorectal Homogenate and Partially Purified AFP in Benign and Malignant Colorectal Tumors:

The same steps mentioned in chapter two, the experiment of (Time course of AFP binding in colorectal tissue homogenate) and chapter three, the experiment of (The time course of partially purified AFP) for the colorectal homogenate and partially purified AFP were performed respectively.

Calculation:

1- The thermodynamic parameters of the standard state obtained from Van't Hoff, the values of the natural logarithm of equilibrium constant (affinity constant Ka) obtained at different temperature were plotted against the reciprocal values of the absolute temperature in Kelvin (1/T), according to the following equation:

$$\text{Ln } K_a = \frac{\Delta S^\circ}{R} - \frac{\Delta H^\circ}{RT}$$

Where:

ΔH°= the enthalpy change of the standard state .

ΔS° = the entropy change of the standard state .

R = the gas constant (8.31414 J.mol^{-1}. K^{-1}).

ΔH° value obtained from the slope , the linear relationship of the plot .

The change in Gibbs free energy of the standard state ΔG° was obtained from the following equation:

$$\Delta G^\circ = - \, R \, T \ln Ka$$

Where Ka is the affinity constant, while the standard state entropy change was obtained from [244]:

$$\Delta S^\circ = \frac{\Delta H^\circ - \Delta G^\circ}{T}$$

2- The thermodynamic parameters of the transition state were obtained from Arrhenius plot of $\ln K_{+1}$ values against $1/T$ values that gave a linear relationship according to the following equation:

$$\text{Ln } K_{+1} = \text{Ln } A - \frac{Ea}{RT}$$

Where A = Arrhenius constant, some times called frequency factor or per exponential factor.

The value of apparent energy of activation (Ea) of the binding reaction can be determined from the slope of the straight line. The enthalpy of transition state ΔH^* was obtained from:

$$\Delta H^* = Ea \, - \, RT$$

Transition state free energy change ΔG^* is calculated from the following equation:

$$\Delta G^* \; = \text{-} \, RT \ln K_{+1} + RT \ln \frac{kT}{h}$$

Where K and h were Boltzman and Plank's constant, which equal (1.38×10^{-23} J.deg^{-1}. K^{-1}), (6.62×10^{-34} J. sec) respectively.

The change in entropy of the transition state ΔS^* was calculated from the following equation:

$$\Delta S^* \; = \frac{\Delta H^* - \Delta G^*}{T}$$

Results and Discussion:

Determination of Affinity Constant (Ka) and the Maximal Binding Capacity (B$_{max}$) of AFP in colorectal Tissue Homogenate Associated with 125I- AFP Antibody

B_{max} and the affinity constant Ka of the binding to ^{125}I-anti AFP Antibody was measured. The experiment was carried out at the optimum conditions that were obtained in previous experiments. Scatchard plot analysis gave straight line as shown in figure (4 – 1, A, B), and the parameters obtained from Scatchard plot are shown in table (4 – 3).

Table (4 – 3): The Kinetic parameters of AFP binding to its 125I- anti AFP Antibody in colorectal tissue homogenate

Temp. °C	K_a x 1010 M-1		K_d x 10 – 10 M		B$_{max}$ Pmol.mg-1 protein	
	Benign	Colorectal Cancer	Benign	Colorectal Cancer	Benign	Colorectal Cancer
4	0.0522	0.0578	16.115	15.142	27.25	46.25
25	0.0543	0.0532	15.064	16.554	29.14	43.28
37	0.0518	0.0514	16.546	18.556	25.56	41.35

| 45 | 0.0458 | 0.0509 | | 17.308 | 19.485 | | 24.35 | 38.64 |

The values Ka and maximal binding capacity (Bmax) were calculated from Scatchard plot at four different temperatures .

It is clear from table (4 – 3) that the affinity constant (K_a) is depended on the type of the tumor (i. e, benign and malignant) and on the temperature. The highest value of Ka occurred at 25°C in the case of Benign but in the case of colorectal cancer, the optimum value of Ka was at 4°C, while the lowest values of Ka in the both groups of colorectal tissue homogenate were at 45°C. The kinetic and affinity constant were different due to the differences in one or more amino acid present in epitope domain [245].

Scatchard plots analysis gave straight lines, indicating that probably only one species of binding site is presented or more but with the same affinity.

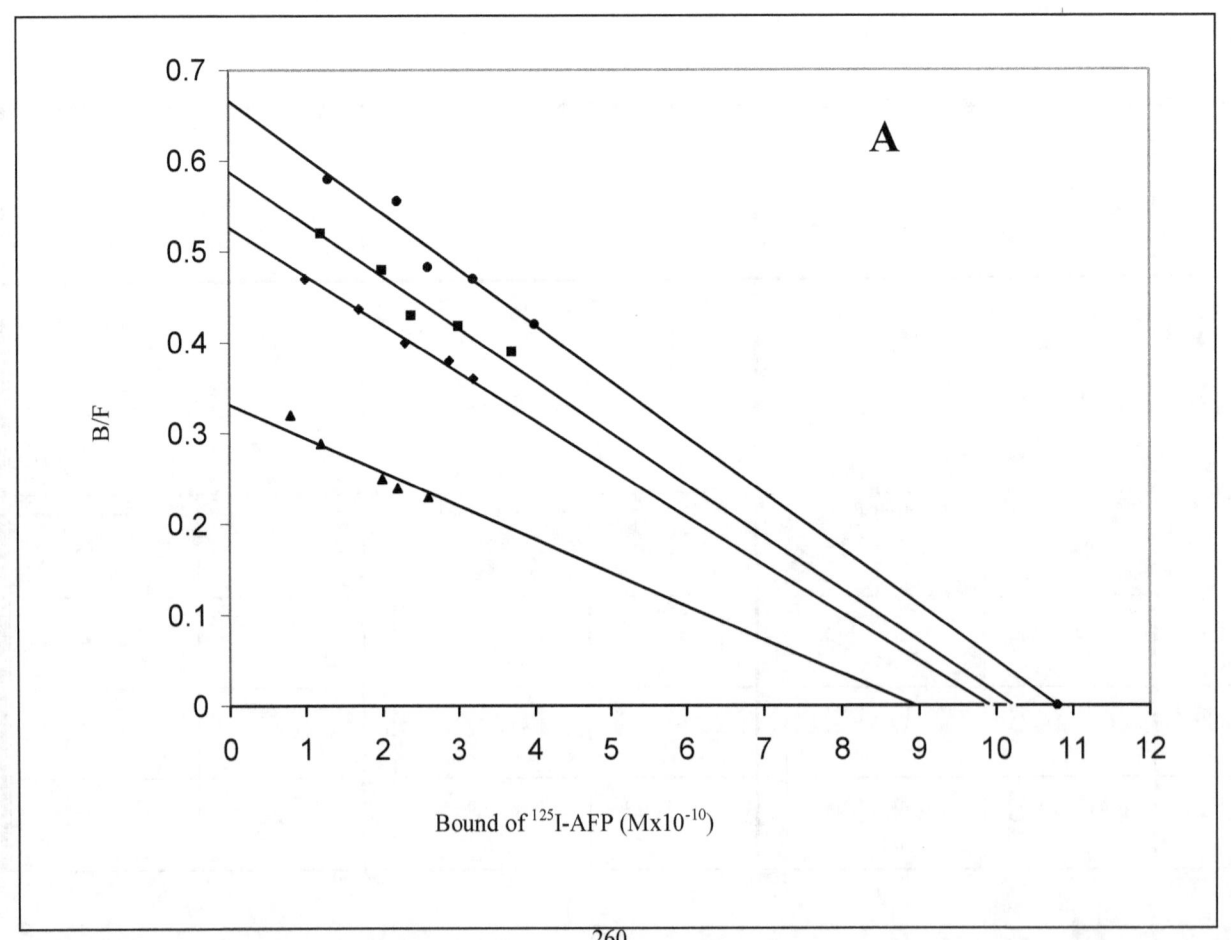

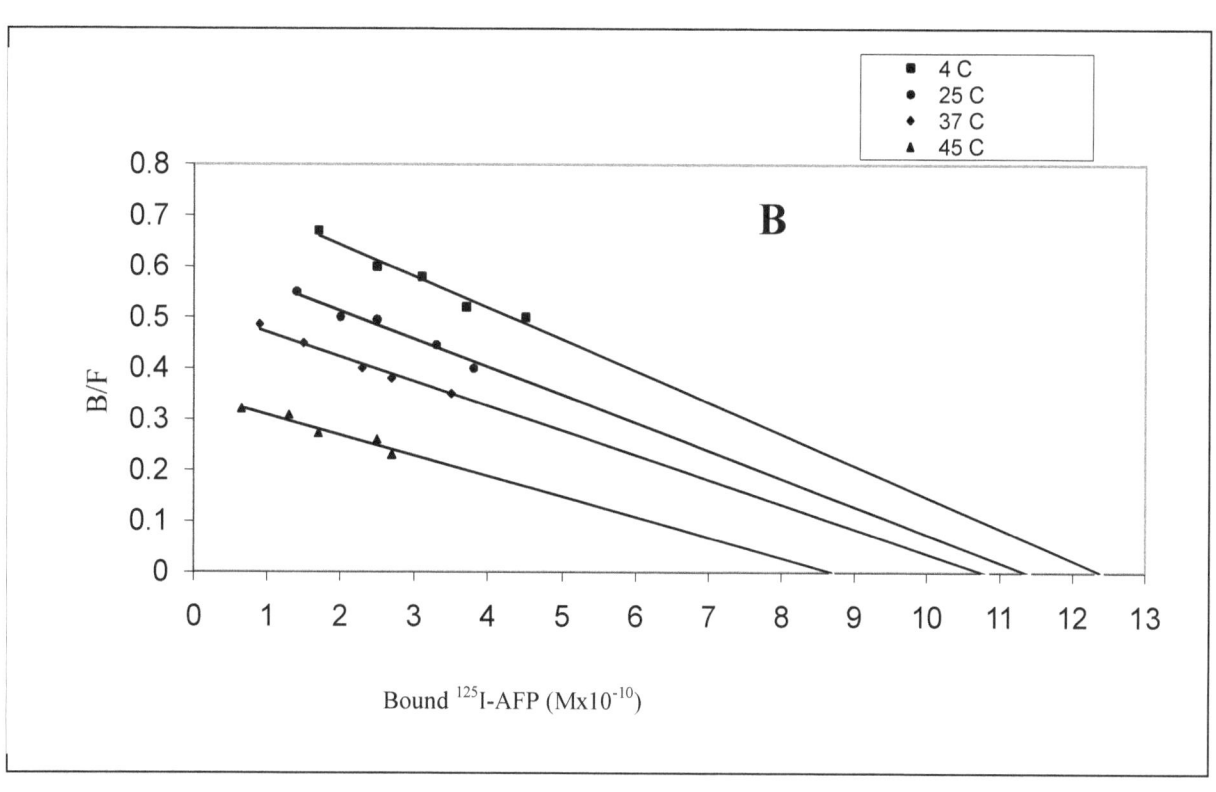

Legend:
- ■ 4 C
- ● 25 C
- ♦ 37 C
- ▲ 45 C

B

B/F (y-axis)

Bound ^{125}I-AFP (Mx10^{-10})

Figure (4-1): Scatchard plot of 125I – anti AFP binding to AFP in

(A) Benign tumor
(B) Malignant tumor
(All other details are explained in the text).

The simplest proposed model representing the binding of ^{125}I – anti AFP Antibody with AFP could be expressed by the following equation

$$^{125}\text{I - Ab} + \text{Ag (AFP)} \underset{k_{-1}}{\overset{k_{+1}}{\rightleftharpoons}} \left[^{125}\text{I - Ab - Ag}\right]$$

Where:

K_{+1}: is the rate of the association of ^{125}I – anti AFP Antibody with AFP.

K_{-1}: is the rate of reverse reaction of dissociation of the complex formed under the same condition.

At equilibrium:

$$K_a = \frac{[^{125}\text{I - anti AFP antibody / AFP}]}{[^{125}\text{I - anti AFP antibody}][\text{AFP}]} \quad\dots\dots\dots\dots\dots\dots(1)$$

$$K_d = \frac{[^{125}\text{I - anti AFP antibody}][\text{AFP}]}{[^{125}\text{I - anti AFP antibody / AFP}]} \quad\dots\dots\dots\dots\dots\dots(2)$$

Thus, $$K_a = \frac{1}{K_d} = \frac{K_{+1}}{K_{-1}} \quad\dots\dots\dots\dots\dots\dots\dots\dots\dots\dots\dots(3)$$

Where:

K_a: is the equilibrium constant of the association (affinity constant).

K_d: is the equilibrium constant of the dissociation of ^{125}I – anti AFP Antibody / AFP) complex.

The reorder of ^{125}I – anti AFP Antibody to AFP was determined by using the following equation (246):

$$\ln[\text{Ab Ag}]e \; \frac{[\text{Ab}]t - [\text{Ab Ag}]t\,[\text{Ab Ag}]e / \text{Ag}]t}{[\text{Ab}]t\,[\text{Ab Ag}]e - [\text{Ab Ag}]e} = K_{+1}\,t \; \frac{[\text{Ab}] + [\text{Ag}]t - [\text{Ab Ag}]e}{[\text{AbAg}]e} \quad\dots(4)$$

Where:

K_{+1}: is the kinetic association constant.

$[\text{AbAg}]e$: is the concentration of (^{125}I – anti AFP/ AFP) complex formed at equilibrium.

$[\text{AbAg}]t$: is the concentration of (^{125}I – anti AFP / AFP) complex after time (t).

$[\text{Ab}]t$: is the total concentration of ^{125}I – anti AFP Antibody.

$[\text{Ag}]t$: is the total concentration of AFP.

Equation (4) represents the second order kinetics, but the percent of binding was in some cases, small and most labeled Antibody remains free and only small fraction binds even at equilibrium, i.e., $[Ab]t \gg [AbAg]e$

Thus:

$$[Ab]\, t \gg \frac{[Ab\ Ag]t\, [AbAg]e}{[Ag]t}$$

So that the following equation could be used in order to fit the pseudo-first order kinetics[247].

On the other hand figures (4 – 2, 4 – 3 and 4 – 4) show the plots of $\ln \dfrac{[Ab\ Ag]e}{[Ab\ Ag]e - [Ab\ Ag]t}$

against time (t) gave a straight line with a slope equal to the observed value of the first rate constant (K_{obs}) in min^{-1}. The rate of constant (K_{+1}) in $mg^{-1}.ml.min$ was calculated at five different temperatures using the following equation[248].

$$K_{obs} = K_{+1}\ \frac{^{125}I\text{ - anti AFP}]t\, [AFP]t}{[^{125}I\text{ - anti AFP} / \text{AFP}]e} \quad \ldots\ldots\ldots\ldots\ldots\ldots(6)$$

Also, the value of K_{-1} at four temperature were calculated using equation (3), whereas the half life time of association (t 1/ 2)$_{ass.}$, which represented the time needed for the formation of half amount of the complex at equilibrium was determined from the concentration of the complex at equilibrium and the time course curve. The half-life time of dissociation (t 1/2)$_{diss}$, was calculated from the following equation.

$$(t\ 1/2\ diss) = \ln \frac{2}{K_{1}} = \frac{0.693}{K_{-1}}$$

The values of $K_{obs.}$, K_{+1}, K_{-1}, (t $_{1/2}$) ass., and (t $_{1/2}$) diss., at different temperature are summarized in table (4 – 4). The values in this table show the highest rate for the association reaction K_{+1}, in benign colorectal tumors occurred at 25°C and at 37°C in malignant colorectal tumors, while the lowest rate occurred at 45°C, so the reaction rate is a temperature dependent, while the rate constant for the reverse reaction K-1 which refers to the rate of dissociation of ^{125}I– anti AFP Antibody from its AFP is temperature independent.

Table (4-4): The effect of temperatures on the kinetic parameters of 125I-anti AFP antibody binding with AFP in homogenates of colorectal tumors.

Temperature °C	$K_{obs.}$ (min.−1)	$K_{+1} \times 10^6$ (M-1.min.-1)	$K_{-1} \times 10^{-4}$ (min-1.)	$(t_{1/2})$ ass. (min.)	$(t_{1/2})$ diss. (min)
Benign					
4	0.0118	3.22	55.47	81	118
25	0.0158	4.08	64.76	43	104.28
37	0.102	2.428	51.32	68	128
45	0.0049	2.049	50.44	52	144
Malignant					
4	0.0163	4.301	67.63	40	99.66
25	0.0112	2.163	43.03	53	142.82
37	0.0091	1.746	38.58	65	156.74
45	0.0058	1.398	35.74	50	179.88

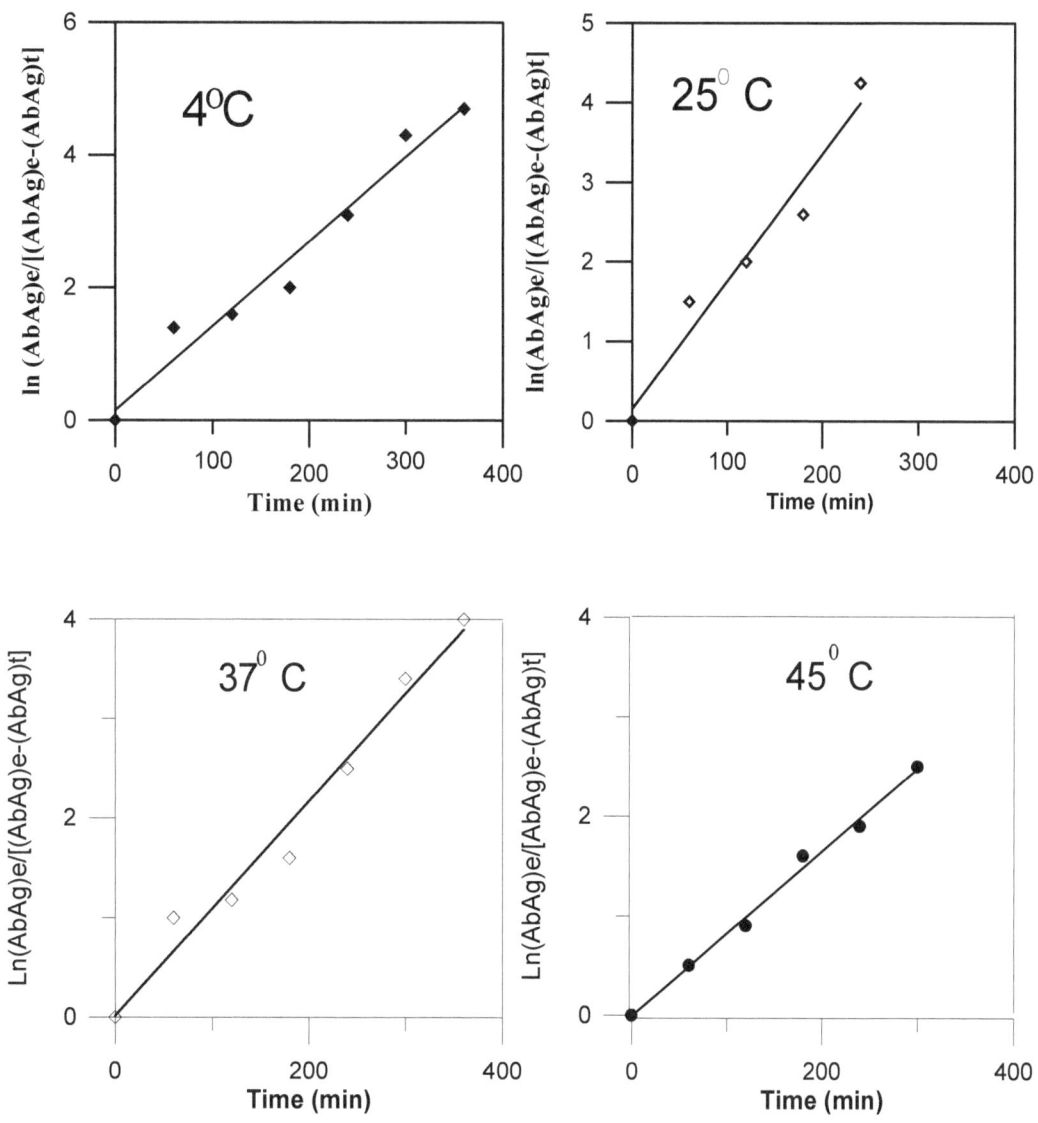

Figure (4-2): Kinetic of ^{125}I-anti AFP Antibody binding to AFP in Benign colorectal tumor. (All other details are explained in the text)

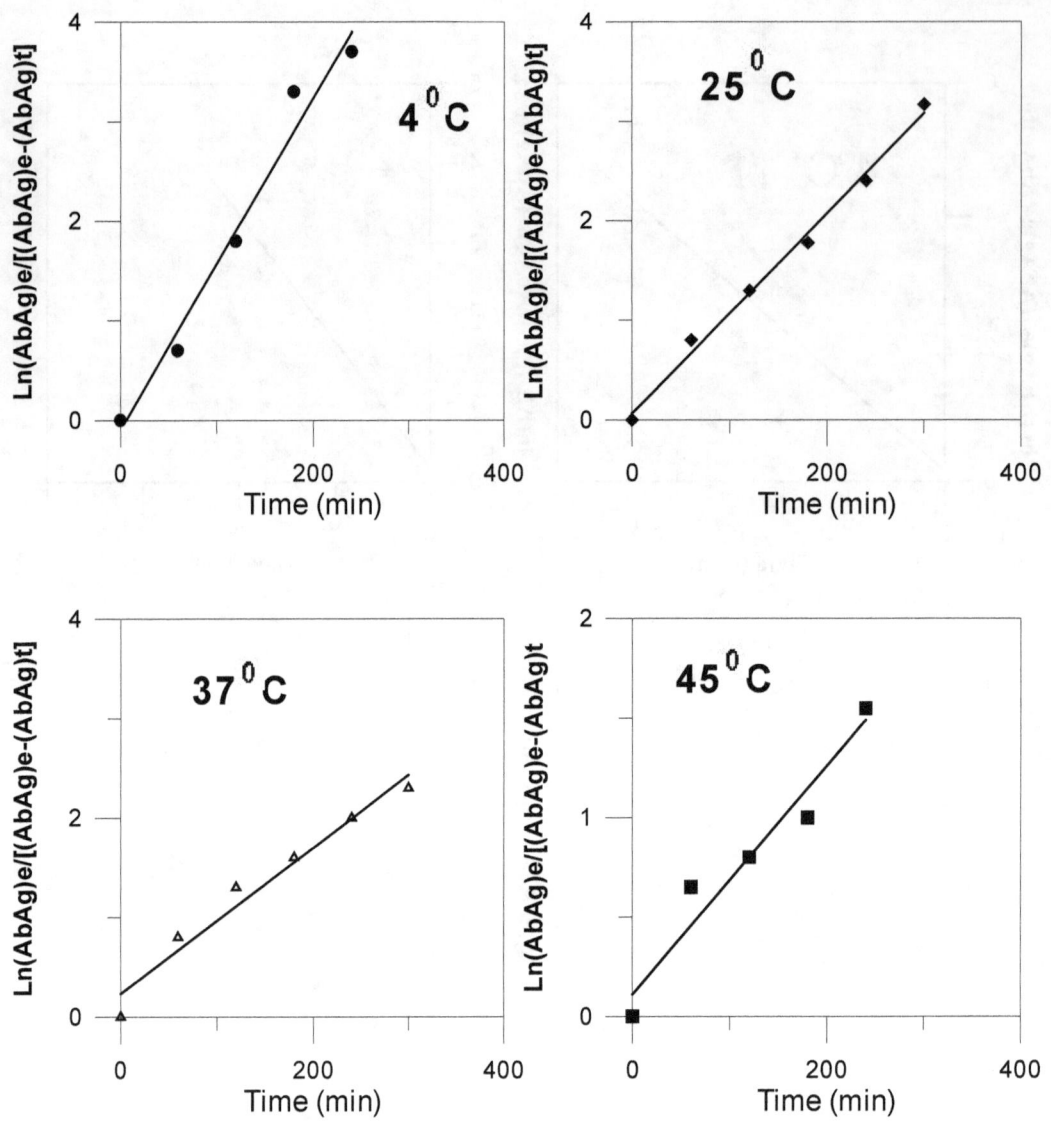

Figure (4-3): Kinetic of ^{125}I-Anti AFP antibody binding to AFP in Malignant colorectal tumor. (All other details are explained in the text).

Kinetic of the Binding of 125I – anti AFP Antibody to partially – purified AFP from colorectal tumors Homogenate:

266

Scatchard plot analysis gave a straight line as shown in fig (4-4) for each partially purified AFP from benign and malignant homogenates of colorectal tumors at each temperature which indicates that the presence of single class of binding sites or more with the same affinity. These results are summarized in Table (4 –5).

Commonly, it can conclude that the crude AFP had lower affinity to bind with its Antibody than the partial – purified AFP in both isolated Antigens.

Table (4 – 5): The Kinetic parameters of 125I- anti AFP Antibody binding to its partially – purified AFP in colorectal tissue homogenate

Temp. °C	$K_a \times 10^{10}$ M-1		$K_d \times 10 - 10$ M		B_{max} Pmol.mg-1 protein	
	Benign	Colorectal Cancer	Benign	Colorectal Cancer	Benign	Colorectal Cancer
4	0.2849	0.2621	36.401	40.823	35.34	46.66
25	0.2382	0.3179	38.078	33.312	31.83	55.24
37	0.2986	0.2545	33.141	42.867	43.12	40.11
45	0.2871	0.2766	35.453	36.419	38.83	48.12

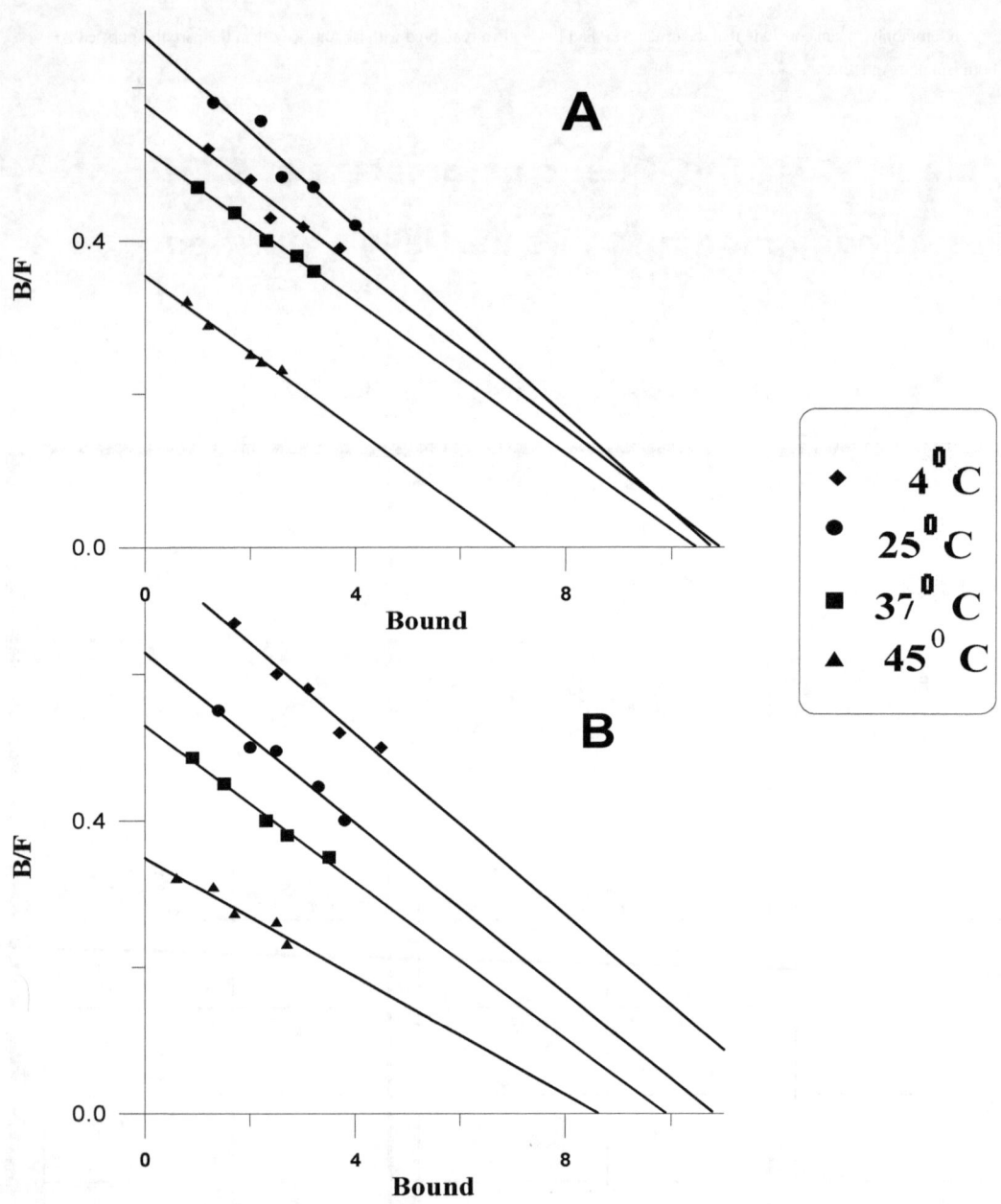

Figure (4-4): Scatchard plots of ^{125}I-anti AFP binding to the partially purified AFP antigen in:

(A)   Benign colorectal tumor

(B)   Malignant colorectal tumor

(All other details are explained in the text)

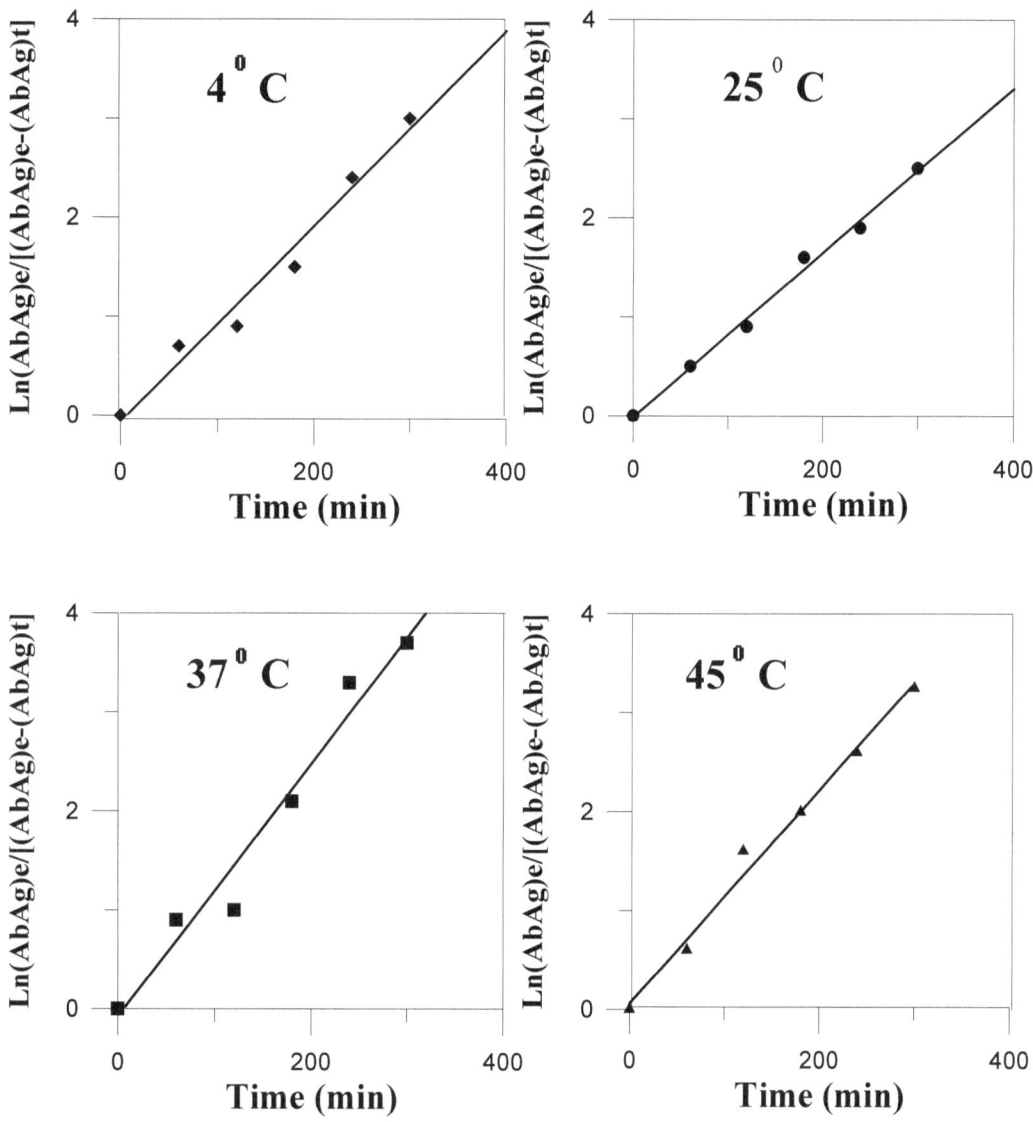

Figure (4-5): Kinetic of ^{125}I-anti AFP antibody binding to partially purified AFP from benign colorectal tumor homogenate.

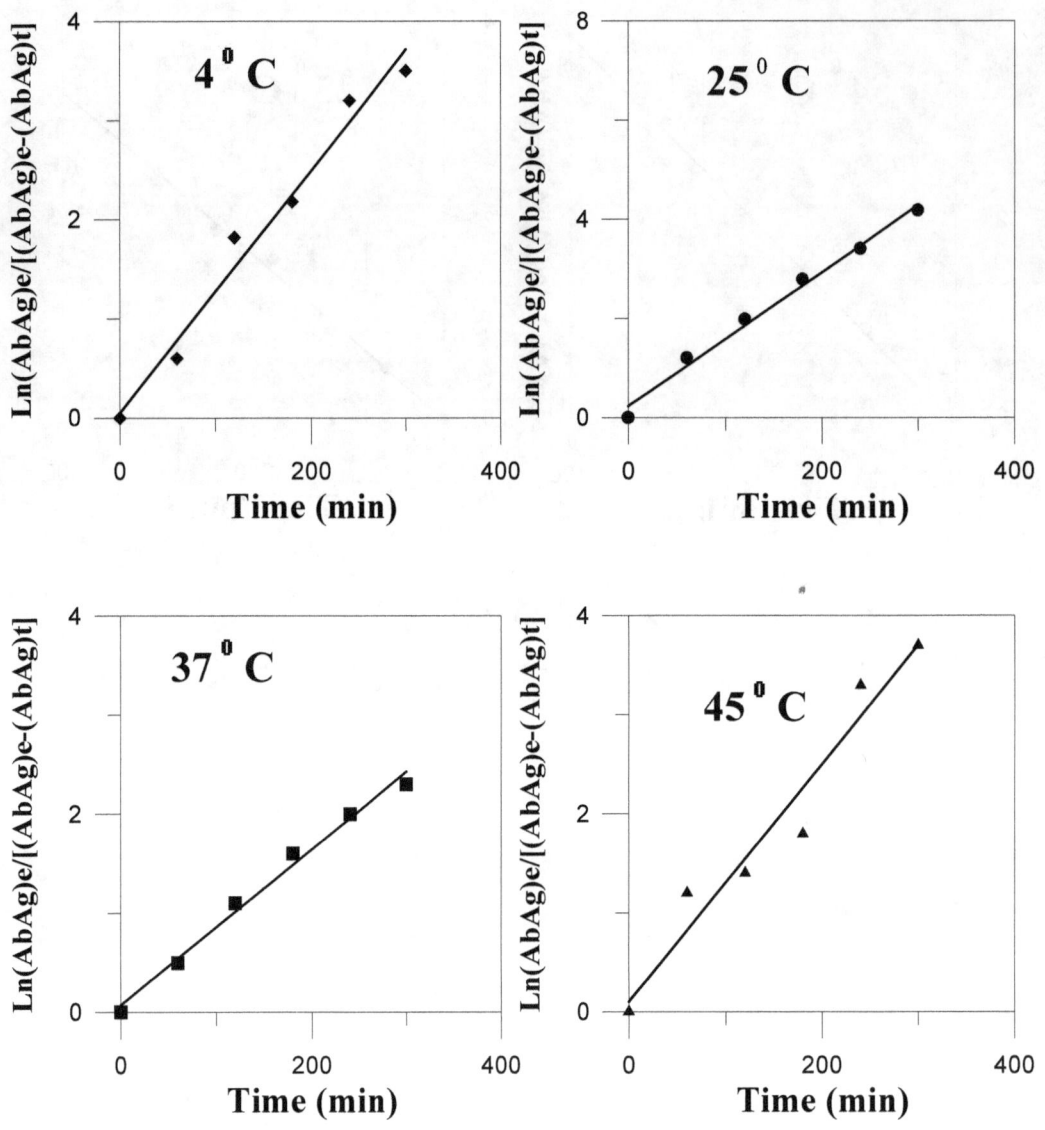

Figure (4-6): kinetic of ^{125}I-anti AFP antibody binding to partially purified AFP from malignant colorectal tumor homogenate.

The thermodynamic studies of interaction of ^{125}I-anti AFP antibody with AFP in colorectal tumor tissues

(A) **Thermodynamic parameters of standard state:**

Figure (4-7) represents the dependence of the equilibrium binding constant (i.e., affinity constant) for the binding of ^{125}I-anti AFP antibody to its AFP of benign and malignant colorectal tumor homogenates on the temperature (Van't Hoff Plot).

The results obtained from Van't Hoff plot revealed that ΔH^o in general had small values and their positive sign ascertain that the reaction was nearly endothermic. The small positive value of ΔH^o may indicate a favorable interaction between ^{125}I-anti AFP antibody to both AFP in malignant breast tumor homogenate and partially purified AFP respectively.

The favorable interactions include the non-covalent interaction, which are fundamentally electrostatic in nature such as charge-charge, charge-dipole, dipole-dipole, charge-induced dipole, dipole-induced dipole interactions, and hydrogen bonds. The sum of these types of interactions can yield some stabilization to the folded structure of the complex [249].

The negative values of ΔG° reflect the stability of the complex hence, the high affinity of the reactants. So, the negative values of ΔG° showed that the overall reaction was energetically favorable in the direction of complex formation.

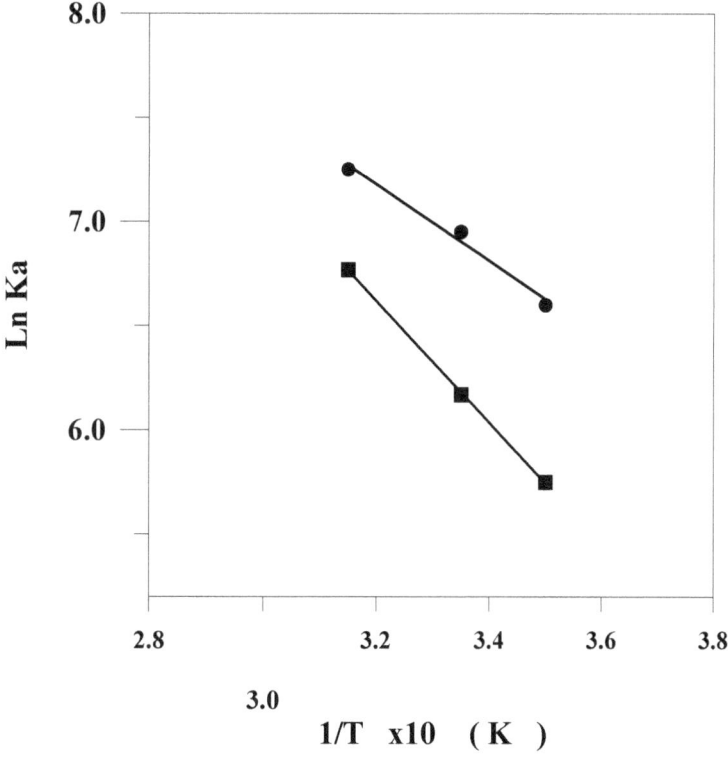

Figure (4-7): Van't Hoff plot for the binding of [125]I-

anti AFP antibody with AFP of (■) benign colorectal tumor homogenates and (●) malignant colorectal tumor homogenates. (All details are explained in the text)

The high negative values of ΔG° for the binding reactions are controlled by small negative and high positive ΔS° values as shown in table (4-6). So, our system is characterized by the contribution of ΔH° and ΔS° to the stability of the complex formed.

This system is characterized by the sole contribution of ΔS° to the stability of the complex formed; while ΔH° has little or no effect [250]. The values of positive ΔS° suggest the binding spontaneity was entropically driven. Entropy was the driven force for the occurrence of the binding, this indicates that the hydrophobic interactions played an important role in stabilizing complex [251].

Table (4-6): Thermodynamic parameters of standard state for the binding of ^{125}I-anti AFP antibody with AFP of colorectal tumor homogenates.

Temperature °C	ΔH° KJ. mol^{-1}	ΔG° KJ. mol^{-1}	ΔS° KJ. mol^{-1}. K^{-1}
Benign			
5	7.528	-18.45	69.28
25	7.528	-19.74	68.42
37	7.528	-20.97	69.73
42	7.528	-18.71	63.52

Malignant			
5	8.554	-20.56	59.35
25	8.554	-21.46	58.34
37	8.554	-22.58	60.24
42	8.554	-20.65	55.48

(B)　　Thermodynamic parameters of transition state:

The transition state theory proposes that the association of two substances to form the final product proceeds through the formation of an activated complex (transition state). Consequently, the interaction of ^{125}I – anti AFP Antibody with colorectal tissue homogenate can be represented as follows

$$^{125}I \quad Ab + Ag \rightarrow [^{125}I - Ab - Ag] * \rightarrow [^{125}I - Ab - Ag]$$
$$\text{labeled} \quad AFP \rightarrow \text{an activated} \rightarrow \text{Final product.}$$
$$\text{antibody} \qquad\qquad \text{complex}$$

The thermodynamic parameters of the transition state (ΔH^*, ΔG^*, and ΔS^*) could be determined from Arrhenius equation and the Kinetic constants.

Figure (4 – 8) shows the Arrhenius plot of ln K_{+1} versus 1/T values. The slop of the straight line represents the activation energy (E_a).

Table (4 – 7) shows the values of thermodynamic parameters of transition state of colorectal tissue homogenate Antigens from malignant and benign colorectal tumor at different temperature. The values of activation energy represents the required energy to overcome the energy barrier of the transition state for the formation of ^{125}I – anti AFP Antibody / AFP complex.

Also the value of activation energy is in accordance with the high positive values of ΔG^*, which indicates that the formation of the activated complex is a non – spontaneous process and required a lot of energy (equal to E_a) to overcome the transition state energy barrier and giving the final product, whereas the high negative ΔS^* revealed that the activated complex had a more structure than the reactants.

From the results obtained for the thermodynamic parameters in the transition state, it can conclude that the positive values of ΔH^* and high positive values of ΔG^* are favorable to overcome the energy barrier of the transition state, the high negative values of ΔS^* mean more arranged structure for the activated complex. The positive values of ΔG^* is mainly attributed to the decrease in the entropy of the transition state. In addition the positive value of ΔH^* shows that the heat content of the activated complex is more than that of isolated species [252,253].

The values of the thermodynamic parameters of the binding reaction, gave an over all idea about the nature of forces that regulate the formation of complex.

The formation of a complex occurs in two steps, the first is the stabilization of the complex by hydrophobic interaction and the second is the stabilization by short-range interactions, such as electrostatic interaction, hydrogen bonding, and Van der Waals' interactions.

Hydrophobic interactions contribute to the complex stability via high positive entropy change ($\Delta S^* > 0$), while electrostatic interactions, hydrogen bonding and Van der Waals interactions contribute to the stability of the complex via negative entropy change ($\Delta S^* < 0$) [254,255].

The thermodynamic data indicated that the binding of ^{125}I – anti AFP Antibody to AFP in colorectal tissue homogenate are entropy driven in agreement with the concept that hydrophobic interactions play an important role in (^{125}I – anti AFP Antibody / AFP) interactions [254,255].

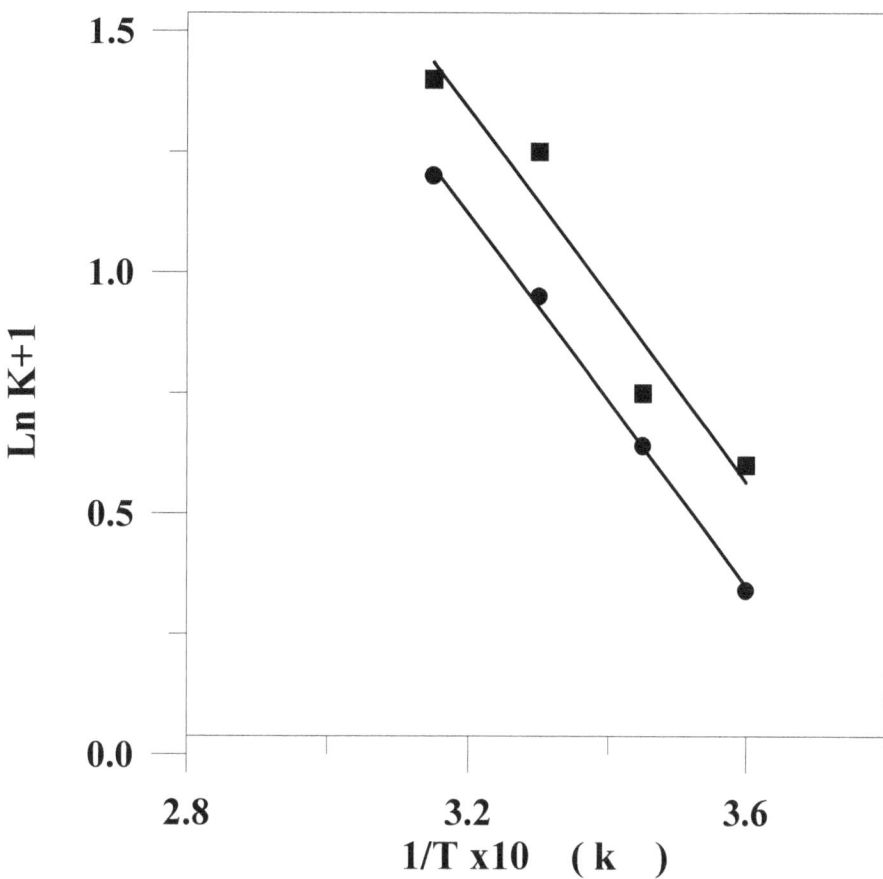

Figure (4-8): Arrhenius plot for binding of ^{125}I-anti AFP antibody with AFP in (■) Benign colorectal tumor homogenates and (●) malignant colorectal tumor homogenates. (All details are explained in the text).

Table (4-7): Thermodynamic parameters of transition state for binding of ^{125}I-anti AFP antibody with AFP from benign and malignant colorectal tumor homogenates

Temperature °C	E_a (KJ.mol^{-1})	ΔH^* (KJ. mol^{-1})	ΔG^* (KJ. mol^{-1})	ΔS^* (KJ.mol^{-1}.K^{-1})
Benign				
5	11.55	9.88	66.69	-210.11
25	11.55	9.72	70.87	-211.24
37	11.55	9.62	73.24	-211.55

42	11.55	9.55	77.05	-212.26
Malignant				
5	7.5	10.24	68.69	-204.35
25	7.5	10.23	7545	-206.34
37	7.5	10.72	78.33	-205.44
42	7.5	1036	83.11	-209.65

Chapter Five

Abstract:

Gel filtration technique was used to separate ^{125}I-anti AFP antibody bound to human AFP from unbound (free) ^{125}I-anti AFP antibody.

The characterization of human-AFP, anti-AFP, and (AFP/Anti-AFP) complex were carried out through the ultraviolet (U.V) spectroscopic studies.

Factors affecting the absorption properties of the molecules under investigation in this chapter such as pH, solvent polarity (solvent perturbation technique, urea and KCl) has been studied.

The spectrophotometric pH titration for h-AFP, anti AFP, and (AFP/anti-AFP) complex showed that pKa for tyrosine was 9.5, 10.2, and 9.9, while for histidine was 5.7, 6.0, and 5.9 respectively.

Introduction:

The ultraviolet absorption of protein solutions in the regions 250 to 310 nm are contributed by phenylalanyl, tyrosyl, and tryptophanyl residues, at shorter wavelengths; the contributions come from other groups such as histidyl residues and the peptide bond [256]. Changes in the environment of these chromophores can lead to alteration in the absorption spectrum, and conformational changes of its chromophoric groups [257]. A variety of environmental changes (e.g. pH, temperature) can affect the absorption spectrum, if the ground and excited states, the altered spectrum of the chromophore can be shifted to longer (red shift) or shorter (blue shift) wavelengths. The shift may or may not be accompanied by a change in intensity of spectrum [258,259].

Molecules absorb light, the efficiency of absorption depend on both the structure and environment of the molecules making absorption spectroscopy a useful tool for characterization of both small and large molecule [260].

Ultraviolet spectral method remains one of the most important methods in immunology for the study of antibody structure and specific ligand binding [261].

The solvent perturbation technique was introduced by Herskovits and Laskowaki [262], in this procedure, chromophores, which are in contact with the solvent, are deliberately perturbed by addition on another solvent.

The most common uses of spectroscopic technique in biochemistry employ the ultraviolet region spectrum [263]. Although it might appear that changes in the U.V. absorption of protein side chains could produce only very limited amount of information, the range of studies carried out using ultraviolet absorption proves to be very wide [264].

This chapter deals with ultraviolet spectrophotometry as tool to study the factors affecting the absorbance at the region of U.V. using human-AFP, anti AFP, and (AFP/anti AFP) complex.

Materials and Methods

5.1 Chemicals and Instruments

5.1.1 Chemicals:

Dimethyl sulphoxide (DMSO), Ethanol, ethylene glycol, and Glycerol were obtained from Fluka Company-Switzerland. Other chemicals and reagents used in the experiments of this chapter were mentioned in chapter two.

The instruments used in this chapter were Zhimadzu double beam U.V-Vis. Spectrophotometer type 160, and other instruments mentioned in chapter two.

5.2 Methods:

5.2.1   Gel Filtration Technique for Separation Free and Bound 125I-Anti AFP Antibody:

5.2.1.1   Preparation of The Column:

The dimensions of the column were 0.9x27 chosen according to the equation in section (3.4).

5.2.1.2   Preparation of The Gel and Determination of Void Volume:

Sephdex G-200 was used to separate free and bound ^{125}I-anti AFP antibody. The gel was prepared as mentioned in section (3.4). The void volume was determined as in section (3.5) and found to be (10 ml).

5.2.1.3   Separation Procedure of (125I-Anti AFP Antibody/ AFP) Complex:

1-   Standard of h-AFP was reacted with ^{125}I-anti AFP antibody at the kit condition (equipped with the kit). At the end of incubation, 720µl of the mixture was applied to surface of the gel, equilibrated with Tris buffer (0.05M, pH 7.2) with flow rate of 4ml/hr and 1ml fractions were collected.

2-   The radioactivity of each fraction was counted in gamma counter.

3-   Two hundred microliter of ^{125}I-anti AFP antibody was completed to 1 ml with Tris buffer (0.05M, pH 7.2), then 720µl from this volume was applied to the column as mentioned above and 2 were repeated.

Solutions:

Tris buffer (0.05 M, pH 7.2) was prepared as described in section (3.4).

Calculations:

1-   Radioactivity (c.p.m.) of each eluted fraction was plotted against the fraction number.

2-   The spectroscopic measurements gave profile of two peaks. The first peak represents the ^{125}I-anti AFP antibody bound to AFP, while the second peak represents ^{125}I-anti AFP antibody the unbound.

5.2.2   The U.V Spectra of (125I-Anti AFP Antibody / AFP) Complex:

The gel filtration experiment in section (5.2) gave two peaks profile. The fractions under each peak were pooled and the absorption spectrum was scanned in U.V region using 0.5cm cuvette against Tris buffer (0.05M, pH 7.2) in reference beam.

5.2.3.1 h-AFP and Anti AFP:

A volume of 20 µl each of human AFP standard and Anti AFP each were completed to 1 ml with Tris buffer pH 7.2. The samples were transferred to 0.5 cm cuvett in the sample beam then the absorption spectrum was scanned against the same buffer in reference beam in region of 200-350 nm.

5.2.3.2 (AFP/Anti AFP) Complex:

1-   Twenty microlitter of human AFP standard provided by (AFP-IRMA Kit) was mixed with 80 µl of ^{125}I-anti AFP antibody. The volume of the mixture was completed to 250 µl with Tris buffer (0.05M, pH 7.2).

2-   The mixture was incubated at 25 °C for one hour.

3-   At the end of incubation, the mixture was centrifuged for 30 min. at 4000 r.p.m. at 4 °C to separate the (AFP/anti AFP) complex.

4-   The supernatants were discarded by decanting the assay tube, and the precipitate formed was dissolved in 1ml Tris buffer pH 7.2 then the sample was placed in 0.5cm cuvette in the sample beam against Tris buffer pH 7.2 in the reference beam then used, the absorption spectrum was scanned.

5.2.4   Factors Affecting The Absorption Properties of h-AFP, Anti AFP, and (125I-Anti AFP Antibody/AFP) Complex:

5.2.4.1 The Effect of pH on The U.V Spectrum:
A. h-AFP and Anti-AFP:

A volume of 20 µl each of h-AFP standard and Anti AFP were completed to 1ml at different pH values (2, 4, 7, and 11). The samples were transferred to 0.5cm cuvett in the sample beam, and the buffer at the adjusted pH in reference beam was used, the absorption spectrum was scanned.

B. (AFP/Anti AFP) Complex:

1-   Twenty microlitters of human AFP standard provided by (AFP-IRMA Kit) was mixed with 80 µl of ^{125}I-anti AFP antibody. The volume of the mixture was completed to 250 µl with Tris buffer (0.05M, pH 7.2).

2- The mixture was incubated at 25 °C for one hour.

3- At the end of incubation, the mixture was centrifuged for 30 min. at 4000 r.p.m. at 4 °C to separate the (AFP/anti AFP) complex.

4- The supernatants were discarded by decanting the assay tubes, and the precipitate formed was dissolved in 1 ml with different pH values (2, 4, 7, and 11) then each sample was placed in 0.5cm cuvette in the sample beam. The buffer at the adjusted pH in the reference beam then used, the absorption spectrum was scanned.

Solutions:

1- Tris (basic buffer, 0.05M), was prepared by dissolving 0.788 gm of Tris (hydroxy methyl amino methane hydrochloride), 0.788 gm of EDTA and 0.2 gm of sodium azide in 50 ml of deionized distilled water, the volume of the mixture was completed to 100 ml with deionized distilled water. The pH was adjusted to 4 by adding few drops of NaOH.

2- Tris (acidic buffer, 0.05M), was prepared by dissolving 0.6057 gm of Tris (hydroxy methyl amino methane), 0.186 gm of EDTA and 0.2 gm of sodium azide in 50 ml of deionized distilled water, the volume of the mixture was completed to 100 ml with deionized distilled water. The pH was adjusted to 7, and 9 by adding few drops of (0.01M) HCl, and also drops of NaOH was added to adjust the pH 11.

5.2.4.2   The Effect of Solvent Polarity (Solvent Perturbation) on The U.V Spectrum:

A. h-AFP and Anti-AFP:

Twenty microlitters of each h-AFP standard and Anti AFP were completed to 1 ml with following solvents dissolved in Tris buffer (0.05M, pH 7.2):

20% DMSO

20% Ethanol

20% Glycerol

20% Ethylene glycol

The absorption spectrum of each sample was scanned against the corresponding solvent in reference beam using 0.5cm cuvette.

B. (AFP/Anti AFP) Complex:

The complex was prepared as mentioned in section (5.5.1 B), the complex was dissolved in the same solvent mentioned in section (5.5.2 A). The absorption spectrum of each sample was scanned against the corresponding solvent in the reference beam using 0.5cm cuvette.

5.2.4.3 The Effect of Urea, KCl, and (Urea,KCl) Mixture on The U.V Spectrum:

A. h-AFP and Anti-AFP:

One hundred microlitter of h-AFP standard and Anti AFP were piptted in a set of three tubes. The volumes were completed to 1 ml Tris buffer (0.05M, pH 7.2) contained (KCl (0.03M), Urea (8M), and mixture of 1:1 of both KCl (0.03M) and Urea (8M) respectively. Each sample was placed in 0.5cm cuvette in the sample beam and the buffer at the sample pH in the presence of the same salt in reference beam then used. The absorption spectrum of each sample was scanned.

B. (AFP/Anti AFP) Complex:

The complex was prepared as mentioned in section (5.5.1 B), and dissolved in the same solution mentioned in section (5.5.3 A). The same procedure was repeated as in section (5.5.3 A).

Solutions:

1- KCl (0.03M) was prepared by dissolving 0.118 gm of the salt in a final volume of 50 ml of Tris buffer (0.05 M, pH 7.2).

2- Urea (8M) was prepared by dissolving 24.024 gm of urea in a final volume of 50 ml of Tris buffer (0.05 M, pH 7.2).

5.2.4.4 Spectrophotometric pH Titration:

A. h-AFP and Anti-AFP:

One hundred microlitter each of h-AFP standard and Anti AFP
- were piptted in a set of tubes. The volumes were completed to 1 ml with

Tris buffer (0.05M) at pH ranging (2-6).

- Each sample was placed in 0.5cm cuvette in the sample beam and the buffer at the sample pH in reference beam then used. The maximum absorption of each sample was measured at 211 nm.

- Another set each of human AFP standard and Anti AFP were piptted in a set of tubes. The volumes were completed to 1 ml with Tris buffer (0.05M) at pH ranging (7-11).

- Step 2 was repeated as above and the maximum absorbance of each sample was measured at 295nm.

- The absorbance of λmax at each pH values was plotted versus the corresponding pH for each h-AFP, and anti AFP.

B. (AFP/Anti AFP) Complex:

1- The complex (AFP/anti AFP) was prepared as mentioned in section (5.5.1 B).

2- A set of the complexes was dissolved in a final volume of 1 ml with buffer at pH range (7-11). The maximum absorbance of each sample was measured at 295nm.

3- The absorbance of λmax at each pH values was plotted versus the corresponding pH.

4- Another set of the complexes was dissolved in a final volume of 1 ml with buffer at pH range (2-6). The maximum absorbance of each sample was measured at 211nm.

5- The absorbance of λmax at each pH values was plotted versus the corresponding pH.

Results and Discussion:

Gel Filtration Technique for Separation of Free and Bound ^{125}I-Anti AFP Antibody:

Figure (5.1) shows the results of gel filtration technique for the (AFP/anti AFP) complex. The complex was prepared from standard h-AFP equipped with kit, where this protein was highly pure. The experiment revealed two peaks profile. The first peak represents the (AFP/^{125}I-anti AFP antibody) complex, while the second peak represents to the unbound (free) ^{125}I-anti AFP antibody.

Figure (5.2) shows the gel flirtation profile of ^{125}I-anti AFP antibody, the results revealed only one peak in the same position of the second peak of the figure (5.1).

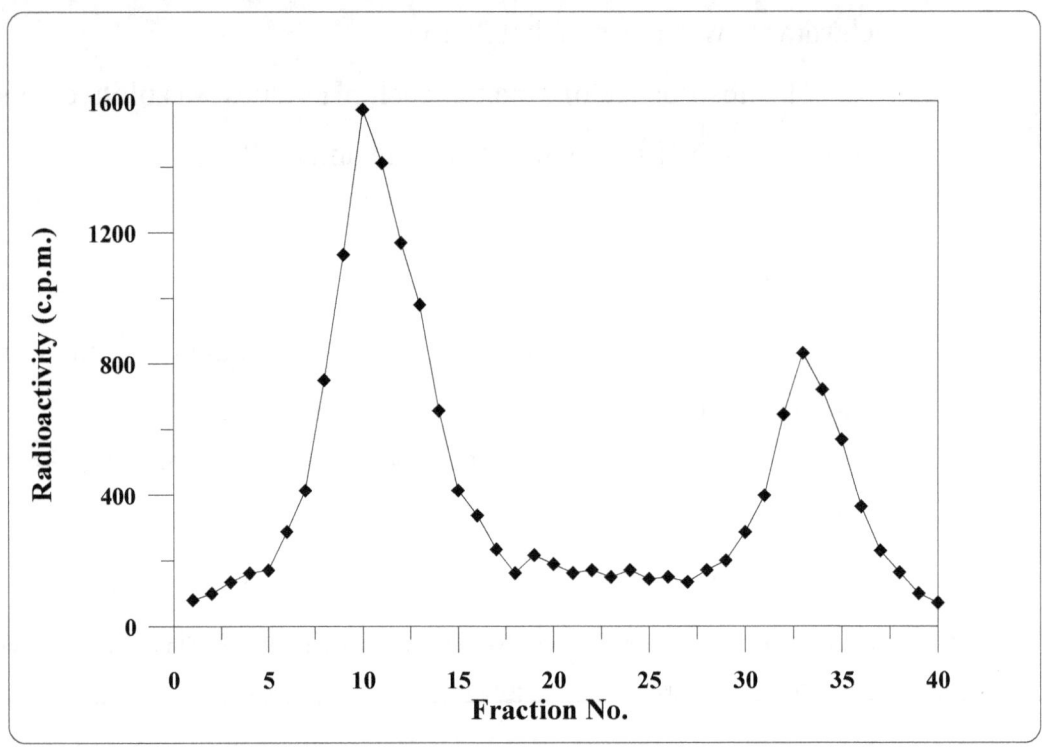

Figure (5-1): The elution profile of the prepared standard complex.
(All details are explained in the text)

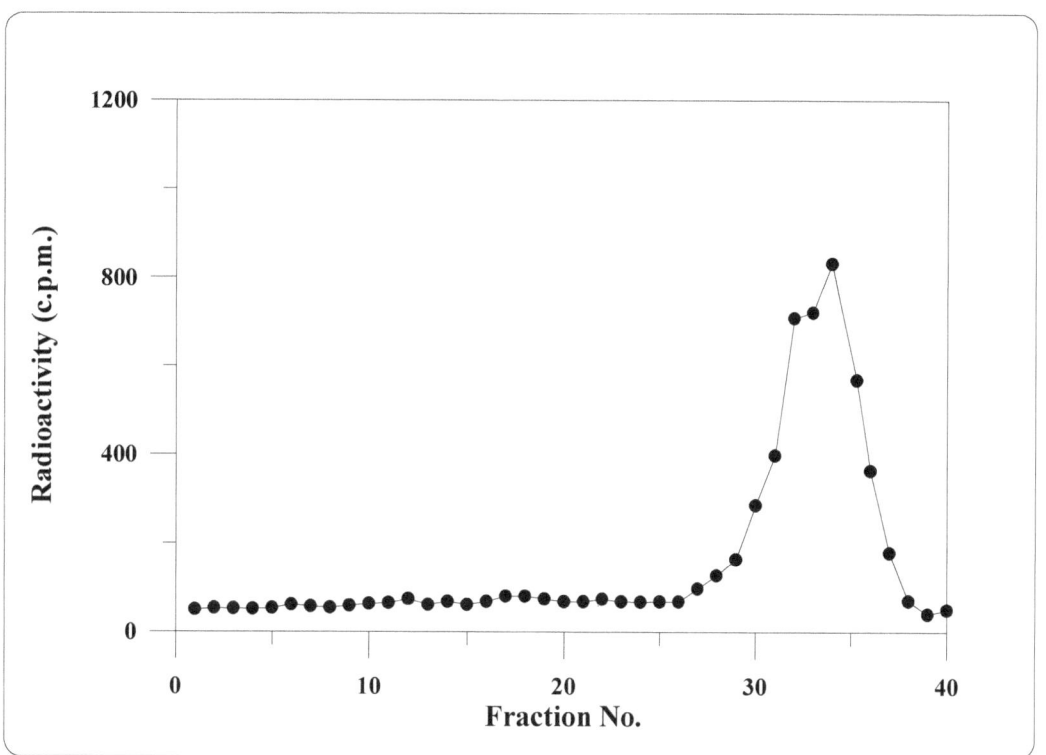

Figure (5-2): The elution profile of ^{125}I-anti AFP antibody. (All details are explained in the text)

Spectroscopic Studies of h-AFP, Anti AFP, and (AFP/Anti AFP) Complex:

The U.V spectrum of h-AFP, anti AFP, and (AFP/anti AFP) complex were scanned to determine their maximum wavelength, and the alteration in the U.V. spectra as a result of their interaction.

The U.V Spectra of h-AFP:

Figure (5.3) illustrated the U.V spectra of h-AFP provided with kit at pH 7.2. As shown, the spectrum consisted of one broad peak at 219 nm represents the amide bond with contribution of tyrosyl residue [259].

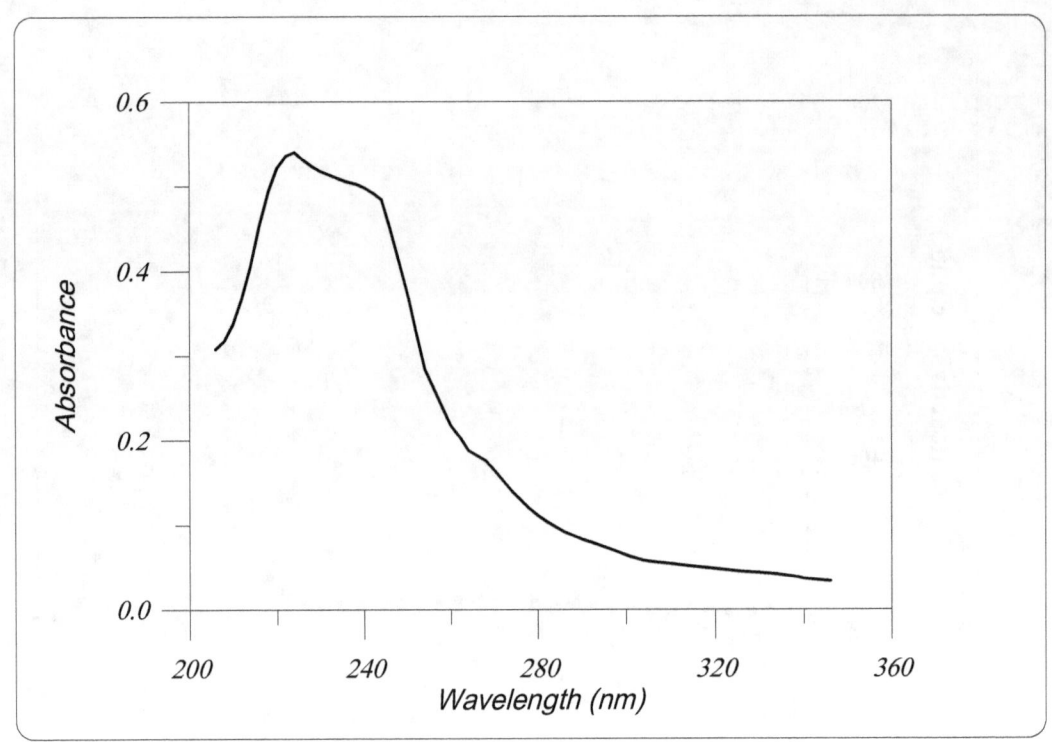

Figure (5-3): The U.V Spectra of h-AFP. (All details are explained in the text)

The U.V Spectra of Anti AFP:

Figure (5.4) illustrated the U.V spectra of anti AFP at pH 7.2. The spectrum shows that the λmax is consisted of two peaks at 218 and 280 nm. The first peak represents the amide bond which assigned to tyrosyl residue, while the second peak represents to the side chain chromophore of tryptophyl residue [260].

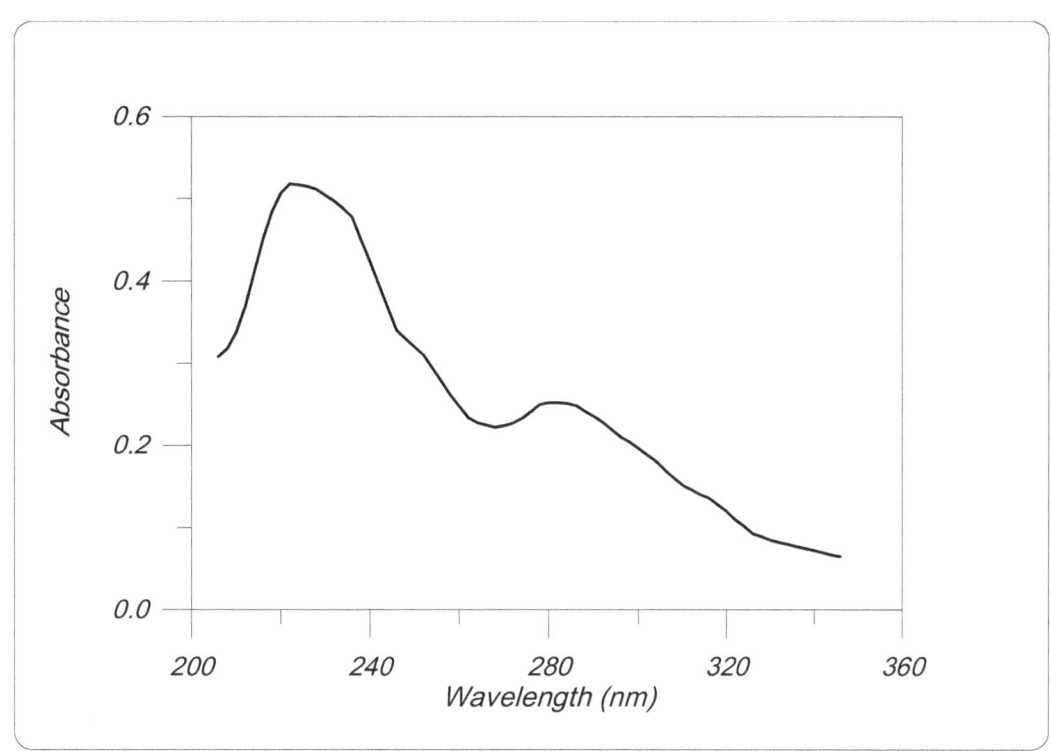

Figure (5-4): The U.V spectra of anti AFP. (All other details are explained in the text)

The U.V Spectra of (AFP/Anti AFP) Complex:

The U.V spectra of (AFP/anti AFP) complex appeared two peaks at 218.2 and 279 nm respectively as shown in figure (5.5). These two peaks were characteristic of the amide bond, which assigned to tyrosyl residue and tryptophyl residue respectively.

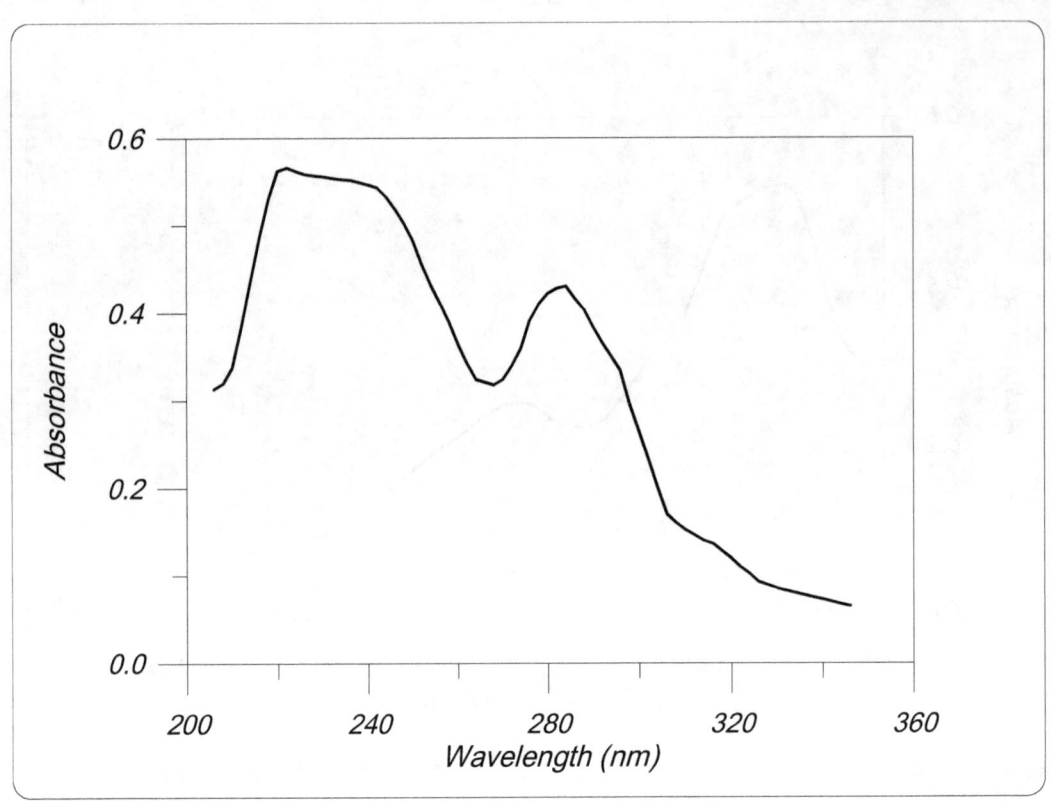

Figure (5-5): The U.V Spectra of (AFP/anti AFP) complex. (All details are explained in the text)

The value of λmax for the spectrum of h-AFP, anti AFP, and (AFP/anti AFP) complex are illustrated in table (5.1).

Table (5.1): The λ_{max} of the U.V. spectrum of h-AFP, anti AFP, and (AFP/anti AFP) complex.

Sample	λmax (nm)
h-AFP	219
Anti AFP	218, 280
(AFP/anti AFP) complex	218.2, 279.0

Factors Affecting The Absorbance of h-AFP, Anti AFP, and (AFP/ Anti AFP) Complex:

The absorbance of a chromophore is primary determined by the chemical structure of the molecule. However, a large number of environmental factors produce detectable change in the λ_{max} [260].

The Effect of pH:

The pH of the solvent determines the ionization state of the ionizable chromophore in the protein molecule. Table (5.2) shows the effect of increasing pH on the λ_{max} value of h-AFP, anti-AFP, and (AFP/anti-AFP) complex.

Table (5.2): The effect of different pH on h-AFP, anti-AFP, and (AFP/anti-AFP) complex.

H	λ_{max}					
	h-AFP		Anti-AFP		AFP/Anti-AFP complex	
	217	277	208	272.2	220	273
	217	275.2	208	270	227	274
	221.4	275.3	214.2	—	224	272
1	227.5	274.4	221.1	—	227	274

The λ_{max} was measured at different range of pH (2, 4, 7, and 11). From this table, two peaks were almost observed at different pH. There were a significant changes in λ_{max1} compared with λ_{max1} in table (5.1). There was disappeared of λ_{max2} at pH 4, 7, and 11 of anti AFP antibody molecule, this is may be due to conformational changes of this molecule [263].

It was found that h-AFP has a remarkably hydrophillic exposed molecular surface at neutral pH and possesses extensive hydrophobic binding sites located in crevices. Conformational changes occur in h-AFP in the acid and alkaline pH regions; extensive, hydrophobic areas of AFP are exposed by both acid and alkaline transitions [265]. At low pH the AFP molecule is transformed into the molten globule state [186]. In the case of tyrosine, it was found that the –OH group can dissociate a proton at high pH, producing a red shift. Below pH 9, where the –OH group is unionized, it could be as a donor in hydrogen bond with a suitable solvent or solute acceptor. It has even been suggested that the π-electron system of the benzene ring may serve as a weak proton acceptor, so that the –OH group might form a hydrogen bond with the tyrosine ring, hence this would also produce a red shift [259,266].

The Effect of Solvent Polarity (solvent perturbation) on The U.V Spectrum of h-AFP, Anti-AFP, and (AFP/Anti-AFP) Complex:

The spectra of chromophores depend on the polarity of their environment, therefore the determination of whether an amino acid is internal or external by measuring the spectra of a protein in a polar and nonpolar solvent is called the *solvent-perturbation method* [260]. The nonpolar solvent itself does not introduce conformational changes. In fact, proteins are rarely studied in completely nonpolar solvents because most proteins are either insoluble or denatured in these solvents [260].

Table (5.3) shows the effect of different solvent on λ_{max} of h-AFP, anti AFP, and (AFP/anti AFP) complex. There was a shift in λ_{max1} toward longer wavelength (red shift) in the presence of 20% ethanol for h-AFP, anti AFP, and (AFP/anti AFP) from 219, 218 and 218.2 nm to 222, 221 and 221.5 nm in each molecule.

Ethylene glycol showed no significant changes in λ_{max1} in both h-AFP and anti AFP, while for (AFP/anti AFP) complex there was an alteration from 218.2 to 221.4 nm. Also λ_{max2} appeared in h-AFP, while a decrease in λ_{max2} from 279 to 273 nm was occurred in the complex.

In the case of 20% glycerol; there was a red shift in λ_{max1} for h-AFP and the complex suggested that the buried tyrosyl regions of the molecules are more readily disrupted [266]. For anti AFP, λ_{max1} was disappeared and no significant change in λ_{max2}, while λ_{max2} appeared in h-AFP which the assigned to tyrosine.

The effect of 20% DMSO showed a red shift in λ_{max1} for the three molecules under investigation and no significant change in λ_{max2} for anti AFP and the complex which is related to tyrosin and phenyl alanine and there was appearance of λ_{max2} in h-AFP.

The several spectral changes obtained in the presence of the perturbants, like the alteration of λ_{max} position and disappearance or appearance of new chromophore on the surface of the molecules. These chromophores were embedded in an interior region of the protein in the absence of the solvent [267, 268]. One of the main assumptions of the solvent perturbation was that the perturbing solvent dose not alter the conformation of the protein but solvent alter the peak positions and intensities by altering the energy and probability of electronic transitions changes in permanent dipole moment during excitation, i.e. the dipole hydrogen bonding which will tend to produce either a short wave or long wave shift depending on the electronic transition and weather the solute is hydrogen donor or hydrogen acceptor [269-271].

Table (5.3): The effect of solvent polarity, solvent perturbation on h-AFP, anti-AFP, and (AFP/anti-AFP) complex on U.V spectrum.

Solvent	λ_{max}					
	h-AFP		Anti-AFP		AFP/Anti-AFP complex	
20% Ethanol	222	274	221	277	220.5	276
20% Ethylene glycol	222	275.6	218.8	278	221.4	273
20% glycerol	227	278	—	277	227	271
20% DMSO	243.2	277	234.6	278	244	274.6

The Effect of Urea, KCl, and (Urea, KCl) Mixture on The U.V Spectrum of h-AFP, Anti-AFP, and (AFP/Anti-AFP) complex:

The effect of urea (8M), KCl (0.03M), and a mixture of 1:1 of urea (8M):KCl (0.03M) on the λ_{max} of h-AFP, anti-AFP, and (AFP/antiAFP) complex were examined. The values of λmax are illustrated in table (5.4).

Comparing the values of λ_{max} of these molecule which obtained in the absence of urea or KCl as in table (5.1) with those obtained in presence of 8M urea in table (5.4), it is clear that there was a significant red shift in λ_{max1} of the polypeptide bond from 219, 218, and 218.2 nm to 228, 224.8, and 227 nm for h-AFP, anti-AFP and (AFP/anti AFP) Complex spectrums respectively. The red shift is due to intramolecular hydrogen bonding between the oxygen of the amide group and the solvent [259].

When KCl (0.03M) used, there was no alteration in the position of the λmax1 for anti AFP while there was a red shift for h-AFP and (AFP/anti AFP) complex. Such a red or blue shift can arise by introducing positive or

292

negative charges near the chromophore, which might interact directly with π-electron system of the benzene ring [259].

Table (5.4): The effect of urea, KCl, (urea,KCl) mixture on the λmax of

h-AFP, anti-AFP, and (AFP/anti-AFP) complex

Solvent	Λmax					
	h-AFP		Anti-AFP		AFP/Anti-AFP complex	
Urea (8M)	25	71	24.8	76	227	269
KCL (0.03M)	27	71	19	78	224 .2	272
(Urea: KCL) mixture	24.8	72	26	75	228	270

When a mixture of urea (8M): KCl (0.03M) was used, there were a significant red shift in λ_{max} (219, 218, and 218.2 nm) to λ_{max} (224.8, 226, and 228 nm) for the molecules under investigation. The shift caused by the mixture may be due to the effect of urea not by KCl. It was found that some of the changes in the absorption may be produced by change in the $n \rightarrow \pi^*$ absorption of polypeptide bonds in the protein either because of change in their geometrical arrangement, or because an environment changes [264].

Spectrophotometric pH Titration:

Many Studies of protein structure require the determination of *pK* values for proton dissociation from ionizable amino acid side chains, because these values give an indication of amino acid location in the protein. This can often be done spectrophotometrically because dissociation often changes the spectrum of one of the chromophores. The observation of tyrosine dissociation was performed by measuring the absorption at 295nm (λ_{max} for the ionized form of tyrosine, and the observation of histidine dissociation was carried out by measuring the observation at all 211nm).

Figure (5.6) shows the pH titration curve of h-AFP, anti-AFP, and (AFP/anti AFP) complex for tyrosine. It was shown that the pKa values of tyrosyle residue were (9.5), (10.2), and (9.9) respectively.

Figure (5.7) illustrated the pH titration of h-AFP, anti-AFP, and (AFP/anti AFP) complex for histidine. It was shown that the pKa values of histidyl residue were (5.7), (6.0), and (5.9) respectively.

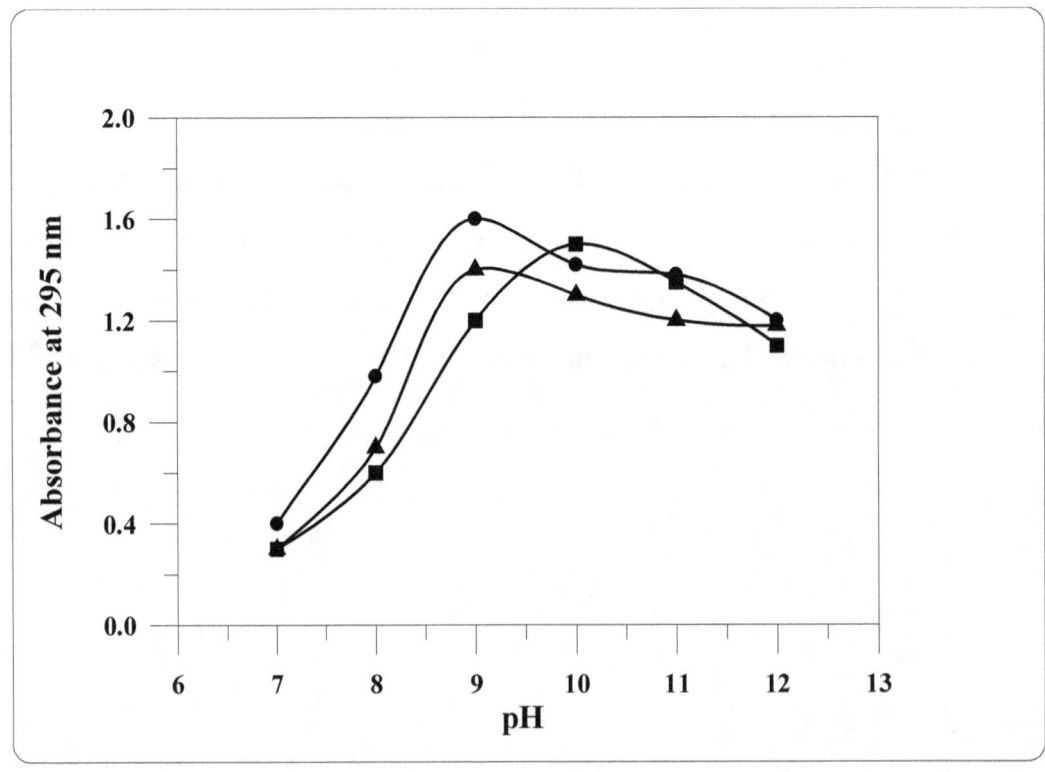

Figure (5-6): Spectrophotometric pH titration (tyrosine) of:

(●) h- AFP, (■) anti-AFP, and (▲) (AFP/anti AFP) complex for tyrosine. (All details are explained in the text)

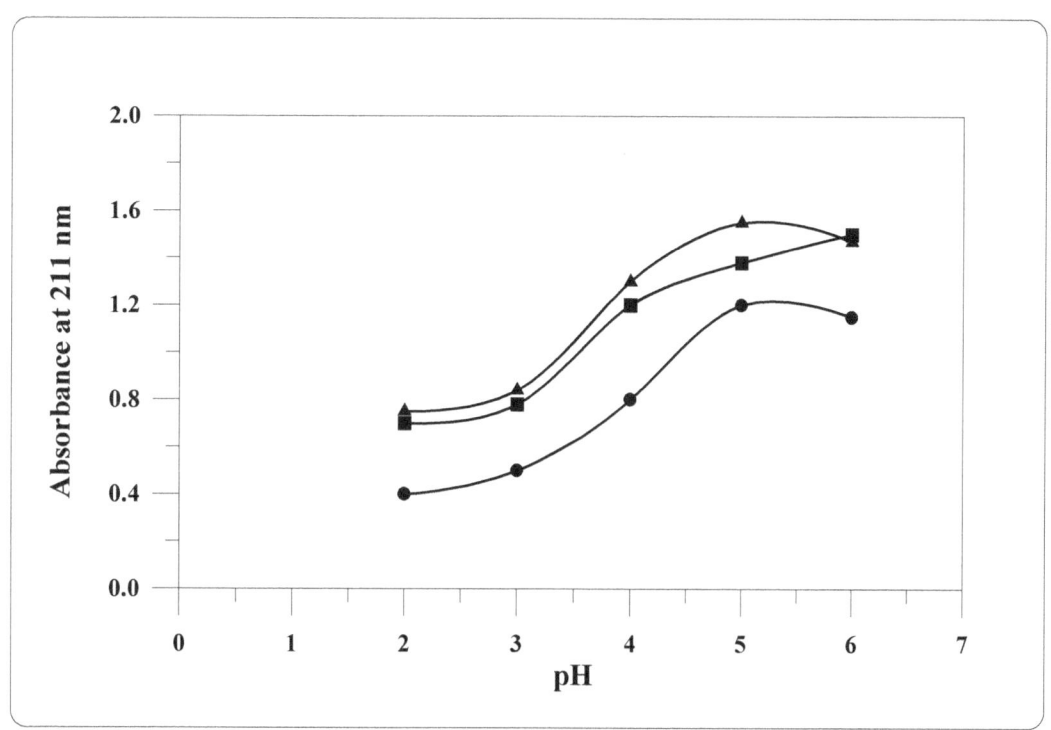

Figure (5-7): Spectrophotometric pH titration (histidine) of:

(●) h- AFP, (■) anti-AFP, and (▲) (AFP/anti AFP) complex for tyrosine. (All details are explained in the text)

Future Work

The following works are suggested for the future:

1. the application of the developed method for the assessment of AFP in other tumors tissues such as pancreas, ovary, etc.

2. application of the developed method for assessment of new markers as tumor-associated antigen (CA27-29 MSA, MCA),galectin-4, tissue polypeptide antigen (TPA), tissue polypeptide specific antigen (TPS), Cathepsin-D,and c-erbB-2 oncoprotein in colorectal tumor.

3. purification and molecular characterization of AFP from different tumor tissues.

4. spectroscopic studies of isolated AFP from different tissues such as pancreas, ovary, etc.

References

1. Berkow R., Fletcher A.J., Beers M.H., "**The Merck Manual of Diagnosis and Therapy**", 16th ed., Merck & Co., Inc. Rahway, N.J. U.S.A., 1998, p. 795.

2. Berkow R., Fletcher A.J., Beers M.H., "**The Merck Manual of Diagnosis and Therapy**", 16th ed., Merck & Co., Inc. Rahway, N.J. U.S.A., 1998, p. 794.

3. Seely R.R., Stephens T.D., Tate P., "**Anatomy and Physiology**", 4th ed., W.C.B. McGraw-Hill, 1998, p. 943.

4. Berkow R., Fletcher A.J., Beers M.H., "**The Merck Manual of Diagnosis and Therapy**", 16th ed., Merck & Co., Inc. Rahway, N.J. U.S.A., 1998, p. 806.

5. Seely R.R., Stephens T.D., Tate P., "**Anatomy and Physiology**", 4th ed., W.C.B. McGraw-Hill, 1998, p. 966.

6. Porth C.M., "**Pathophysiology Concepts of Altered Health States**", 4th ed., J.B. Lippincott Company, 1994, pp. 756, 757.

7. Berkow R., Fletcher A.J., Beers M.H., "**The Merck Manual of Diagnosis and Therapy**", 16th ed., Merck & Co., Inc. Rahway, N.J. U.S.A., 1998, p. 811.

8. Damjanov I., Linder J., "**Anderson's Pathology**", 10th ed., Mosby-A Times Mirror Company, 1996, p. 2304.

9. Cotran R.S., Kumar V., and Collins T., "**Robbins Pathologic Basis of Disease**", 6th ed., W.B. Sunder Company, 1999, p. 856.

10. Tiernery L.M., McPhee S.J., and Papadakis M.A., "**Current Medical Diagnosis and Treatment**", 38th ed., New York; Appleton & Lange, 1999, p. 653.

11. Damjanov I., Linder J., "**Anderson's Pathology**", 10th ed., Mosby-A Times Mirror Company, 1996, p. 2307.

12. Cotran R.S., Kumar V., and Collins T., "**Robbins Pathologic Basis of Disease**", 6th ed., W.B. Sunder Company, 1999, p. 865.

13. Rosai U., "**Ackerman's Surgical Pathology**", 8th ed., Mosby-A Times Mirror Company, 1996, p. 1287.

14. Markowitz, A. J.; Winawer, G. J.; (1997). J. Am. Cancer Soci. , **47** (2): 93.

15. Nagel, H.; Bahlo, M.; Klapdor, R.; et al., (1999). Am. Heart. J., **137** (6): 1044.

16. Berkow R., Fletcher A.J., Beers M.H., "**The Merck Manual of Diagnosis and Therapy**", 16th ed., Merck & Co., Inc. Rahway, N.J. U.S.A., 1998, p. 1680.

17. Bombi, J. A.; (1988). Cancer. , **61**: 1472.

18. Rickert, R. R.; Aucrbach, O.; Garfinkel, L.; et al., (1979). Cancer. , **43**: 1847.

19. O᾽Brien, M. G.; Winawer, S. J.; Zauber, A. G.; et al., (1990). , Gastroenterology. , **98**: 371.

20. Winawer, S. J.; Zauber, A. G.; O᾽Brien, M. G.; et al., Cancer. , **70**: 1236.

21. Match, W.; Demling. ; Hermanek, P.; (1986). Endoscopy. , **18**: 17.

22. Lawrence, J. Bronelt. ; (1999) "**Clinical Practice of Gastroentrology**", vol. 1 2nd. ed., Philadelphia: Cuuent Medicine Inc., Chap. 66, pp. 588 – 595 & Chap. 48, pp. 752 – 776.

23. Joseph, R.; Bertino. ; (1996) "Encyclopedia **of Cancer**". Vol. 1. 1 st. ed., California, Academic Press. pp. 441.

24. Mayers, M. A.; (1998). " **Neoplasms of the Digestive Tract: Imaging Staging & Management.** ", Philadelphia: Lippincott – Raven Publishers, Chap. 19. pp. 203.

25. Berkow R., Fletcher A.J., Beers M.H., "**The Merck Manual of Diagnosis and Therapy**", 16th ed., Merck & Co., Inc. Rahway, N.J. U.S.A., 1998, p. 1677.

26. Chow, W.; Dwvesas. , (1993). Cancer. , **71**: 3819.

27. Bostick, R. M.; Slattery, M. L.; Potter, J. D; et al., (1993). Epidemilol. Rev.**, 15**: 499.

28. Christopher, H.; Edwin, R. Ch.; John, A. H.; Nicholas, A. B.; (1999) " **Davidson's Principles and Practice of Medicine** " 8 th . ed., U. K., Harcourt Brace & Company Limited. Chap 9. pp. 671.

29. Ministry of Health (1989). **Results of Iraqi Cancer Registry.** (1986 – 1988).

30. Ministry of Health (1993). **Results of Iraqi Cancer Registry.** (1989 – 1992).

31. Bos, J. L.; Fearon, E. R.; Hamilton, S. R.; (1987). Nature. , **327**: 293.

32. Muto, T.; Bussey, HJ. R.; Morson, B. C.; (1985). , Cancer. , **36**: 2251.

33. Kato, H.; Tamaik, K.; Morioka, M.; Najai, T.; et al., (1999) Cancer., **54** (3): 1544.

34. O'Brien, M. J.; O'kean, J. C.; Zauber, A.; Gottlieb, L. S. ; et al ., (1992). Cancer. , **70**: 1317.

35. Harnden, D. G.; (1984). Carcinogenesis. , **5**: 1535.

36. Hamilton, S. R.; (1993). Gastroenterology**. , 105**: 3.

37. Couture, J.; Swallow, C.; Redston, M.; Gallinger, S.; et al., (1997). Cancer. , **21** (5): 233.

38. Bishop, J. M.; (1987). Science. , **233**: 305.

39. Kelin, G.; (1981). Nature. , **294**: 313.

40. Watson, F. R.; Kyle, K.; Turnbull, R. B.; et al., (1987). Ann. Surg.**, 166**: 420.

41. Warell, D. A.; Leadingham, J. G. G.; Weatherall, D. J.; (1988). "Oxford **Text Book of Medicine**". , Vol . 1, London: English Language Book Society Oxford University Press. pp. 12, 146, 157.

42. Winawer, G. J.; Markowitz, A. J.; (1997). J. Am. Cancer. Soc.**, 47**, (2): 93.

43. Mach, J. P.; Deperthes, D.; Finnern, R.; Couty Jouve, S. ; Houimel , M . ; (1999). **, http: // www. isrec. ch / reports /mach. asp. , 09 / 24 / 1999, pp. 1 – 6.**

44. Yactayos. ; Foultier, M. T.; Patrice, T.; et al., (1990). Dig. Dis. Sci.**, 35**: 545.

45. Arroyop, J.; Stern, J. D.; Unger, S. W.; et al., (1990). Am. Surgeon. , **56**: 153.

46. Wiand, H. S.; Martenson, J.; O 'Connell, M.; (1994). **N.** Eng. J. Med., **331**: 502.

47. Macmillan, W. E.; Wolberg, W. H.; Welling, P. G.; (1978) Cancer Res., **38**: 3479.

48. Ullman, B.; Lee, M.; Martin, D. W. Jr.; Santi, D. V.; (1978). Proc. Natl. Acad. Sci., **35**: 123.

49. Horton, J.; Olson, K. B.; Sullivan, J.; Reilly, C.; et al., (1970). Ann. Intern. Med., **73**: 897.

50. Siefert, P.; Baker, L. H.; Reed, M. L.; Vaitkevicius, V.; (1975). Cancer. , **35**: 123.

51. Ansfiel, D. F.; Klotz, J.; Nealon, T.; Ramirez, G.; et al., (1977). Cancer. , **39**: 34.

52. Erlichman, C.; Fine S.; Wong, A.; Elhakim, T.; (1988). J. Clin. Oncol. , **6**: 469.

53. Lokich J. J.; Anlgren, J. D.; Gullo, J. J.; Phillips, J. A.; Fryer, J. G.; (1989). J. Clin. Oncol**. 7**: 425.

54. Leichman, C. G.; Leichman, L.; Spears, C. P.; Rosen, P. J. ; et al ., (1990). Pharmacol. , **26**: 57.

55. Poplin, E. A.; Kraut, M.; Baker, L.; Brodfuehrer, J.; Vaitkevicius V.; (1991). Cancer. , **67**: 367.

56. Inglis, J. K.; (1989) "A **Text Book of Human Biology**". 3 rd. ed., New York: Pergamon Press, Chap. 7, pp. 114 – 115.

57. Gylys, B. A.; Wedding, M. F.; (1988). "Medical **Terminology a Systems Approach".** 2 nd. Ed. Phiiladelphia: F. A. Davis Company. Chap. 6 pp. 92.

58. Ferguson, J. E.; Hulse, P.; Jayson, G.; Lorigan, P. ; Scarffe, J. H . ; (1995). B. J. Cancer. , **72**: 193.

59. Presant, C. A.; Wolf, W.; Waluch, V.; Wiseman, C. L.; Weitz, I.; and Shani, J.; (2000). J. Am. Soc. Clin. Oncol. , **8**: 107.

60. O 'Dwyer, P. J.; Panal, A. R.; Weiner, L. M.; Comis, A. L. ; (1989) . Proc. Am. Soc. Clin. Oncol. , **8**: 107.

61. Holland, J. F.; Bast, R. C.; Morton, D. L.; (1997). "**Cancer Medicine**". , 4 th . Ed. London: Willians & Wilkins. Vol. 2, Section XXXI, Chap. 121. pp. 2029 – 2072.

62. Moertl, C. G.; (1988). J. Clin. Oncol. , **6**: 934.

63. Bleiberg, H.; Rougier, P.; Wilker, H. J.; (1998). "**Management of Colorectal Cancer**". London: Martin Dunitz Ltd. Chap. 3. pp. 35 – 54.

64. Pamies R.J. and Crawford D.R. ;Med. Clinc. N. Amer.; 1996; **80:**185-199.

65. Burtis, C. A; Edward, R.; Ashwood, E. R.; (1999) "Tetiz **Text Book of Clinical Chemistry**". 3 rd. ed., Philadelphia, W. B. Saunders Co. Chap. 23. pp. 722 – 748.

66. Burtis, C. A; Edward, R.; Ashwood, E. R.; (1999) "Tetiz **Text Book of Clinical Chemistry**". 3 rd. ed., Philadelphia, W. B. Saunders Co. Chap. 23. pp. 754-756.

67. Daivid, W.; (2001) "Immunoassay **Hand Book**". 2 nd. ed. U. K. nature publishing Co. Chap. 63. pp. 635 – 662.

68. Rubin, P. "**Clinical Oncology: A Multidisciplinary Approach for Physicians and Students**" 7th ed. Philadelphia, W.B. Saunders Co.; pp. 791 (1993).

69. The American Society of Clinical Oncology. "J. Clin. Oncol." **14**: 2843-2877 (1996).

70. http://Nauvoo.byu.edu//Academy/microbiol.Tumor/table/htm/10/1/1999.

71. http://www.med.uni.muenchen.de/egtm/detail/4htm.

72. Koprwsk, H.; Herlyn, M.; Steplewise, Z.; Sears, H. F.; (1981). Science, **212**: 53.

73. Magnant, J. L.; Brockhus, M.; Smith, D. F.; Blaszrzyk, M.; et al., (1981). Science, **212**: 55.

74. Diezn. ; Cerdan, F. J.; Pollan, M.; Maestro, M. L.; Ortega, M. D.; et al., (1994). Anticancer Res. **14**: 2819.

75. Carpelan, H. M.; Haglund, C.; Kusela, P.; Jarvinen, H.; Roberts, P. J.; (1995). Br. J. Cancer, **71**: 868.

76. Doos, W. G.; Wolff, W. I.; Shinya, H.; et al. (1975). Cancer, **36**: 1997.

77. Stevens, D. P.; Mackay, I. R.; (1973). Lancet, **2**: 1238.

78. Chu, T. M.; (1973). J .Natl. Cancer. Inst., **51**: 1119.

79. Vincent, R. G.; Chu, T. M.; (1973). Thorac Cardiovase Surg., **66**: 320.

80. Dittrich , C .; Jakes , Z .; Havelec .; et al ., (1985). Cancer, **8**: 181.

81. Ona, S. W.; Zamcheck, N.; Blair, P.; et al., (1973). Cancer, **31**: 324.

82. Van Nagell, J. R.; Donaldson, E. S.; Wood E. G. ; et al . , (1978). Cancer, **42**: 1527.

83. Goldenberg, D. M.; (1988). Arch. Pathol. Lab. Med., **112**: 580.

84. Wanebo, H. J.; Rao, B.; Pinsky, C. M.; et al., (1978). N. Engl. J. Med., **229**: 448.

85. Campo, E.; Munoz, J.; Miquel, R.; Palacin, A.; et al., (1994). Am. J. Pathol. , **145**: 301.

86. Jessup, J. M.; (1994). Am. J. Pathol. , **145**: 253.

87. Walach, N.; Guterman, A.; Zaidman, J. L.; et al., (1991). Tummori. , **77**: 164.

88. Verazin G.; Riley, W. M.; Gregory, J.; et al., (1991). Dis. Colon Rectum. , **33**: 139.

89. Barillari, P.; Rammacciato G.; De Angelis, R.; et al., (1989). Int. J. Colorectal. Bis. , **4**: 230.

90. Derimanov S.G.; Nucl. Med. Commun. ; 1984 Mar.; **5**(3): 169.

91. Kiyasu Y., Stao M., Sakai K., *et al.*; Gan-No-Rinsho. ; 1983 May; 29(5): 454. (Medline, abstr.).

92. Webb A,Scott-Mackei P, Cunningham D. et al. Ann Onco. 1995; **6**:581.

93. Ishiguro T., Sugitachi I., Sakaguchi H., and Itani S.; Cancer; 1985; **55**:156.

94. Zinser J.W.;Rev. Gastroenterol. Mex.; 1997 Jul-Sep.; 62(3): 145.

95. Mizejewski G.J.; J. Theor. Biol.; 1995; **176**:103.

96. Bagshawe K. D.; **Medical Oncology (Medical aspects of Malignant Disease)**; Blackwell Scientific Publication; London; 1975; p: 257.

97. Smith J. A., Fracis T. I., Edington G. M., and Williams A. O.; Br.J.Cancer; 1971; **25**:337.

98. Delves P. J., and Roitt I. M.; **Encyclopedia of Immunology**; 2nd ed.; Acadimic Press Limited; New York; 1998; p: 798.

99. Burtis C.A., and Ashwood E.R.; **Tietz Fundamentals of Clinical Chemistry**; 4th ed.; W.B. Saunders Company; London; p: 344.

100. Kaneko J.J., Harvey J.W., and Bruss M.L.; **Clinical Biochemistry of Domestic Animals**; 5th ed.; Acadimic Press; New York; 1997; pp: 762, 122.

101. Nowak T., and Handford A.G.; **Essentials of Pathophysiology (Concepts and Applications for Health Care Professionals)**; 2nd ed.; The McGraw-Hill Companies, Inc.; New York; 1999; p: 148.

102. McLeod, JF.; Cooke, NE. ; J. Biol. Chem.; 1989; **264**: 21760-21769.

103. Lichenstein, HS.; Lyons, DE.; Wurfel, MM.; Johnson, DA.; McGinley, MD.; Leidle, JC.; Trollinger, DB.; Mayer, JP.; Wright, SD.; Zukowski, MM.; J. Biol. Chem.; 1994; **269**: 18149-18154.

104. Mizejewski, G.J.; J. Theo. Bio.; 1995; **176**: 103-113.

105. Mizejewski, G.J.; Exp. Bio. Med.; 2001; **226**: 377-408.

106. Carter, Dc.; He, XM.; Science; 1990; 302-304 . [Medline].

107. Luft, A.J.; Lorscheider, F.L.; Biochemistry; 1983; **22**: 5978-5981.

108. Yang, F.;Luna, V.J.; McAnelly, R.D.; Neberhaus, K.H.; Cupples, R.I.; Bowman, B.H. ; Nucleic Acids Res.; 1985; **13**: 8007-8017.

109. Deutsch H.F.; Adv. Cancer Res.; 1991; **56**: 253.

110. Burtis C.A., and Ashwood E.R.; **Tietz Fundamentals of Clinical Chemistry**; 4th ed.; W.B. Saunders Company; London; p: 263.

111. Mizejewski, G.J.; Critical Reviews in Eukayotic Gene Expression; 1995; **5**(3 and 4): 281.

112. Terry, W.D.; Nature; 1976; **260**: 804-805.

113. Alpert, E.; Drysdale, J.W.; Isselbacher, K.J; Schu, P.H.; J. Biol. Chem.; 1972; **247**: 3792-3798.

114. Hirai, H., Nishi, S.; Watabe, H.; Tsukada, Y.; Cancer Res.; 1973; **14**: 19-34.

115. Krusius, T.; Ruoslahti, E.; J. Biol. Chem.; 1982; **257**: 3453-3457.

116. Gillespie, JR.; Uversky, VN. ; Biochim. Biophys. Acta; 2000; **1480**: 41-56.

117. Ruoslahti, E.; Seppälä, M.; Int. J. Cancer; 1971; **7**, 218-225.

118. Ruoslahti, E.; Seppälä, M.; Pihko, H.; Vuopio, P.; Int. J. Cancer; 1971; **8**, 283-288.

119. Ruoslahti, E.; Engvall, E.; Pekkala, A.; Seppälä, M.; Int. J. Cancer; 1978; **22**: 515-520.

120. Taketa, K.; Hepatol.; 1990; **12**: 1420-1432.

121. Tamaoki, T.; Morinaga, T.; Sakai, M.; Protheroe, G.; Urano, Y.; Ann. N. Y. Acad. Sci.; 1983; **417**: 13-20.

122. Uversky, V.N.; Narizhneva, N.V.; Ivanova, T.V.; Tomashevsk, A.Y.; Biochemistry; 1997; **36**: 13638-13645.

123. Haefliger D., Moskaitis J.E., Shoenberg D.R., and Wahli W.; J. Mol. Evol.; 1989; **29**: 344.

124. Mcleod J.F., and Cooke N.E.; J. Bilo. Chem.; 1989; **246**: 21760.

125. Mizejewski, G.J.; BioEssays; 1993 May; **15**(5): 427.

126. Yang F., Luna V.J., McAnelly R.D.,et al.; Nucleic Acid Research ; 1985 ; **13**(22) : 8007.

127. Goeken N.E., and Thompson J.S.; J. Immunol.; 1977; **119**: 139.

128. Ruoslathi E., and Terry W.D.; Nature; 1976; **260**: 804.

129. Morinaga T., Sakai M., Wegmann T.G., and Tamaoki T.; Proc. Natl. Acad. Sci. U.S.A.; 1983; **89**: 4604.

130. Gorin M.B., Cooper D.L., Eiferman F., et al.; J. Biol. Chem.; 1981; **256**(4): 1954.

131. Lester E.P., Miller J.B., and Yachnin S.; Proc. Natl. Acad. Sci. U.S.A.; 1976; **73**: 4645.

132. Hirai H., Nishi S., Watabe H., and Tsukada Y.; Cancer Res.; 1973; **14**: 19.

133. Kioussis D., Eiferman F., Gorin M.B., et al.; J. Biol. Chem.; 1981; **256**(4): 1960.

134. Dudich, I.; Tokhtamysheva,N.; Semenkova, L.; Dudich, E.; Hellman, J.; Korpela, T.; **Biochemistry**; 1999; **38**: 10406-10414. [Abstract]

135. Morinaga, T.; Sakai, M.; Wegmann, T. G.; Tamaok, T.; Proc. Natl. Acad. Sci. U.S.A.; 1983; 80: 4604-4608.

136. Mizejewski, G.J.; Proc. Soc. Exp. Bio. Med.; 1997; **226**: 333-362.

137. Mizejewski, G.J.; **Bioessays**; 1993; **15**: 427-432. [Medline]

138. Yoshima, H.; Mizuochi, T.; Ishii, M.; Kobata, A.; Cancer Res.; 1980; **40**: 4276-4281.

139. Yakimenko, E.F.; Karamova, E.R.; Goussev, A.I.; Hilgers, J.; Abelev, G.I.; Yazova, A.K.; Tumor Biol.; 1998; 19: 301-309.

140. Nustad, K.; Paus, E.; Kierulf, B.; Bormer, O.P.; Tumor Biol.; 1998; **19**: 293-300.

141. Luft, A.J.; Lorscheider, F.L.; Biochemistry; 1983; **22**: 5978-5981.

142. Mizejewski, G.J.; Crit. Rev. Eukaryot Gene Expr,; 1995; **5**: 281-316.

143. Mizejewski, G.J.; Life Science; 1995; **56**(1): 1.

144. Mizejewski, G.J.; P. S. E. B. M.; 1997; **215**: 333. [Minireview]

145. Smally J. R., and Sarcione E. J. ; Biochem Biophys Res. Commun.; 1980; **92**(4): 1429.

146. Mizejewski, G.J., and Porter I. H.; **Alpha-Fetoprotein and Congenital Disorders**; Academic Press; Orlando; 1985; p: 5.

147. Cradall B.F.; Alpha-fetoprotein A Review; CRC. Crit. Rev. Clin. Lab. Sci.; 1981; **15**: 127.

148. Norgaad-Pedresen B., and Axelsen N. H.; Clin. Chem. Acta.; 1976; **71**: 343.

149. Marrink J.; Scand. J. Immunol.; 1978; **8**: 309.

150. Sacione E. J., and Smalley J. R.; Cancer Res.; 1976; **36**: 3203.

151. Meng Sen, L.I.; Ping Feng, L.I.; Fei Yi Yang; Shi Peng, H.E.; Guo Guang, D.U.; Gang, L.I.; Cell Res.; 2002; **12**: 151-156.

152. Deutsch, H.F.; Adv. Cancer Res.**; 1991; 56**: 253-312.

153. Abelev, G.I.; Eraiser, T.L. ; Semin. Cancer Biol.; 1999; **9**: 95-107.

154. Leffert, H.L.; Sell, S.J.; J. Cell Biol.; 1974; 61: 823-829.

155. Sell, S.; Skelly, H.; Leffert, H.L.; Muller-Eberhard, U.; Kida, S.; Ann. N.Y. Acad. Sci.; 1975; **259**:45-58 (1975).

156. Keel, B.A.; Eddy, K.B.; Cho, S.; May, J.V.; Endocrinol.; 1991; **129**: 217-225.

157. Mizejewski, G.J.; MacColl, R.; Mol. Cancer Ther.; 2003; **2**: 1243-1255.

158. Toder, V.; Blank, M.; Gold-Gefter, L.; Nebel, L.; Placenta; 1983; **4**: 79-86.

159. Mizejewski, G.J.; Jacobson, H.I. **"Biological Activates of Alpha-fetoprotein"** Boca Raton, F.L.; CRC Press; 1987; Vol. II: pp. 162-179.

160. Leal, J.A.; May, J.V.; Keel, B.A.; Endocrinol.; 1980; **126**: 669-671.

161. Mizejewski, G.J.; Warner, A.S.; J. Reprod. Fertil.; 1989; **85**: 177-185.

162. Mizejewski, G.J.; Keenan, J.F.; Setty, R.P.; Biol. Reprod.; 1990; **42**: 887-898.

163. Dudich, E.; Semenkova, L.; Gorbatova, E.; Dudich, I.; Khromykh, L.; Tatulov, E.; Grechko, G.; Sukhikh,G.; Tumor Biol.; 1998; **19**: 30-40.

164. MacLusky, N.J.; Naftolin, F.; Science; 1981; **211**: 1294-1302.

165. David, C.P.; Merle, A.E.; Harold, F.D.; J. Biol. Chem.; 1978; **253**: 2114-2119.

166. Aoyagi, Y.; Ikenaka, T.; Ichida, F.; Cancer Res.; 1979; **39**: 3571-3574.

167. Crandall, B.F.; CRC. Rev. Clin. Lab. Sci.; 1981; **15**: 127-185.

168. Gillespine, J. R.; Uversky, V.N.; Biochem. Biophys. Acta; 2000; **1480**: 41-56 (2000).

169. Mizejewski, G.J.; Porter, I.H. Eds. **"Alpha-Fetoprotein and Congential Disorder"** Orlando: Academic Press; 1985; pp. 5-34.

170. Nikolic, J.A.; Glas. Srp. Akad. Nauka.; 1992; **42**: 57-73. [Abstract]

171. Shirabe, K.; Takenaka, K.; Gion, T.; Shimada, M.; Fujiwara, Y.; Sugimachi, K.; J. Surg. Oncol.; 1997; **64**: 143-146.

172. Bidart, J.; Thuillier, F.; Augereau, C.; Chalas, J.; Daver, A.; Jacob, N.; Labrousse, F.; Voitot, H. ; Clin. Chem.; 1999; **45**: 1695-1707.

173. Tsuchida, Y.; Kaneko, M.; Yokomori, K.; Saito, S.; Urano, Y.; Endo, Y.; Asaka, T.; Takeuchi, T.; J. Pediatr. Surg.; 1978; **13**: 25-29 (1978). [Abstract]

174. Smith, J.B.; O'Neil, R.; Am. J. Med.; 1971; **51**: 767-771.

175. Germa-Lluch, J.R.; Begent, R.H.; Bagshawe, K.D.; Br. J. Cancer; 1980; **42**: 850-855.

176. Gerl, L.; Lamerz, R.; Mann, K.; Clemm, C.; Willmanns, W.; Anticancer Res.; 1997; **17**: 3047-3049.

177. Sell, S. ; Cancer Res.; 1973; **33**: 1010-1015.

178. Kohn, J.; Weaver, P.C.; Lancet; 1974; **304**: 334-337.

179. Hunter, W.M.; Budd, P.S.; J. Immuno. Meth.; 1981; **45**: 255-273.

180. www.diakey.com.

181. Masopust, J.; Kityier, K.; Rádl, J.; Koutecký, J.; Kotál, L.; Int. J. Cancer; 1968; **3**: 364-373.

182. Watabe, H.; Int. J. Cancer; 1974; **13**: 377-388.

183. Wu, J.T.; Knight, J.A.; Clin. Chem.; 1985; **31**: 1692-1697.

184. Katsumata, Y.; Sato, M.; Tamaki, K.; Rsutsumi, H.; Yada, S.; Oya, M. ; J. Forensic Sci.; 1985; **30**: 1210-1215. (Abstract)

185. Kang, Y.; Matsuura, E.; Sakamoto, T.; Sakai, M.; Nishi, S.; Tumor Biol.; 2001; **22**: 254-261.

186. Uversky, V.N.; Kirkitadze, M.D.; Narizhneva, N.V.; Potekhin, S.A.; Tomashevski, A.Y.; FEBS letters; 1995; **364**: 165-167.

187. Kodama, T.; kameya, T.; Hirota, T.; Shikosato, Y.; Ohkura, H.; Mukojima, T.; Kitaoka, H.; Cancer; 1981; **48**: 1647-1655.

188. Morimptp, H.; Tanigawa, N.; Inoue, H.; Muraoka, R.; Hosokawa, Y.; Hattori, T.; Cancer; 1988; **61**: 84-88.

189. Dempo, K.N.; Chisaka, N.; Yoshida, Y.; Kaneko, A.; Onoe, T.; Cancer Res.; 1975; **35**: 1282-1287.

190. Xing Wang, W.; Jin Hui, Y.; Ru Gang, Z.; Li Xia., G.; Yong, X.; Hong, X.; World J. Gastroenterol.; 2001; **7**: 345-351.

191. Burg-Kurland, C.L.; Purnell, D.M.; Combs, J.W.; Hillman, E.A.; Harris, C.C.; Trump, B.F.; Cancer Res.; 1986; **46**: 2936-2943.

192. Taketa, K.; Okada, S.; Win, N.; Hlaing, N.K.T.; Win, K.M.; Acta Med. Okayama; 2002; **56**: 317-320.

193. Bei, R.; Budillon, A.; Reale, M.G.; et al. ; Cancer Res.; 1999; **59**: 5471-5474.

194. Amato, R.; Kim, E.E.; Prow, D.; Andreopoulos, D.; Kasi, L.P. ; J. Cancer Res. Clin. Oncol.; 2000; **126**: 161-167.

195. Brochert A., and Staatz W. D.; Neuron; 1991; **7**: 231-237.

196. Alpert, E.; Pinn, V.W.; Isselbacher, K.H.; New Engl. J. Med.; 1971; **285**: 1058-1059.

197. McIntire, K.R.; Waldmann, T.A.; Moertel, C.G.; Vl Go.; Cancer Res.; 1975; **35**: 991-996.

198. Jones, E.A.; Clement-Jones, M.; James, O.F.; Wilson, D.I. ; J. Anat. ; 2001 ; **198**: 555-559. [Abstract]

199. Mizejewski, G.J.;Exp. Biol. Med. ; 2001 ; **226**(5) : 377-408.

200. Leather A. J., Gallegos N. C., Kocjan G., and Hu W.; Br. J. Surg.; 1993; **80**: 777-780.

201. Adam B.; Br. J. Cancer; 1997; 76: 628-633.

202. Janson, J. C.; and Pyden, L.; " **Protein Purification (Principles , High – Resolution , Methods and Applications) "** . , 2 nd . ed.; John Willey and Sons, Inc; New York; 1998; PP. 30, 79.

203. Al – Khayt, T. H.; (1991). **"Molecular Characterization of Prolactin Receptors in Human Prostate"**. Ph. D Thesis Supervised by Al – Mudhaffar S. A.; College of Science, Baghdad Univ.

204. Chamberlain, J.; Jargarince, N.; Ofher, P.; Biochem. J.; 1966; **99**: 610.

205. Farrant, T. J.; (1997). **"Practical Statistics for the Analytical Scientist"**. L G C. pp. 16, 49.

206. Scott-Mackie P, Cunningham D, et al. Ann Oncol.; 1995; **6**:581.

207. Deutch H. F.; Adv. Cancer Res.; 1991; **56**: 253-312.

208. Roitt, I.; Brostoff, J.; Male, D.; **"Immunology"**; 5th ed.; Iondon Mosby Philadelphia st. Louis; 1998.

209. Changux, J. P.; Mol. Pharmacol. ; 1966 ; **2** : 369.

210. Bryant, N. J.; (1986). "**Laboratory Immunology and Serology**" 2. nd. ed , Philadelphia, W. B. Saunders Co. Chap. 5. pp. 49 – 52.

211. Helen, Ch.; Mansel, H.; Siraj, M.; Neil, S.; (1999). "**Essential of Clinical Immunology**". 4 th. ed., U .K., Blackwell Science Ltd., Chap. 19 pp. 314 – 321.

212. Dadliker, W. B.; and Satussure, V. A.; Immu. Chem.; 1970; **7**: 799.

213. Steiner A. L., Kipnis D. M., and Utiger R. ; Proc. Nat. Acad. Sci. USA; 1969; **64**: 367.

214. Segal, I. H.; (1975). "**Enzyme Kinetics: Behavior and Analysis of Rapid Equilibrium and Steady – State Enzyme System**" John Wiley and Sons, New York. , pp . 100.

215. Dixon, M.; and Webb, E.; (1979) "**Enzyms**". 3 rd .ed., London, Long man Group Limiteded; pp. 273.

216. Devilin, T. M.; (1986) "**Text Book of Biochemistry with Clinical Correlation**". 2 nd. ed., New York, Inc., John Wily and Sons. , pp 273.

217. Price, N .C.; and Stevens, L. (1989); "**Fundamentals of Enzymology**". 2 nd. ed. New York, Oxford University Press. , pp. 125.

218. Melander, W.; Horvath, C.; Arch. Biochem. Biophys.; 1977; **183**: 200.

219. Edigington T. S.; J.Immunol.; 1971; **106**: 673-689.

220. William, E. P.; (1998). "**Fundamental Immunology**". 4 th. ed., Philadelphia, Lippicott. Raven, Chap 4, pp. 75 – 110.

221. Mellor. ; Maley. ; **Nature.;** 1974; **159**: 370.

222. Williams, R. J. P.; (1959). "**The Enzymes**" 2 nd. ed**.,** New York, Academic Press, Vol. I, pp. 391.

223. Al – Gurnawi, Z. A.; (1999). "**Physical and Chemical Properties of Prostate Specific Antigen in Some Prostate Diseases**". M.Sc. Thesis Supervised by Al – Mudhaffar S. A.; College of Science, Baghdad Univ.

224. Karamova E. R., Yazova A. K., and Goussev A. I.; Tumor Biol.; 1998; **61**: 310-317.

225. Ruoslaht E., and Seppala M.; Int. J. Cancer; 1971; **8**: 283-288.

226. Mizejweski G. J.; Experimental Biology and Medicine; 2001; **226**: 377-408.

227. Al – Rubae 'I, S. H.; (2002). "**Biochemical Characterization of CA 15 – 3 in Sera and Tissues of Breast Tumors**". Ph. D Thesis Supervised by Al – Mudhaffar S. A.; College of Science, Al – Mustansiriyah Univ.

228. Friefelder D.; **Physical Biochemistry, Application to Biochemistry, Molecular Biology**; 2nd ed., San Fracisco; W. H. Freeman & Company; 1982; chapter 14: pp. 494-591.

229. Brostoff, J.; and Male, D.; (1994). "**Clinical Immunology, an Illustrated Outline**". Philadelphia. , Mosby. , Chap. 8, pp. 112.

230. Shiu, R. P. C.; Friesen, H. G.; J. Biol. Chem.; 1974; **249**: 7902.

231. Al – Kazzaz, F. F.; (2000). "**Molecular Characterization of Carcinoembryonic Antigen (CEA) in Some Colorectal Tumors**". Ph. DThesis Supervised by Al – Mudhaffar S. A.; College of Science, Al – Mustansiriyah Univ.

232. Moro R., Tamaoki T., Wegmann T., et al.; Tumor Biol.; 1993; **14**(2): 116.

233. Mizejewski G.J.; J. Theor. Biol.; 1995; **176**:103.

234. Naval J., Villacampa M. J., Goguel A. F., and Uriel J. ; Proc. Natl. Acad. Sci. USA ; 1985 ; **82** : 3301.

235. Kanevsky V. Y., Pozdnyakova L. P., Aksenova O. A., et al.; Biochem. Molecul. Biol. Int.; 1997; **41**(6): 1143.

236. Uriel J., Trojan J., Moro R., et al. ; Ann. N. Y. Acad. Sci. ; 1983 ; **417** : 321.

237. Wiseman, T.; Williston, S.; Brandts, J. F.; Lung – Nan, L.; Anal. Biochem. ; 1989; **179**: 131.

238. Wibdenmeyer, J. A.; Schuck, P.; and SmithGill, S. J.; J. Biol. Chem.; 1999; **274**: 26838.

239. Sundberg, E. J.; et al ; Biochem. ; 2000 ; **39**: 15375.

240. Gomez, J and Freire. ; J. Mol. Biol.; 1995; **252**: 337.

241. Myszaka, D. G.; et al. ; Proc. Natl. Acad. Sci. USA.; 2000 ; **97**: 9026.

242. Wibdenmeyer, J. A.; Schuck, P.; and SmithGill, S. J.; J. Biol. Chem.; 1999; **274**: 26838.

243. Kalstan, M. B. ; Onyekwere , O .; Sidransky , D .; Vogelstein , B .; Craig , R . W.; Cancer Res.; 1991; **51**: 6304.

244. Adams, A.; Karrott, D.; Biochem. Biophys. Res. Commun.; 1985; **128**: 2.

245. Jose, M.; Tomothy. A. S.; Joann, A. K. B.; et al. ; J. Virol.; 1998 ; **72** (7): 6244.

246. Weiland, G. A.; Molinof, P. B.; Life Sci.; 1981; 29: 313.

247. Seely, G. A.; Wang, W. Y.; Sathanick, H. A.; Biochem. Biophy. Acta.; 1980; **632**: 535.

248. Segel, I. H.; **"Biochemical Calculation"**; 3 rd. ed. John Willey & Sons, Inc.; 1979: pp. 311.

249. Waelbroeck, M.; Van-Obberghen, E.; De-Meytes, P.; J. Biol. Chem.; 1979; **254**: 7736-7740.

250. Nemethy, G.; Scheraga, H.A.; J. Phys. Chem.; 1962; **66**: 1775-1780.

251. Haro, L.S.; Talamantes, F.J.; Mol. Cell Endocrinol.; 1985; **43**: 199-204.

252. Brown, E. M.; Hauser, D.; Troxler, F.; et al. ; J. Biol. Chem.; 1976; **251**: 1232.

253. Villacampa, M. J.; Moro, R.; Uriel, J.; et al. ; Cancer Res. ; 1984 ; **44**: 5314.

254. Stull, J. T.; and Blumenthal, D. K.; Biochem.; 1982; **21**: 2386.

255. Storm, D. I.; Wierman, E. M.; Laport, D. C.; et al.; Biochem.; 1980; **19**: 3814.

256. Devlin, T.M. **"Text Book of Biochemistry with Clinical Correlation"** 2nd ed. John Wiley and Sons, Inc, New York; 1986; pp.125, 66.

257. Loskowski, M.; Leach, S.J.; Scheraga, H.A.; J. Am. Chem. Soc.; 1960; **5**: 71.

258. Scheraga, H.A. **"Protein Structure"** New York, Academic Press; 1961; Chap.VI, pp. 175-287.

259. Leach, S.J.; Scheraga, H.A.; J. Biol. Chem.; 1960; **235**: 2827-2829.

260. Freifrlder, D. **"Physical Biochemistry: Physical Application to Biochemistry Molecular Biology"**. 2nd. Ed., San Francisco: W. H. Freeman & Company; 1982; Chap. 14, pp. 494-591.

261. Chanse, M. W.; Williams, C. A. **"Methods in Immunology and Immuno- Chemistry"**. New York: Academic Press.; 1968; Vol. II, Chap. 10, pp. 163.

262. Laskowski, M. J.; Herskovits, T. T.; J. Biol. Chem.; 1960; **235**: 56-57.

263. Mathews, Ch K.; Holde, K. E. **"Biochemistry"**. Callifornia. The Benjamin/ Cummings Publishing Co.; 1990; Chap. 6: pp. 191.

264. Leach, S. J. **"Physical Principles and Technique of Protein Chemistry"**. Part A. 5th. Ed., London: Academic Press.; 1969; Chap. 3, pp. 102-170.

265. Zizkovsky, V.; Strop, P. Korcakova, J. Havranova, M.; Mikes, F.; Ann. N.Y. Acad. Sc.; 1983; **417**: 49-56.

266. Martin, R.B.; Edsall, J.T.; Wetlaufer, D.B.; Hollingworth, B.R.; J. Biol. Chem.; 1958; **233**: 1421-1428.

267. Herskovits, T.T.; J. Biol. Chem.; 1965; **240**: 628-638.

268. Herskovits, T.T.; Sorensen, M.; Biochemistry; 1968; **7**: 2523.

269. Nagacura, S.; Baba, H.; J. Am. Chem. Soc.; 1953; **74**: 5693.

270. Pimentel, G.C.; J. Am. Chem. Soc.; 1957; **79**: 3323.

271. Brealy, G.J.; Kaska, M.; J. Am. Chem. Soc.; 1955; **77**: 4462.

272. Kaplan L.A. and Pesce A.J., "**Clinical Chemistry, Theory, Analysis, and Correlation**", 2nd ed., The C.V. Mosby Company, 1989, p. 612.

Tumor Markers in Breast Tumors (Sialic Acids)

Prof. Dr.sami A. Al-Mudhaffar
Dr. Ammar Khuait

contents

chapter one
BREAST CANCER:

The breast cancer is the commonest tumor in women, but it may also occur in the male, accounting for almost 20% of all malignancies [1]. Over half a million women develop breast cancer every year [2]. Breast cancer has an enormous impact on the individual patient, and it often strikes in the prime of life. There is no proved method of primary prevention. Treatment may be physically disfiguring and emotionally disruptive. Breast cancer has an unpredictable course and the risk of metastases continues for 20 years or more. When breast cancer; results in death this is often after a prolonged, painful, and disabling period of disease [2].

The incidence of and mortality from breast cancer vary greatly around the world. In the late 1980s mortality in Britain was not only higher than in most other countries in Western Europe, it was among the highest in the world [3].

Incidence in Britain, however, was similar to that on other western European countries. Incidence has been rising in many parts of the world, including the United States, Canada, Europe, the Nordic countries, Singapore, and Japan, where the rates are the lowest in the world. Much of this rise may have resulted from increased diagnostic activity, and will accelerate with the introduction of screening. The corollary of high mortality coupled with average incidence in Britain is that survival is worse than elsewhere in Europe [3]. As well as the probability of developing the disease increases through out the life and the mean age of women suffering from breast cancer is 60-61. The disease is more common in whites than nonwhites. The incidence of the disease among nonwhites women is mostly blacks however, is increasing specially in a younger women [4].

1.1.1 Clinical Types of Carcinoma of the Breast:
It is difficult to give any one growth a distinct classification, and clinical significance may be minimal. However, because the tumor may present in such a wide variety of ways it is relevant to recognize certain clinical types [21].

1. Invasive duct carcinoma: It is the commonest form and is met with principally in middle-aged or elderly women.

2. Medullary carcinoma: accounts for about 5 per cent of all breast cancer and affects a somewhat earlier age group than the average.

3. Colloid (mucinous) carcinoma and tubular carcinoma: These two variants account for about 5 per cent of all invasive duct cancers. They appear as well defined masses more common in the elderly patient.

4. Inflammatory carcinoma: Is a fortunately rare, highly aggressive cancer seen usually during pregnancy and lactation but may occur at any age unassociated with these events.

5. Paget's carcinoma: It is not common (about 1% of all breast cancers), usually multicentric in the nipple and breast ducts, but it is important because it appears innocuous.

6. 'Pseudo' lipomatous carcinoma: True lipoma of the breast is extremely rare. However, a carcinoma may sometimes develop a covering of the soft breast and subcutaneous tissue around itself to mimic a lipoma.

Etiology of Breast Cancer:

There are different causes of breast cancer, the most important of these causes are:

• **Genetic predisposition:** Breast cancer has a strong genetic component; there are at least two breast cancer genes, **BRCA1** and **BRCA2** [5]. But these two genes probably play little role in most breast cancers [6]. Indeed, 90%-95% of "sporadic" breast cancers are unlikely to be due to inherited susceptibility [6].

• **Endogenous hormones:** Particularly prolactine and oestrogens probably have an important role in the development of breast cancers [7]. However, other hormones such as growth hormone, thyroxin and progesterone also play a role, and insulin and cortisol are required for lactation. Androgens have an inhibitory effect on breast development and lactation [8].

• **Viruses and other carcinogens** [9]:

Other factors associated with increased risk of breast cancer in women such as[10,11]:

1. The age, (the mean and the median age is 60-61 years).

2. Family history, (first-degree relatives).

3. Environmental and diet.

4. Fibrocystic disease probably does predispose to cancer.

5. Ionizing radiation in large doses, as with repeated chest fluoroscopy is associated with carcinoma of the breast.

6. Others organ cancer, such as ovary and uterine corpus.

7. A women who has had cancer in one breast is at increased risk of developing cancer in the opposite breast.

8. Alcohol consumption probably increases the risk of breast cancer slightly.

1.1.3 Early Detection and Diagnosis of Breast Cancer:

It is necessary to early detection of breast cancer before it has spread to axillary lumph nodes. A solitary lump in the breast must be regarded with suspicion until proved benign. Hardness tethering to skin or deep tissues, skin ulceration, nipple retraction or the presence of enlarged axillary nodes are all features pointing towards malignancy.

Mammography using radiological methods, thermography and needle biopsy are used in establishing diagnosis. Various systems of staging have been devised. They all grade the extent on criteria such as localization within the breast, extension to adjacent groups of lymph nodes, involvement of skin and deep structures, and metastases to distant sites. The aim of each classification is to indicate prognosis principally to provide an objective way of comparing methods of treatment.

Histology of the tumor is of prognostic significance. Recently there has been considerable interest in tumor oestrogen receptors as a guide to the suitability of oestrogen of anti-oestrogen therapy. Tumors that lack these receptors are unlikely to respond to hormone treatment[12].

1.1.3.1. Staging of Breast Cancer (4,21):

TNM classification: The international Union against Cancer has recommended a staging system known as **TNM** (tumor, nodes, and metastases). The details of a method of clinical staging in relation to the TNM classification is given in (Table 1.1).

Table (1.1): Staging of breast cancer by clinical examination.

Stage	Clinical extent
I	Tumor < 2 cm in diameter Nodes, if present, not felt to contain metastases without distant metastases
II	Tumor < 5 cm in diameter Nodes, if palpable, not fixed without distant metastases

III	Tumor > 5 cm or Tumor any size with invasion of skin or attached to chest wall Nodes in supraclavicular area without distant metastases
IV	With distant metastases

1.1.4 Treatment and Surgery of Breast Cancer:

Treatment may be curative or palliative. Curative treatment is advised for stage I and stage II, while treatment can be palliative for patients in stage III and IV and for previously treated patients whom develop distant metastases [13].

The extent of disease and its invasiveness are the major factors influenced the primary therapy [14].

Several approaches have been suggested to the plan of treatment of breast cancer [15]. However, radical breast surgery is less widely used than formerly.

Now endocrine and chemotherapy are increasing in popularity as adjuvants to the basic treatment. Schedules are continually evolving and there are no strict guidelines a part from reserving oophorectomy for premenopausal patients.

The presence of oestrogen receptors in the tumor increases the likelihood of response to castration and indeed to adrenalectomy. This latter manoeuvre is of greatest benefit to the premenopausal group who have responded to oophorectomy and then relapsed. Hypophysectomy is also practiced but less so than the other endocrine a blations possibly owing to the more restricted availability of the necessary surgical expertise [16].

1.1.4.1. Hormonal Therapy:

Oestrogens are the most commonly used hormonal agents and are particularly effective in postmenopausal women with oestrogen receptors in the tumor cell. Androgens have been employed but are currently unpopular because of their virilizing effects. Progestogens and corticosteroids have also been used but are generally thought to be of only minor value. Anti-oestrogens, of which the most used in tamoxifen, induce remissions in the majority of oestrogen receptor-positive patients both pre-and postmenopausal [17,18].

1.1.4.2. Chemotherapy:

It is used in metastatic breast cancer [19]. Cytotoxic therapy is usually reserved for advanced cases. The view that micrometastases are already present in many patients who on clinical grounds had been thought to have early local disease has widened the indications for chemotherapy, so that some centers are giving drugs with the primary excision whether or not there is distant spread. The list of agents employed is ever increasing and includes alkylating agents, antitumor antibiotics, antimetabolites, and the vinca alkaloids. These drugs may be given in various combinations, and may be combined with hormones as well as with surgery and radiotherapy [20].

1.1.4.3. Radiotherapy:

It is usually used after surgery (Particularly for tumors in stage I and II), palliative radiotherapy may be advised for locally advanced cancers with distant metastasis in order to control ulceration, pain and other manifestations in the breast and regional nodes. Irradiation of the breast and chest wall and the axillary, internal mammary, and supraclavicular nodes should be undertaken in an attempt to cure locally advanced and inoperable lesions when there is no evidence of distant metastases. Radiotherapy is especially useful in the treatment of the isolated bony metastasis and chest wall recurrences [21,22].

1.1.5 Benign Breast Disease:

It is present with one or more of four symptoms and signs. They may describe a discrete lump, an area of lumpiness, nipple discharge or a pain in the breast. This disease represent (Fibrocystic Disease or Mammary Dysplasia, Fibroadenoma and Duct papilloma) [23].

1.1.5.1. Mammary Dysplasia (Fibrocystic Disease):

The disorder, also known as fibrocystic disease or chronic cystic mastitis, is the most frequent lesion of the breast. It is common in women 30-50 years of age, but rare in postmenopausal women; this suggests that it is related to ovarian activity. Estrogen hormone is considered a causative factor. The microscopic findings of fibrocystic disease include cysts (gross and microscopic), papillomatosis, adenosis, fibrosis, and ductal epithelial hyperplasia. Mammary dysplasia may produce an a symptomatic lump in the breast that is discovered by accident, but pain or tenderness often calls attention to the mass. There may be discharge from the nipple. Fluctuation in size and rapid appearance or disappearance of a breast tumor are common in cystic disease [24,25].

1.1.5.2. Fibroadenoma:

This common benign neoplasm occurs most frequently in young women, usually within 20 years after puberty. It is somewhat more frequent and tends to occur at an earlier age in black than in white women.

Multiple tumors in one or both breasts are found in 10-15% of patients. The typical fibroadenoma is around, firm, discrete, relatively movable, nontender mass 1-5 cm in diameter. The tumor is usually discovered accidentally [26,27].

1.1.5.3. Duct papilloma:
It is usually occurs in women between 35 and 50 years of age, but rare before the age of 25. In the majority of cases, bright red blood or a serosanguineous discharge or, less often, a dark blood-stained discharge from the nipple is the only symptoms. On examination, a cystic swelling can sometimes be felt beneath the areola; pressure upon it will cause a discharge from the orifice of the affected duct on the nipple. Available evidence suggests that these lesions rarely undergo malignant transformation [21,28].

1.1.6 Biochemical Aspects of Breast Cancers:
Through the literature survey, it has been found that there are a several types of biomolecules may be vary in concentration and nature between the normal cells and tumor cells.
These differences include: enzyme activity, hormone levels and their receptor between normal and cancerous tissues [29].
Several enzymes and their isoenzymes are studied in breast cancer patients, they include: ribonuclease, sialyltransferase and galactosialytransferase and lactate dehydrogenase. Most of these enzymes revealed variations of clinical significance [30,31].
Also hormones investigated in breast cancer patients, they include: FSH and LH, progesterone, oestradiol and HCG. These hormones showed marked
variations and may used in the clinical evaluation of the disease [32]. Many properties of mammalian cells are expressed at, or mediated through, the cell surface. Among these properties are those, which distinguish a malignant cell from a normal cell. As neoplastic changes are expressed at the cell surface, altered surface characteristics are essential for the abnormal growth and behavior of malignant cells [33]. Sialic acid is thought to be important in determining the surface properties of cells and has been implicated in cellular invasiveness, adhesion and immunogenicity [34].

1.2 Tumor markers:
Definition:
Several definitions of tumor markers exist in the literature [35]. A restricted and classical definition of tumor markers pertains to the measurement of certain analytes either using chemical, biochemical or immunochemical methods in conveniently obtainable body fluids such as urine and blood. However, this narrow definition excludes the important class of tissue and cellular markers. Not all tumor markers are secreted into body fluids such as (**CA170, CA174**) membrane bound antigens and several intracellular antigens including aberrant nucleic acid sequences. However, at the fifth international conference on tumor markers (Stockholm, 1988) a consensus definition of one category of tumor markers was adopted: "Biochemical tumor markers are substances developed in tumor cells and secreted into body fluids in which they can be quantitated by non-invasive analysis. Because of a correlation between marker concentration and active tumor mass, tumor markers are useful in the management of cancer patients. Markers, which are available for most cancer cases, are additional, valuable tools in patient prognosis, surveillance and therapy monitoring whereas they are presently not applicable for screening. Serodiagnostic measurements of markers should emphasize relative trends instead of absolute values and cut-off levels". The general potential uses of tumor markers include: Screening, diagnosis, differential diagnosis and classification, staging and grading, prognosis and monitoring treatment [36].
1.2.1 Measurement of Tumor Markers:
Different assay techniques have been used for the measurement of tumor markers. The in vitro measurement of tumor markers is in two realms: histochemical and body fluid analysis. Current probes for such measurements are largely based on monoclonal antibodies and polyclonal antibodies although some assays based on the total absence of immune reactions still exists. While immunohistochemical analysis of tumor markers generally are qualitative, the availability of image analysis techniques, software and digital signal acquisition methods provide a degree of quantitation of the intensity of staining and the degree of heterogenicity of tumor marker expression in a tumor mass. Almost all in vitro immunoassays for body fluid analytes are quantitative [36].
Today, most immunoassays of tumor markers are based on monoclonal antibodies using the dual monoclonal sandwich assay technique.
Further refinements in assay technology include the development of homogeneous assays wherein there is no separation of the bound and free reactants: new non-isotopic signal generation such as time resolved, fluorescence, bioluminescence and chemiluminescence and dye based particles[37].
Recently a new technique called immuno-PCR (polymerase chain reaction) has been introduced that uses a short piece of DNA as the tag in an immunoassay[38].
The in vivo identification of tumor mass needs a special mention here. Anti-tumor polyclonal antibodies and now monoclonal antibodies have been radiolabeled with a variely of diagnostic isotopes such as ^{131}I, [111]In

and ^{99m}Tc and infused into cancer patients for in situ detection by and external gamma camera of primary and recurrent metastasis exploiting tumor markers as targets. Although this technique of scintigraphy is largely a qualitative in vivo measurement of the tumor marker, some estimate of the quantitative uptake in tumor markers has been made for dosimetry and eventual radiotherapeutic applications [39].

1.2.2 Classification of Tumor Markers:

Tumor markers can be broadly classified into tumor specific antigens and tumor-associated markers. Most tumor markers were often heralded as highly tumor specific, but subsequent studies demonstrated their presence in normal tissues of the adult or in various stages of ontogeny.

As a result, very few tumor-specific antigens can be recognized. The idiotypes of immunoglobulins of B cell tumors and certain neo-antigens of virus induced tumors are two examples that are strictly tumor specific. The vast majority of tumor markers are in reality tumor-associated antigens and can be classified into two types based on their size. The low-molecular weight tumor markers (- < 1000 Daltons) include some nucleosides, lipid associated sialic acid, polyamines, pseudouridine, pigment derivatives, and other metabolites. The macromolecular tumor antigens are the most important sub-type useful in the clinical management of cancer patients.

The large cancer antigens are either enzymes, growth factors, hormones, receptors, biological response modifiers, oncogenes and their products, or glucoconjugates which include glycoproteins and glycolipids. The classification of tumor markers into its various categories are summarized in (Table 1.2) [40].

Table (1.2): Classification of tumor markers.

Category	Examples
I. Tumor –specific markers	B-cell tumor immunoglobuline idiotype: virus induced antigens e.g.: SV40T antigen, T-cell receptor of T-cell leukemia.
II. Tumor-associated markers **Low molecular weight markers**	Polyamines, nucleoside derivatives; sialic acid (lipid associated) Vanillylmandelic acid and catechoamine. Van metabolites.
Macromolecular markers **Enzymes, Isoenzymes**	Placental alkaline phosphatase: prostate specific antigen, prostatic acid phosphatase, Thymidine kinase: Neuorone specific enolase. HCG: 11.2: EGI: estrogen and progesterone receptors.
Hormones cyctokines, growth factors soluble and receptors.	C-myc: src: ras: crb: neu: sis. CEA, AFP, lewis X; lewis Y.
Oncogenes and oncoproteins	CA125; CA15-3; SIA; CA19-9; TAG72; Matrix protein; N-CAM.
Oncofetal proteins	Lewis X; Lewis Y; GM2; GD2; CA19-9 glycolipid Philadelphia chromosome; pre-cancerous cells in PAP smear
Complex glycoconjugates: **i- Glycoproteins and glycosamino-** **glycans.** **ii- Glycolipids** **6. Cellular markers**	

Some of Tumor Markers Used in Breast Cancer:

Tumor markers may be used to indicate the risk presence status, or future behavior of cancer. However, it is more valuable to use tumor markers for discrimination between malignant and benign tumors. New markers are frequently introduced into clinical practice without rigorous analysis, with the assumption that any information available to the clinician will help the patient[41].

1.3.1 CA 15-3 As A Marker For Breast Cancer:
The CA15-3 test measures the serum level of a mucin-like membrane glycoprotein, which is shed from tumor cells into the bloodstream. The CA15-3 epitope is recognized by two monoclonal antibodies in a double-determinant or sandwich radioimmunoassay.

1.3.2 CEA As A Marker For Breast Cancer:
CEA belongs to family of cell-surface glycoproteins with increased expression found in a variety of malignancies, including breast cancer. CEA is not recommended for screening, diagnosis, staging, or routine surveillance of breast cancer patients following primary therapy. Routine use of CEA for monitoring response of metastatic disease to treatment is not recommended. But in the absence of readily measurable disease, an increasing CEA level may be used to suggest treatment failure.

1.3.3 Estrogen Receptors and Progesterone Receptors As Markers For Breast Cancer:
The estrogen and progesterone receptors are intracellular receptors that are measured directly in tumor tissue. These receptors are polypeptides that bind their respective hormones translocate to the nucleus [42,43], and induce specific gene expression. There are three domains on these polypeptides: a C-terminal hormone binding domain, a central DNA binding domain, and an N-terminal domain that is important for transcription [44].
When the hormone binds to its respective receptor, the DNA binding domain is modified so it binds to DNA quite avidly and initiates transcription. This process clearly affects the growth of the cell. However, the intricacies of how the hormone and receptor complex affect cell growth are only partially understood[45,46].

1.3.4 DNA Flow Cytometrically Derived Parameters As Markers For Breast Cancer:
DNA diploid tumors are those in which a single peak containing an amount of DNA similar to normal control cells is generated by flow cytometry. DNA aneuploid tumors have additional peaks on DNA

histogram. Presumably representing cells containing more or less nucleic acid than is found in 46 normal chromosomes [47].

1.3.5 C-erbB-2 (HER-2/neu) As Marker For Breast Cancer:

The C-erbB-2 gene encodes a transmembrane tyrosine kinase that is the receptor for a family of peptide hormones.

1.3.6 P53 As A Marker For Breast Cancer:

P53 is a tumor suppressor gene on the short arm of chromosome 17 that encodes a protein that is important in the regulation of cell division. The P53 gene product appears to regulate transcription of several other genes. The full role of P53 in the normal and neoplastic cell is unknown. There is evidence that the gene product is important in preventing the division of cells containing damaged DNA. P53 gene deletion or mutation is a frequent event along with other molecular abnormalities in colorectal carcinogenesis.

1.3.7 Cathepsin-D As A Marker For Breast Cancer:

In 1979, a glycoprotein was discovered in the culture medium of hormone-dependent human breast cancer. It had a dependence on estrogens in that it could be increased by estrogens and inhibited by antiestrogens. It was discovered to be a 52-Kd protein, which is a precursor to a lysosomal acidic protease. Cathepsin-D, this proteolytic enzyme can react against basement membranes. Cathepsin-D also has mitogenic activity on MCF-7 cells that are estrogen-depleted. Further studies showed that cathepsin-D was relatively low in resting mammary cells but was elevated in malignant and benign proliferative breast diseases. These findings raised the suspicion that the Cathepsin-D could both promote abnormal growth of cells as well as contribute to the metastatic potential of malignant cells through its disruption of the basement membrane and therefore might be a marker for a poor prognosis in breast cancer [48].

1.4 The Sialic Acids:

Definition:

The term sialic acids represent a group of nine-carbon sugar collectively called neuraminic acid. Sialic acids, are the N-acetyl or N-glycolyl derivatives of the parent D-neuraminic acid (Fig. 1.1). They contain carboxyl, acetamide, ketone, deoxy group, as well as hydroxyl group [49,50].

($C_9H_{17}O_8N$) D-Neuraminic acid

($C_{11}H_{19}O_9N$) N-Acetyl Neuraminic acid

($C_{11}H_{19}O_{10}N$) N-glycolyl Neuraminic acid

Fig. (1.1): Structure of the Sialic acids

Some of these sialic acids are known to carry o-acetyl substituents located at carbon No. 4 (C_4) and/or at the various position of the polyhydroxy side chain which are carbon No. 7, 8 and 9 (C_7, C_8, and C_9), (Fig. 1.2) [51].

C_6 = unsubstituted sialic acids

Fig. (1.2): Structure of O-acetyl substitute sialic acids.

Sialic acids are usually located terminally and they are non-reducing residue of the carbohydrate prosthetic group of numerous glycoproteins and glycolipids [52].

1.4.1 Types of Sialic Acids:

Ten sialic acids were isolated, they differ in having various degrees of O-acetylation and N-glycolyation [53]. (Nana) which is N-acetyl neuraminic acid is the most common in mucous glycoproteins and in glycolipids [53]. The major derivatives of neuraminic acid as components of glycoproteins and glycolipids are N-acetylated or N-glycoloylated and frequently also O-acetylated [54]. O-acetyl groups were found at C-4, C-7, C-8 and most frequently at C-9 of Neu5Ac or Neu5Gc; in some cases O-lactoyl groups are also present at C-9 [55]. Mono-and oligo-O-acetylated sialic acids were isolated in pure form e.g. from glycoproteins of submandibular glands and horse erythrocyte membranes [56]. In rat, rabbit and mouse erythrocyte membranes 9-O-acetylated sialic acids, and in horse erythrocytes 4-O-acetylated neuraminic acid derivatives were identified [57]. It is difficult to identify the particular type of sialic acids present in certain tissue because, the two common analytical methods used, the Erlich and the Resorcinol method, are carried out in strong acids and under such condition the acyl groups are removed [53].

1.4.2 Physical and Chemical Properties of Sialic Acids:

314

Purified sialic acids are colourless; it is only after heating at 100°C for several hours that they turn yellowish. They do not melt, but decompose with discoloration over a range of several degrees, discoloration usually preceding decomposition. The sialic acids are rather strong acids (α-keto-acids) with a pK_a value of 2.6 [58].

The sialic acids and methoxyneuraminic acid are easily soluble in water. Bovine O,N-diacetylneuraminic acid dissolves readily in methanol, the other acids are only sparingly soluble in methanol. All sialic acids and methoxyneuraminic acid are insoluble in ether and light petroleum. The sialic acids are very unstable to both acid and alkali; aqueous solutions decompose already when kept for some time at room temperature due to their acidity. All sialic acids, but not methoxyneuraminic acid, reduce Fehling's solution on heating. The sialic acids consume in the oxidation by hypoiodite an amount of iodine varying with time between 33 and 90% of that calculated for the presence of one aldehyde group per molecule [53].

1.4.3 Methods for Determination of Sialic Acid:

A variety of procedures have been used for the measurement of total sialic acid. These can be broadly classified as colorimetric, fluorometric, enzymatic and highly sensitive high performance liquid chromatographic (HPLC) procedures[58].

1.4.3.1. Colorimetric procedures:

Two classical procedures have stood the test of time. One uses resorcinol and the other uses periodic and thiobarbituric acids. The resorcinol based assay uses heat and strong acid to hydrolyze glycosidic bonds. The released free sialic acids are reacted with resorcinol and copper ions to give a colored compound, which is extracted and measured at 580 nm. While the procedure described by Warren is typical of the periodic and thiobarbituric acid procedure, which measures only free sialic acid that is released after an initial hydrolysis step. In this procedure, formyl pyruvic acid formed as a result of periodic acid oxidation of free sialic acid is reacted with thiobarbituric acid to yield a red color, which is measured at 549 nm [59].

1.4.3.2. Fluorometric procedures:

In a typical and more specific assay formaldehyde that is formed upon oxidation of free sialic acid by periodic acid is reacted with acetyl acetone. The yellow product is excited at 410 nm and the resulting fluorescence is measured at 510 nm [60].

1.4.3.3. Enzymatic procedures:

Enzymatic assays are based on conversion of free sialic acids released by the enzyme neuraminidase to pyruvate and acetyl monnosamine with the aid of the enzyme acetyl neuraminic acid pyruvatelyase or neuraminic acid (NANA) aldolase. The resulting pyruvate can be coupled to the lactate dehydrogenase NADH system to measure the oxidation of NADH to NAD at 340 nm. Alternatively pyruvate can be coupled to pyruvate oxidase, flavine adenine dinucleotide (FAD), and thiamine pyrophosphate (TPP) to form hydrogen peroxide, which in turn is coupled to peroxidase in presence of 4-aminoantipyrine and a toluidine derivative to form a red chromogen, which is measured at 550 nm. The reactions associated with the NANA-aldolase-pyruvate oxidase-peroxidase system [61], as shown below:

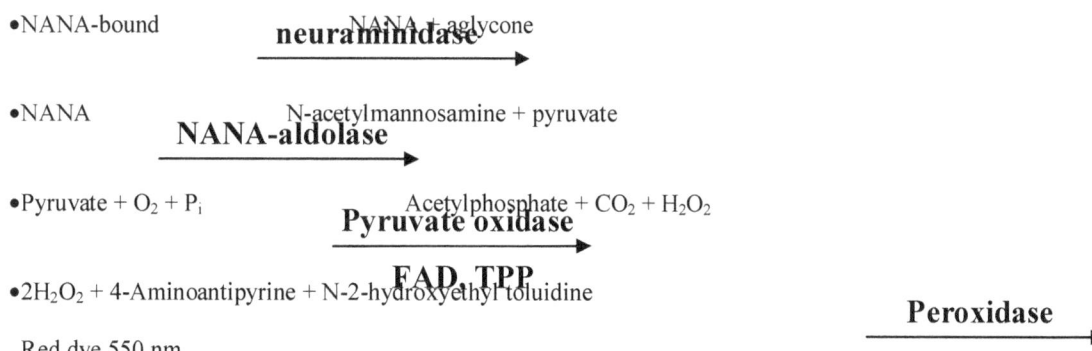

Red dye 550 nm
Reactions involved in a typical enzymatic assay for the measurement of sialic acid.

1.4.3.4. High performance liquid chromatographic (HPLC) procedures:

The HPLC procedures provide the ultimate sensitivity. In one such procedure, sialic acid released from the sample by acid hydrolysis is converted to highly fluorescent derivatives by reacting with a fluorogenic agent for alpha-keto acids such as 1,2-diamino 4,5-methylene dioxybenzene in dilute sulfuric acid. The fluorescent derivatives are separated on an octadecyl (C18) bonded silica column using a reverse phase solvent system. The chromatographic step takes only 12 minutes allowing detection of levels as low as 25 femtomoles (F.mol) or 7.7 picograms (pg) of N-acetylmeuraminic acid and 23 F.mol or 7.5 (pg) of Nglycolylneuraminic acid, in an injection volume as small as 10 microliter. The procedure is capable of analyzing precisely sialic acids in a 5 μL of serum sample [62].

1.4.4 Functions of Sialic Acids:

Sialic acids are widely distributed in the human body and they perform important functions [63]. The plasma glycoproteins are soluble and they have a hydrophilic character, it is believed that these two characters are due to the presence of sialic acids as a terminal sugar residues in these plasma glycoproteins [64]. In addition to the role of sialic acids in protection of the plasma glycoproteins from splitting by proteolytic enzymes, they also play a role in the turn over of some plasma glyscoproteins; for example ceruloplasmin, which its removal from the circulation is believed to occur when sialic acids has been removed from its molecule; so it will be recognized by the liver cell plasma membrane [65].

The human red blood cell is studied with nearly 20 million molecules of sialic acid on the outer cell membrane which contributes to its electronegative charge, and by cell to cell repulsion prevents red blood cell from aggregating. Owing to its negative charge, sialic acid can bind positively charged molecules and thus play a role in the transport of such molecules [58]. RBC aging process is due to the removal of sialic acids from its plasma membrane, so RBC will be taken up by the endothelial system to be destroyed [66]. Similarly, the injection into rabbits of a desialylated preparation of ceruloplasmin resulted in the removal of the desialoglycoprotein from the circulation within a few minutes after injection, in contrast to the intact glycoprotein, which exhibited a normal survival time [66]. Sialic acids are believed to decrease the immunogenicity of the glycoproteins, for example, human orosomucoid (which contain 11.8% NANA) was found to be a poor immunogenic in the rabbit but the removal of NANA by neuraminidase enhance the immunogenic properties of this glycoprotein at least five folds [67]. In the same way the removal sialic acids from feutin enhances its combination with the antibodies. In RBC, the removal of sialic acids from its surface enhances the binding of influenza viruses to the RBC. Thus, sialic acid seems to protect self protein from self immune response and unfortunately it also seems to mask a foreign protein from rejection caused by host immunogenicity [67].

The presence of sialic acids in the sera of patients with various tumors inhibit the attack of immune lymphocytes to the tumor, so they cause masking to the tumor cells. However, the increased amount of sialic acid on the tumor cell surface can, by increasing adhesiveness, contribute to the formation of larger tumor emboli. Metastatic spread is also facilitated by sialic acid molecules increasing the adherence of tumor cells to vascular endothelium at secondary sites of implantation and by increasing the ability to aggregate platelets [68].

In glycohormones sialic acids perform important function, and their biological activity affected by the removal of sialic acids, example of these hormones are Human Chorionic Gonadotropin (HCG) and Follicular-Stimulating Hormone (FSH), this effect is presumably because sialic acids removal destroys the ability of these hormones to reach the target cells, so sialic acids seem to posses a hormone-receptor recognition function [69,70].

In mucin, sialic acids known to give the mucin it's viscosity and protect it from proteolytic splitting. Colonic mucin are predominantly neuraminidase resistant and contain O-acetyl substituent in the polyhydroxy side chain of the sialic acids, and it was found that the reduction in this substituent associated with diseases of the colon like ulcerative colitis and Crohn's disease [71,72].

1.4.5 Role of Sialic Acids in Disease Conditions:

McNeil et al [73], stated that the determination of sialic acids in serum is a sensitive, accurate method of estimating the glycoproteins content of the blood. Thus, sialic acids level increase in the same diseases that cause glycoproteins level to increase. Carter and Martin [74] demonstrated an increase in the level of sialic acids in serum of the patients with rheumatic arthritis, bacterial infection, cirrhosis, myeloma and macroglobulinanemia. They suggested that this increase is due to increased production of abnormal glycoproteins with normal sialic acids.

Elevated serum sialic acids has been shown to increase in patients with variety of malignant diseases including, melanoma, breast cancer and lung cancer [75-77]. Furthermore, patients with metastatic cancer had significantly higher sialic acids level than cancer patients without metastatic [78]. Increase in the level of serum sialic acids is usually due to bound sialic acids, which is bound to lipids or proteins, causing increase concentration of total serum sialic acids [79].

Brozmanova and Skrovina [80], studied serum sialic acids level in patients with bon tumors. They observed significant elevation among sarcoma patients as compared to those with benign tumors, therefore sialic acids level was suggested to be useful for comparison and diagnosis of bone tumors.

Serial determination of serum sialic acids has found to be useful monitor of tumor burden [76], and it had been shown to be proportional to the tumor stage [81-84].

Again the relation between sialic acids, tumor stage and clinical course indicates that serum sialic acids analysis could prove clinically important as a monitor of tumor burden in diseased patients, specially as sialic acids correlate both with disease progression and remission, and it may provide the earliest indication of tumor response to drugs [76]. One of the interesting facts about sialic acids is that it tend to increase in smokers and with age [85]. The study of side chain O-acetyl substitution of sialic acids is important, it has a

316

value in identifying mucin producing metastasis arising from carcinoma of the colon [86], important in studies of perianal Paget's diseases, in distinguishing between carcinoma arising from the anal gland and those arising from the rectal epithelium [71].

In addition, reduction in the side chain O-acetyl substitution pattern of these sialic acids has been found to be associated with colonic malignancy, Crohn's disease and ulcerative colitis [87]. However, detailed studies about the biological significance of O-acetyl groups in sialic acids have shown that the degree of O-acetylation and the position of O-acetyl groups in sialic acids play a significant role in the action of sialidases. O-Acetyl groups at the sialic acid side chain reduce the rate of enzymatic hydrolysis, while a corresponding residue at C-4 prevents the action of all sialidases tested so far [88]. It was furthermore observed that the degree of 9-O-acetylation exerts an effect on immunological and complement reactions [89]. Therefore, the distribution of N-acetyl and/or N-glycolyl, with or without O-acetyl substituent on different tissue or secretion differ appreciably and may express the special function of these sialic acids in that tissue, for example the presence of O-acetylated sialic acids confers some protection to the epithelial surface of colon against the fecal stream (Bacteria and Enzymes) and it was suggested that the substitution with O-acetyl at C-4 position is responsible for sialic acids vibrio cholera neuraminidase resistance[90,91].

1.4.6 Sialic Acid and Cancer:

There has been much discussion of the relation between sialic acid and cancer. The elevation of serum sialic acid levels has been reported in cancer patients [92]. It has also been reported that, in melanoma patients serum sialic acid levels are greater than those in normal donors and the levels are in proportion to tumor burden [93]. The increase of sialylated carbohydrates, such as sialyl Lewis [a] (CA19-9) and sialyl Lewis[x] (CSLEX-1), on the cell surface is generally observed with malignant or transformed cells [94]. Plasma from cancer patients shows an almost uniform elevation in sialyltransferase activity [95]. The elevated sialylated carbohydrate level in the plasma of cancer patients may contribute to the pathological immunodepression by blocking leukocyte interaction with endothelial cell leukocyte adhesion molecule-1 [96]. These sialylated carbohydrates that are increased in cancer patients have a basic oligosaccharide structure in common, with blood group antigens (ABH, Lewis, etc.); thus, it is considered that the former is induced by modification of the latter with cancerization. It is reported that oligosaccharides with the structure of ABH antigenic determinants are present in normal human urine [97]. However, there has been no investigation of the nature of oligosaccharides in the urine of cancer patients. LASA levels have been reported to be useful in monitoring patients with malignant melanoma. In one study when tumor recurrence was correlated with elevated LASA levels, the increased level was found as early as 9.3 months (median value) prior to recurrence [98]. Higher levels of TSA and LASA have been reported in leukemia patients compared to patients with anemia [99]. The TSA levels were significantly higher in acute myeloid leukemia compared to chronic myeloid leukemia and acute lymphatic leukemia patients. The LASA levels were significantly elevated in acute myeloid leukemia patients as compared to other leukemic patients. The sensitivity of sialic acid as a marker for leukemia is high with the sensitivity of LASA approaching 85 percent. The TSA levels in patients with oral and maxilla facial malignancy were reported to be significantly higher in patients with stage III and IV cancer, when compared to patients with stage I and II cancer. During follow-up of response to treatment while TSA levels declined during remission of disease, they became elevated with recurrence and metastasis [100].

1.4.7 Proteins Containing Sialic Acid:

Glycoproteins:

Definition:

Glycoproteins can be best defined as "conjugated proteins containing as prosthetic group one or more heterosaccharides, the latter is usually branched, lacking repeating units and bound covalently to the peptide chain" [101].

1.4.7.1. Chemical characteristics and structure:

There are great variations in chemical and physical properties of glycoproteins according to their location and function [102]. Molecular weight of glycoproteins may vary from 15,000 to over 2 million Dalton, usually contain 15 or fewer sugar units which are attached to the protein back-bone [102].

Glycoproteins are isolated from most organisms including: plants, bacteria, fungi, viruses and animals, and their chemical structure differ in the different organs, for example: the submandibular salivary mucin has a simple composition, it is almost exclusively N-acetylneuraminic acid and N-acetylgalactosamine [63]; while the intestinal glycoproteins are more complex, they have no unique amino acid composition but they do contain a characteristic group of sugars that include D-galactose, L-fucose, N-acetylglucosamine, N-acetylgalactosamine, and furthermore the chemical structure of glycoproteins may differ even in the different parts of the same organ in the same individual, for example, mucin in small intestine contain both neutral fucomucin and acid non-sulphated mucin however, in large intestine acid-sulphated and acid non-sulphated mucin are present [103]. In addition to the variation in glycoproteins structures in the same individual, variation among different strains of the same species has been reported [104]. In general, mucus glycoproteins from various sources share the following features: (i) very high molecular weight, usually in excess of 10^6 Dalton: (ii) approximately 75% carbohydrate and 20% protein core, with the remainder consisting of variable amounts of sulphate and water: and (iii) isoelectric point below 4, due to charged sialic acids and sulphated groups [105].

1.4.7.2. Structure of the Oligosaccharides attached to Glycoproteins:

Nine different sugar residues are generally found in the oligosaccharides chain attached to the protein core, these sugars differ according to their site and function, for example; glucose is found only in collagen, while galactose and mannose are more common and widely distributed. The two most frequently found hexoses are N-acetylgalactosamine and N-acetylglucosamine. Fucose, which is 6-deoxygalactose, is a common constituent, and frequently located at the terminal site in the neutral glycoproteins.

Two pentoses; arabinose and xylose were isolated from different tissues like dermatine. The ninth sugar is the sialic acids, of which N-acetylneuraminic acid (NANA) is an example, these acids are terminally located non-reducing residue of the carbohydrate prosthetic groups in the acidic glycoproteins [102].

On the other hand the N-acetylhexosamine are the most common sugars present at the proximal end of the oligosaccharide side chain, attaching to the protein back-bone, through the amino acids asparagine, serine, therionine, hydroxylysine or hydroxyproline (the first three usually present in mucus glycoproteins, while the last two present in connective tissue glycoproteins).

There are two types of linkage between carbohydrate of the polysaccharides side chain and the above amino acids, these are (i) the O-glycosidic linkage; and (ii) the N-glycosidic linkage (Fig. 1.3) [102,106].

(Serine)

(N-acetylgalactosamine)

(i)

(Asparagine)

(N-acetylglucosamine)

(ii)

Fig. (1.3): The O-glycosidic linkage (i) and the N-glycosidic linkage (ii).

Although it is possible to dissociate the polysaccharides chain complex by changes in pH or ionic strength, yet it is not possible to separate carbohydrate from peptide portion of the glycoproteins without directly degrading the entire molecule, therefore carbohydrates are integral part of the glycoproteins [107

1.4.7.3. Functions of glycoproteins:

Glycoproteins are thought to participate in many important functions. Glycoproteins may serve as a structural molecules in the cell, the major portion of the glycoprotein in the animal cell is associated with the cell surface, and approximately 70% of the sialic acid-containing glycoproteins are found in the surface membrane [108].

318

Glycoproteins also serve as a structural molecules in collagen, elastine, fibrine, and bone matrix. Concerning collagen Herp and Pigmen in 1958 found that rat skin contained an insoluble component; after the collagen fraction had been removed by intensive treatment of the collagen with hot alkali; this component showed to contain the sugar characters of glycoproteins [67]. The other important function of glycoproteins is that they act as lubricants and protective agents, one of the well known examples is mucin; mucin is a glycoproteins containing viscous fluid being continuously secreted by the wet mucosa like respiratory, gastrointestinal and urinary systems. An example of the protective function of the glycoproteins is that of the human fetus, in which the surface epithelium of the stomach secretes neuraminidase resistant acid-mucin to protect the stomach wall from digestion by hydrochloric acid and the digestive enzymes [109].

Not only gastrointestinal mucin have this lubricant and protective function, but other mucin also performs similar function, for example respiratory mucin which protect the respiratory epithelium from the external environment by providing a barrier to the epithelium.

Furthermore, the immunoglobulin content of the respiratory mucin; which is recognized as a first line of defense against infection; is a glycoprotein [110].

Other important function of glycoproteins is to serve as a transport molecules for vitamins, lipids, minerals, trace elements and hormones. Example of those glycoproteins are transcobalamin (bind vitamin B_{12}), prealbumin and transferrin. Prealbumin, which is of a special biological interest because it is responsible for the transport of a hormone and a vitamin together, it appears to contain one binding site for one molecule of thyroxine and it also transports retinol (vitamin A), which is bound indirectly by the retinol-binding protein that forms a protein-portion complex with prealbumin [111,64]. The other example of glycoprotein as a transport molecule is transferrin, which transport iron, it contains 8-9% carbohydrate in its molecule [64].

Glycoproteins could serve as a defense mechanism molecule in the body for example immunoglobulins, histocompatability antigen, complement and interferon. Immunoglobulines form a set of glycoproteins that have the ability to bind other molecules with a high degree of specificity. These molecules are foreign bodies or non-self and they are called antigens [112]. Furthermore, the carbohydrate moieties of glycoproteins which are displayed on the cell surface act as immunogenic determinants (antigenic behaviour of the molecule).

Example of this immunogenic determinant is the blood group antigen; i.e. the four different types of the blood group (A, B, AB and O); which are determined by the sequences and arrangements of the sugar residues and the glycosidic bond. So the type of terminal sugars and their arrangement in the oligosaccharides side chain of glycoproteins, which are present in the RBC membrane, will determine the blood group, and it is specificity of the individual [113,114,115,66].

The other and important function of glycoproteins, is that same hormones are glycoproteins (glycohormones) for example Human Chorionic Gonadotropin (HCG) and Thyroid-Stimulating Hormone (TSH). Human chorionic gonadotropin appears in the urine and serum of women in significant quantities during the first trimester of pregnancy. Thus, pregnancy test based on the estimation of the level of this glycoprotien in urine and serum of pregnant women.

Not only hormones, but glycoproteins may be a part of an enzyme (glycoenzymes); these are widely distributed among animals, plants and microorganisms; and they are biologically active molecules for example proteases, nucleases and clotting factors. The last are essential glycoenzymes present in the blood, they are important in haemostasis (stop bleeding). Some of these clotting factors may be deficit in certain diseases causing defect in haemostasis, an example of such a disease is haemophilia, which is a congenital disease [69]. Glycoproteins may also act as cell attachment and recognition sites, for example cell-cell, virus-cell, bacterial-cell and hormone-receptors. In 1953 Coman, D.R., suggested that malignant cells were less adhesive than their normal counterparts. Adhesion is defined as the attachment of the cell to each other, leading to the formation of tumor or, under normal conditions, to organs. Concerning recognition, the presence of galactose on the serum glycoprotein was necessary for the latter to be taken up by liver cell membrane, which in turn require sialic acids for this function. On the other hand, the circulating glycoproteins rapidly leave the portal system after desialization [108].

In hormone-receptor recognition as a functions of glycoproteins, insulin and glucagon are excellent examples. The receptors of these two hormones have been recognized as glycoprotein [116]. One of the interesting facts in glycoproteins function, is that they act as antifreeze in Antarctic fishes [102].

1.4.7.4. Role of Glycoproteins in Disease Conditions:

Considerable interest has been focused toward the plasma glycoproteins in the past few years, centered largely around demonstration of increased level of glycoproteins in the plasma of individuals suffering from various disease conditions.

Most pathological conditions involve glycoproteins are due to defective degradation of glycoproteins rather than synthesis [65].

In a previous study [117], it was demonstrated that patients suffering from various malignant conditions postulated that, the increase in glycoproteins arises as a result of depolymerization of the ground substance adjacent to the tumor; with subsequent release of the glycoproteins to the circulation, on the other hand, many other workers suggested that this increase is due to tumor destruction, proliferation and repair [118].

However, other workers suggested that the major possibilities of the increased level of glycoproteins in malignant diseases are due to; (i) increase in carbohydrate content of normal glycoproteins, (ii) increase glycoproteins production by the tumor itself, (iii) increase glycoproteins synthesized by the liver or by the lymphoreticular tissue [85].

Glycoproteins, which increase in malignant diseases [119,120], are also known to increase in the acute phase of diseases (acute phase protein or reactant)[85], infections and inflammatory diseases like rheumatic arthritis [117].

It was found that continuous estimation of glycoproteins level in plasma may give a clue about the progress of the disease. The other interesting fact about glycoproteins was that their level is more in smoker than in non-smoker persons[85].

It has been found that, glycoproteins level decreased in vitamin A deficiency, and drugs like Glutamine also cause reduction in glycoproteins level because both affect glycoproteins synthesis [121].

In several congenital diseases, including mucopolysaccharidoses and sphingolipidoses, glycoproteins have been blamed. These diseases are due to defect in glycoproteins catabolism, causing abnormal storage of glycoproteins and increase their accumulation. These diseases are called inbora error of metabolism and affect mostly serum and membrane glycoproteins [65].

It is well known that one of the glycoproteins functions is to determine the blood group of the individual. Furthermore, there are some relation between blood group and susceptibility to certain malignant diseases, for example blood group A individuals are more likely to have salivary gland tumors and carcinoma of the stomach than are individuals who are O, B or AB blood group. This susceptibility is of unknown etiology but it is believed to be due to the alteration in glycoproteins secreted by salivary gland or stomach of those people[63].

Deficiency in certain glycoproteins cause a disease condition, for example deficiency in the clotting factors (which are plasma glycoproteins) cause a defect in clotting mechanism of the affected patient with defect in haemostasis [64].

The tow leading diagnostic markers in human cancer, carcinoembryonic antigen (CEA) and α-fetoprotein are both glycoproteins, they increase in malignant diseases and their level may return to normal after treatment (with surgery or radiotherapy), but their level starts to increase again if the tumor starts regrowth. From this point, it seems that their level is beneficial to estimate the progress and follow up of the disease if estimated in many occasions [122,64].

1.4.8 The Lectins:

Lectins are divalent or multivalent carbohydrate-binding proteins with the ability to agglutinate erythrocyte (RBC of certain blood group), bacteria and other normal and malignant cells [123].

A great variety of carbohydrate-binding proteins (lectins) have been isolated from plants and have proved to be very useful in investigations of glycoproteins, glycolipids and polysaccharides and in studies of cells. Many of these plant lectins are able to agglutinate erythrocytes or other cells and they have therefore often been termed phytohaemagglutinins. Lentil lectin is the haemagglutinaing lectine isolated from common lentil lens culinaris and shows a specific binding affinity towards α-D-dlucose and α-D-mannose residues [124]. The ability of lectins to interact with soluble glycoproteins has been used to isolate and fractionate glycoproteins, this ability is due to binding of lectin to the oligosaccharides moieties of glycoproteins. So it is used in chromatographic techniques for glycoproteins fractionation [125]. Most of the soluble lectins isolated from vertebrate tissues bind β-galactoside. The best studied are group of dimeric protein found in many organism including the electric eel, chicken and man [126].

Another group of vertebrate β-galactoside-binding lectins can be isolated as monomers [127], other soluble lectins are multimeric. The serum of the eel contains lectin composed or have twelve subunit per molecule [128].

1.4.8.1. Biological functions of lectins:

It is reasonable to expect that the known biochemical properties of lectins dictate their endogenous biological function. Some of these properties and the functions that might be inferred from them are summarized in (Table 1.3) [129].

320

Table (1.3): Some common properties of lectins that suggest biological functions.

Property	Function suggested
Specific binding sites: All of one kind Of different kinds	Recognition of complementary oligosaccharide receptors (range of specificities)
2. More than one carbohydrate-binding site	Cross-linking glycoproteins or glycolipids in membranes and/or solution High affinity (multisite) binding to molecules or a cell surface with multiple receptors
3. Agglutinin	Binding cells together: Like cells (promoting adhesion, fusion, etc.) Unlike cells (promoting symbiosis, infection, phagocytosis, etc.)
4. Abundant	Structural rather than catalytic function
5. Generally not integrated in membranes	Relative freedom of movement in or between cellular compartments

One major property of lectins is their specific saccharide-binding sites. In the many lectins that are multimers of an identical subuint these binding sites are the same. In contrast, some lectins are composed of subunits with different binding sites. These include the lectin from the red kidney bean, phaseolus vulgaris. It is composed of two different subunits combined into five different forms of noncovalently bound tetramers [130].

Since the subunits have markedly different specificities for cell surface receptors, each combination could be envisioned to have a different function. For example the homotetramer of one subunit might agglutinate cells with an appropriate receptor, whereas a tetramer that contains only one of these subunits per molecule might inhibit such agglutination.

The common finding of more than one lectin in seeds, slime molds and even vertebrate tissues [131], raises the possibility of concerted specific reactions due to concurrent display of these proteins on a structure like a cell surface. A highly specific interaction directed by the binding properties of more than one type of lectin molecule could result.

The specificity of the binding sites of the lectins suggests that there are endogenous saccharide receptors in the tissues from which they are derived or on other cells or glycoconjugates with which the lectin is specialized to interact. Unfortunately no endogenous receptor for a lectin has yet been unambiguously identified, despite the fact that their carbohydrate-binding sites may be specialized for association with highly specific complex oligosaccharides [132]. The other properties and functions of lectins included that the fact of lectins have more than one carbohydrate-binding site suggests that they could act to cross-link glycoproteins and glycolipids in membranes of the same cell for various organizational purposes. Agglutination activity suggests functions in binding like or different cells for a variety of purposes ranging from morphogenesis in embryos to phagocytosis of one cell by another. The marked abundance of lectins suggests that they play a structural role rather than an enzymatic or catalytic function. The fact that many lectins are readily isolated as water-soluble materials suggests that if they play a role in membrane function it is by association with oligosaccharides on membranes without the constraints imposed by being integrated within the membrane bilayer. It of course remains possible that some functions of lectins may be mediated by properties that have not yet been discovered [133].

1.4.8.2. Applications:

Lectins play an important role in research into a variety of cellular properties and processes.
Their major biological effects, such as cell agglutination and mitogenic stimulation, appear to be mediated initially through interactions at the level of cell surfaces and mimic various physiologically important processes. Since lectins can be obtained in a purified form and show well-defined interaction specificities, they have frequently been used in model systems. Boldt and Coworkers [134], used lentil lectin, to study the

mitogenic responses of various populations of human lymphocytes. Lectins also provide convenient "markers" in cell surface studies. Lentil lectin was used by Scott and Rosenthal[135], to characterize plasma membrane vesicles shed by guinea pig macrophages on exposure to sulphydryl-blocking reagents.

Aim of the thesis

The aim of this thesis concerns with the followings:

❖　　　Determination the serum levels of TSA, LASA, TP and TSA/TP ratio in breast and other cancer patients and then to carry out a comparison of their levels with those of normal individuals and Asthma diseases as a pathological control.

❖　　　Estimation of the levels of mucoid and protein-bound hexose in sera of breast and other cancer patients and to compare these levels with those from normal individuals and Asthma diseases as a pathological control.

❖　　　Measurement of SOD activity in sera of patients with breast cancer to compare these values with those from benign patients and normal healthy controls.

❖　　　Isolation and study the binding characteristics of human cancerous breast lectin to glycoprotein present on the human erythrocyte surface and then determination of the optimum conditions for this binding through the study of the effect of various factors.

❖　　　Purification and identification of lectin from human cancerous breast homogenate, and determination of the molecular weight of purified lectin. However, study the physiochemical properties and the effect of various factors on binding activity for purified lectin.

❖　　　Determination of the kinetic and thermodynamic parameters associated with the binding of lectin to glycoprotein at four temperatures (4, 11, 18 and 25°C).

ABSTRACT

This investigation was used to determine the concentration of total sialic acid (TSA), lipid-associated sialic acid (LASA) and total protein (TP) in sera of normal donors (n = 25), patients with chronic non-malignant diseases (Asthma disease n = 20) as a pathological control, patients with breast cancer (premenopausal n = 40, postmenopausal n = 35 and benign patientsn = 25) and patients with other cancer (endometerial carcinoma n = 15 and sarcoma n = 5).

Data analysis show a significant increase ($P < 0.001$) in the mean values ($\pm$ SD) of TSA and TSA/TP in sera of cancer patients when compared to normal healthy individuals and pathological control (Asthma disease). However, the mean values ($\pm$ SD) of total protein (TP) in sera of breast cancer patients did not show significant differences with respect to normal controls and patients with chronic non-tumoral disease (Asthma disease) except for other cancer patients, which show a significant differences ($P < 0.001$). Also, from results analysis there is a statistically significant rise in the mean values ($\pm$ SD) of LASA in sera of cancer patients when compared to the mean values ($\pm$ SD) with normal controls and Asthma disease.

Chapter Two

Introduction

Many glycoproteins and glycolipids from malignant cells differ in carbohydrate composition from those found in normal cells [136]. Since many of these glycoconjugates contain sialic acid, which can be shed into the circulation, total sialic acid (TSA) is of great interest as a marker of malignancy [137], although it has not been demonstrated to be specific for any type of cancer. There are many controversies regarding the quantitative changes in TSA occurring in cancer patients [138]. It can only be said that increase in TSA is associated with certain diseases, including cancer, and is roughly related to tumor size.

Some authors [139] have based their work upon the fact that alterations in glycolipid metabolism are well documented in many tumors, including human cancers. These authors have reported raised levels of lipid sialic acid (LSA) in sera of patients with various neoplasms, but others [140] have found raised LSA values in sera of patients with acute inflammatory disease.

This has led to the conclusion that the high levels of the so-called LSA in the sera of patients with cancer probably emerge from associated inflammation [141] and, hence, the acute phase reactant glycoproteins mainly build up the LSA fraction. Moreover, Dnistrian et al [142] and Katopodis et al [143] reported that excessive sialic acid levels are found frequently in carcinomas of the breast, pancreas, colon, ovaries and other organs. In actuality, some investigators believe that high levels of sialic acid are markers for neoplasms of these organs, but they are not thought to be very reliable.

In a previous study [144], it was demonstrated that patients with breast cancer had lower neuraminidase levels compared with subjects without a personal or family history for breast cancer. Consequently, it was suggested that inadequate activity of neuraminidase enzyme may be a marker for breast cancer.

The objective of this part was to investigate the diagnostic utility of serum levels of total sialic acid (TSA) and lipid-associated sialic acid (LASA) in breast and other cancer patients.

Materials and Methods:

1. **2.1 Chemicals:**

All chemicals and reagents used in this study were of analytical grade.

Table (2.1) shows the chemicals used and the companies that supplied them.

Chemicals	Company
1. Chloroform, Na, K-tartarate, Folin-Cio calteau, Butyl acetate, HCL, Perchloric acid, H_2SO_4, Benzoic acid, Mannose, Fructose, NBT, Riboflavin, EDTA, NaCl, NaOH, NaCN, $CaCl_2$ and Neuraminidase.	BDH
2. Ethanol, Methanol, Galactose, Orcinol, Tris (hydroxymethylaminomethane) and Bovine serum albumin (BSA).	Fluka
3. Resorcinol, Na_2HPO_4 and Urea.	May and Baker
4. Phosphotungstic acid, $CaCO_3$, $CuSO_4.5H_2O$, KH_2PO_4 and Xylose	Riedel-De Haënag
5. Polyethyleneglycol	Merk
6. Sialic acid and D-Glucuronic acid	Sigma

2. **2.2 Instruments:**

Table (2.2) shows the instruments used and companies that supplied them.

Instruments	Company
1. Spectrophotometer Ultraspec type 4050	LKB
2. PH meter	Pye-Unicam
3. MSE-centrifuge	England
4. Memert water bath	West Germany
5. SM-shaker	England
6. Electrophoresis system type LKB 2117 Multiphor system	LKB

3. **2.3 Patients and Blood Samples:**

Twenty five samples of blood were taken from physically normal volunteers, these samples were used as a control, the control volunteers aged between 25-40 years, and consisted of 20 females and 5 males, who gave no history of previous diseases.

Twenty samples of blood were taken from patients with Asthma diseases were used as a pathological controls.

Twenty samples of blood were taken from patients with other cancer also were used as a pathological controls.

Two groups of breast cancer patients and one group pf patients with benign breast tumors were included in this study.

Group 1 contained 40 premenopausal patients with breast cancer. Group 2 consisted of 35 postmenopausal patients with breast cancer. Group 3 comprised 25 patients with benign breast tumors.

All patients were admitted for treatment to AL-Karama teaching hospital, Saddam Medical City and AL-Saddoon Hospital. Patients suffered from any disease that may interfere with our study were excluded.

The blood samples were collected from these patients, left for 30 minutes at room temperature; blood clots were separated at 3000 RPM for 10 minutes by using centrifuge. Sera were aspirated and stored in caped sterilized tubes at -20°C until time of analysis.

The host information of all patients and normal healthy subjects is summarized in Table (2-3).

Table (2-3): The host information, which are used in this study

Groups	Number	Type of tumor	Metastases	Age (range)
Controls: **Normal healthy** **Non-tumoral disease:**	25	-	-	25-40
Chronic Asthma disease	20	-	-	20-60
Breast cancer: **Premenopausal**	40	38 Infiltrative ductal carcinoma 2 Infiltrative labular carcinoma	Liver and axillary lymphnode	30-42
Postmenopausal	35	34 Infiltrative ductal carcinoma 1 Infiltrative lobular carcinoma	Liver and axillary lymphnode and bone	48-68
Benign	25	24 Fibroadenomas 1 Fibrocystic disease		21-45
Other cancers	20			30-50

4.

5. **2.4 Preparation of Stock Solutions:**

6. Chloroform/methanol (2:1 v/v) solution: (prepared by mixing two volumes of the first to one volume of the second reagent).

7. Phosphotungstic acid solution (1mg/ml): (10 gm of the reagent was dissolved in 10 ml of deionized water then was heated to produce clear solution).

8. Resorcinol stock (2% w/v): (2 gm of solid resorcinol was dissolved in 100 ml of deionized water), prepared daily.

9. Copper sulphate (0.1 M): (prepared by dissolving 4 gm of the salt in 250 ml of deionized water).

10. Resorcinol reagent: (10 ml of 2% w/v stock resorcinol plus 9.75 ml of deionized water plus 0.25 ml of 0.1 M copper sulphate, brought to final volume of 100 ml with concentrated hydrochloric acid), prepared daily.

11. Butyl acetate/methanol (85:15 v/v) reagent: (85 ml of butyl acetate was added to 15 ml of methanol).

12.

13. **2.5 Determination of Biochemical constituents in Sera of Patients with Breast Cancer, Other Cancer and non-Cancer (Asthma disease).**

2.5.1. Estimation of Total Serum Proteins:

The method of Lowry et al [145] was used to estimate serum total proteins, using bovine serum albumin as standard. Protein, concentrations are expressed as g/dL of sera. Fig. (2.1) shows the standard curve of protein, which was constructed by plotting the absorbance at 600 nm against standard protein concentration.

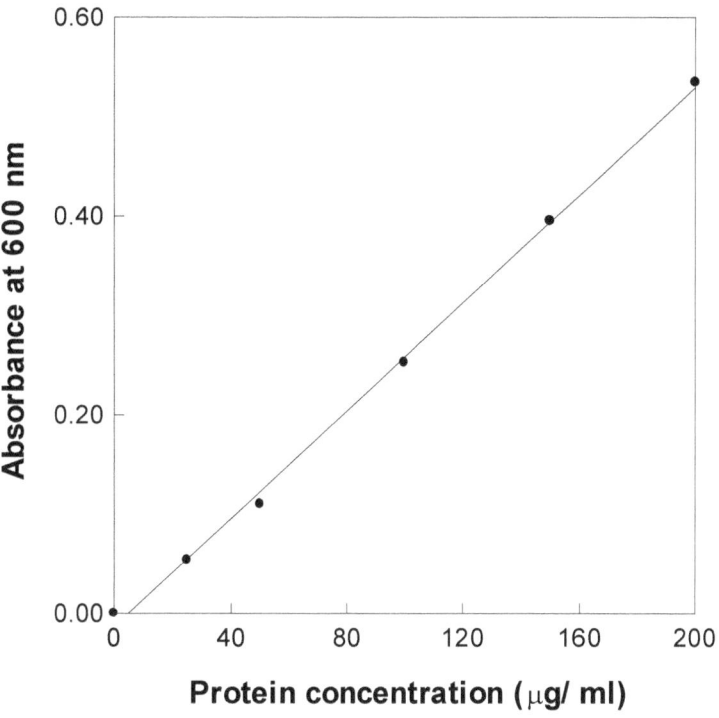

Fig. (2.1): Standard curve of protein determination in human sera by Lowry method.

2.5.2. Estimation of Total Serum Sialic Acid and Lipid-Associated Sialic Acid:

Serum LASA was estimated spectrophotometry by the method of Katopodis et al [143], with slight modification in the volume of sample used. Fifty microliters of serum was placed in test tube (triplicate) with 150 μL deionized water, then vortexed for five seconds and placed on ice. Three ml of cold (4°C) chloroform: methanol (2:1 v/v) was added, then again vortexed and 0.5 ml of cold deionized water was added, then also vortexed after that, centrifuged for 5 minutes at 3000 rpm, at room temperature. One ml of the resulting upper layer was transferred to another tube and 50 μL of phosphotungstic acid (1g/ml) was added, then vortexed and allowed to stand at room temperature for 5 minutes, then centrifuged for 5 minutes at 3000 rpm. The supernatant fluid was removed and the remaining precipitate was dissolved in one ml deionized water.

TSA levels were determined as follows:

Twenty microliters of sera and 980 μL of deionized water were placed in test tube, vortexed and placed on ice.

To each assay tube for LASA and TSA one ml of resorecinol reagent was added, and placed in a 100°C (boiling water bath) for exactly 15 minutes, followed by 10 minutes on a ice bath, two ml of butyl acetate: methanol (85:15 v/v) was added to each tube, then vortexed and centrifuged for 10 minutes at 3000 rpm. The extracted chromophore was read at 580 nm against distilled water.

Standard sialic acid: The standard sialic acid solutions with different concentrations (5, 10, 15, 20, 25, 30, 35, 40, 45) μg/ml were prepared by serial dilutions from a stock standard solution of sialic acid (1000 μg/ml). Fig. (2.2) shows the standard curve of sialic acid which was constructed by plotting the absorbance at 580 nm against the corresponding concentration of standard sialic acid solution, and was used to determine the TSA and LASA levels in the serum samples.

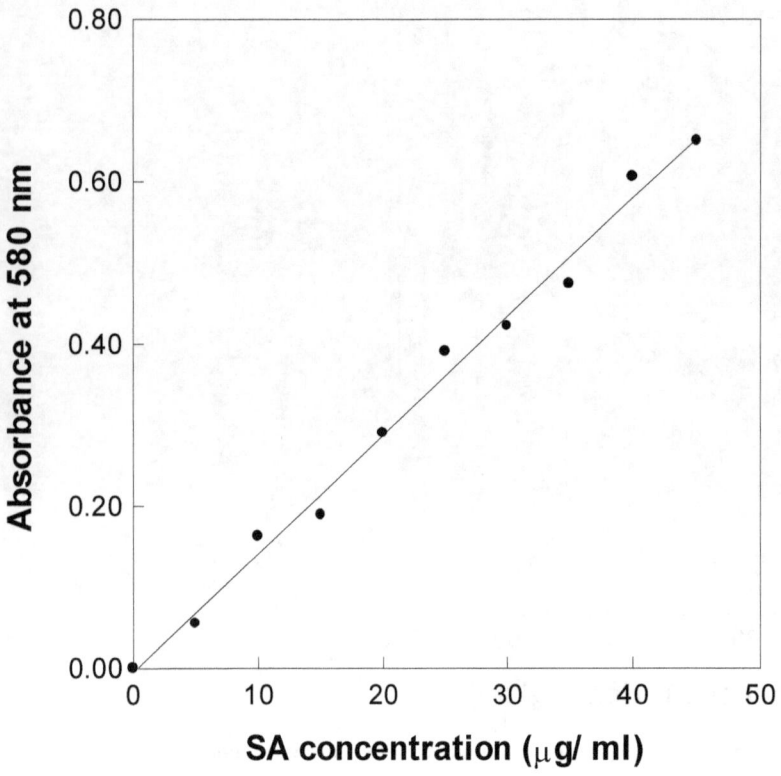

Fig. (2.2): Standard curve for determination of sialic acid concentration in human serum.

14.2.6. STATISTICAL METHODS:

The results for TSA, LASA, TP and TSA/TP ratio were analyzed statistically and values were expressed as mean ±SD. The level of significance was determined by student's t-test [146].

ResultS and Discussion

<u>Determination of Total Sialic Acid (TSA) and Total Protein (TP) Concentrations in Sera of Patients With Breast Cancer, other Cancers and Non-C Individuals:</u>

The individual and mean serum concentrations of TSA and TP for the normal healthy controls, pathological controls (chronic Asthma disease) and patients with breast cancer and other cancers are summarized in table (2-4). This table presents comparisons of the mean values of sera TSA an TP between different groups above, it shows that the TSA levels in breast and other cancer patients were significantly elevated (P < 0.001) as compared to the normal healthy and pathological controls (P < 0.001 and P < 0.01 respectively). However, comparison of serum TSA for different types of breast cancer was studied, the postmenopausal patients show a high level of TSA as compared to the premenopausal patients.

Table (2-4): Comparison of mean values for TSA, LASA and TP in sera of normal, pathological controls and in patients with breast cancer and other cancer.

Groups	No.	TSA (mg/dL) (mean ±SD)	LASA (mg/dL) (mean ±SD)	TP (g/dL) (mean ±SD)	TSA/TP (mg/g) (mean ±SD)
Controls: **Normal healthy**	25	56.7±8.2	17.86±1.59	7.03±0.83	8.27±0.73
Non-tumoral disease: **Chronic Asthma disease**	20	80.8±1.25	21.57±4.02	7.06±0.91	11.54±0.7
Breast cancer: **Premenopausal** **Postmenopausal** **Benign**	40 35 25	100.29±17.38 128.25±21.43 85.6±0.16	32.13±10.01 30.93±10.16 19.29±1.84	7.1±1.02 7.16±1.13 7.08±0.93	14.23±1.45 18.01±1.71 12.28±0.8
Other cancer:					

Endometerial carcinoma	15	103.5±11.5	27.3±7.02	7.91±1.2	13.4±1.7
Sarcoma	5	94.6±9.31	23.92±5.88	8.1±1.03	12.7±1.9

On the other hand, the mean values of total protein (TP) in sera of breast cancer patients did not find any significant differences when compared to the mean values with normal healthy and pathological controls, and on the contrary, the mean values of total protein in sera of other cancer patients were observed a significant differences ($P < 0.01$) than that in breast cancer patients, normal healthy and pathological controls (Asthma disease). It is clear from table (2-4) the mean values of total sialic acid to total protein ratio (TSA/TP) in sera of breast cancer patients (premenopausal and postmenopausal patients) were significantly elevated ($P < 0.01$) as compared to the normal healthy and patients with chronic Asthma disease, whereas, the serum levels of TSA/TP ratio in postmenopausal patients were significantly higher ($P < 0.01$) than those in premenopausal patients with regard to the mean values of TSA/TP ratio in sera of other cancer patients (endometerial carcinoma and sarcoma), show a significantly higher increase than those observed in normal healthy, patients with non-tumoral disease and benign patients ($P < 0.01$), but it is lower than in breast cancer.

Table (2-5) shows the specificity and sensitivity of TSA test in normal controls and in patients with breast cancer, other cancer and non-cancer (Asthma disease). The specificity was calculated as % of patients with cancer who had normal TSA values than normal while the sensitivity was calculated as % of patients with cancer who had higher TSA values than normal. Through table (2-5), the results show that 37 out of the 40 patients with breast cancer (premenopausal patients) 92.5% had elevated values of TSA. Of the groups of 35 patients with breast cancer (postmenopausal), 34 (97.1%) had elevated values for TSA. For benign patients, 22 out of the 25 patients included in group (88%) had elevated values of TSA. But in other cancer, test sensitivity of TSA was varied from 93.3 to 100%. However, in normal controls test specificity of TSA was 84% so that only 16% of those tested were falsely positive. Fig. (2.3) shows the distribution of the individual values of TSA in sera of cancer patients, non-cancer patients (Asthma disease) and normal healthy.

Table (2-5): Specificity and sensitivity of TSA in normal, controls and in patients with breast cancer, other cancer and non-cancer.

Groups	No.	TSA<65 mg/dL (normal)	Specificity % true negative	TSA>65 mg/dL (elevated)	Sensitivity % true positive
Normal	25	21	84	4	16
Breast cancer: Premenopausal Postmenopausal Benign	40 35 25	3 1 3	7.5 2.9 12	37 34 22	92.5 97.1 88
Other cancer: Endometerial carcinoma Sarcoma	15 5	1 0	6.7 0	14 5	93.3 100
Non-cancer: (Asthma disease)	20	4	20	16	80

Note: Sensitivity and specificity of TSA measuring as follows: Sensitivity was calculated as % of patients with cancer who had higher TSA values than controls (normal). Specificity was calculated as % of patients with cancer who had normal TSA values than controls (normal).

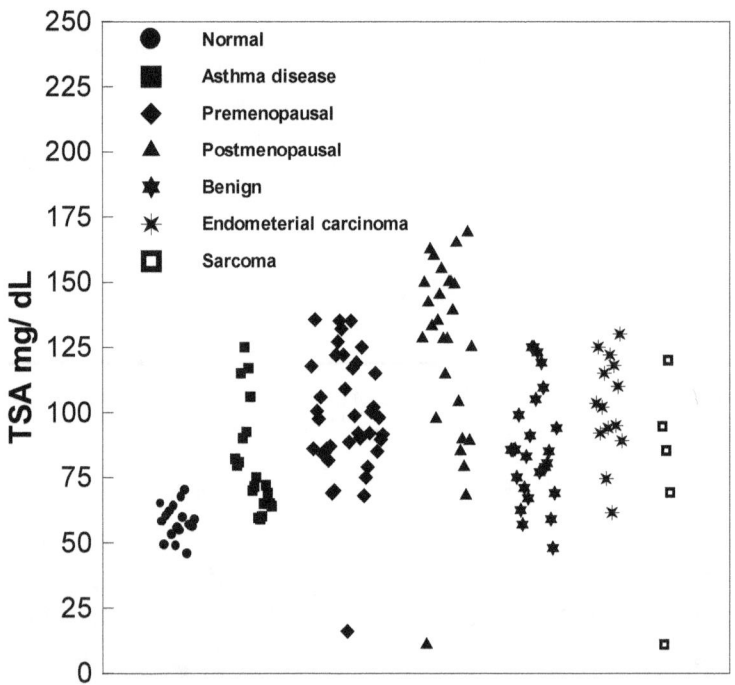

Figure (2.3): Distribution of the individual values of TSA in sera of breast cancer patients, other cancers, non-cancer (Asthma disease) and normal healthy.

Glycoproteins and glycolipids are present in membrane of malignant cells and differ in carbohydrate composition. This contribute to aberrant cell to cell recognition, cell adhesion, antigenicity and the invasiveness demonstrated by malignant cells [147]. Therefore, alterations in important glycoprotein constituents like sialic acid naturally assume importance in malignancy. Variations in serum TSA levels have been found to be useful for diagnosis, staging, prognosis and treatment monitoring of cancer patients

330

[148]. TSA levels when normalized for TP variations and expressed as TSA/TP, become more tumor specific [149]. Increased glycolytic activity has been observed in malignant cells. During neoplastic transformation, the carbohydrate chains in glycolipids and glycoproteins are frequently altered. There is a close relationship between the expression of certain carbohydrate antigens and oncogenesis [150].

In an elegant study examining the significance of the linkage of sialic acid residues in cancer-associated carbohydrate antigens, by using specific monoclonal antibodies, it was demonstrated that not all sialic acids are specific to cancer. Indeed, this study demonstrated that there are significant variations in the cancer specificity depending on the difference in linkage in sialic acid residues [151]. However, the relevance of sialic acid to the tumor cell is apparent from the increased sialylation and sialyltransferase activity observed in many cancer cells [70]. The aberrant glycosylation found in cancer cell membranes is presumably due to the activation of new glycosyl transferases that are characteristic of tumor cells and are absent or present only in small quantities in normal cells [152]. Thus, for instance, a relatively specific sialyltransferase is found to be present by as much as 2.5 to 11 times in greater amounts in transformed cells when compared to control cells [153].

Sialic acid bound to membrane glycoproteins and glycolipids apparently enters the circulation by either shedding or by cell lysis.

Approximately 98 to 99.5 percent of total sialic acid found in serum or plasma is bound to glycoproteins. Only a very small fraction of sialic acid is bound to lipids, which is mainly in the form of gangliosids. Normal levels of total sialic acid in serum are approximately in the range of 51 to 84 mg/dL. In contrast, the contribution of the pure lipid fraction to total sialic acid level is barely in the range of 0.4 to 0.9 mg/dL [154].

Since sialic acid is one of the component of glycoprotein and glycolipid, several investigators and our study concentrate on their levels in the sera or plasma or tissues of patients with cancer diseases. On the other hand, most of these studies have been concerned with TSA level.

Total sialic acid (TSA) level is increased in a variety of tumors, and this level is directly related to cancer burden and disease recurrence [154].

In a recent study on the usefulness of TSA in lung cancer, data obtained in this study show that the mean concentration of TSA was significantly higher in lung cancer patients when compared to benign and normal controls [155].

Erbil et al [75] reported on increased levels of TSA in genitourinary tumors, concluding that serum TSA levels were highly correlated with the stage and grade in patients with advanced urological cancer.

15. Determination of Lipid-Associated Sialic Acid (LASA) Levels in Sera of Patients with Breast Cancer, Other Cancers and non-Cancer individuals:

Serum lipid-associated sialic acid (LASA) levels were determined in normal healthy persons and in patients with non-cancer (Asthma disease) as a pathological controls and patients with breast cancer, other cancers, using the method of Katopodis et al [143].

Table (2-4) shows the individual and mean serum concentrations of LASA in different groups of cancer patients, non-tumoral disease (Asthma disease) and in normal healthy, also this table presents comparisons of the mean values of sera LASA between these groups.

The results in this table reveal an overall elevation in LASA levels for each group of patients with cancer when compared to the normal healthy and pathological controls. This increased LASA values in sera of breast cancer patients except benign patients showed statistically significant differences than those values obtained from sera of normal control, patients with non-malignant diseases and other cancers ($P < 0.001$, $P < 0.01$ and $P < 0.01$) respectively.

It is clear from table (2-4) the level of LASA in both types of breast cancer patients (premenopausal and postmenopausal) show equivalent value.

Table (2-6) represents the percentages of LASA test specificity and sensitivity in normal individuals, patients with breast cancer, other cancers and non-cancer (Asthma disease), using the value of 19 mg/dL as the upper limit of normal.

The results presented in this table show that 35 out of the 40 patients with breast cancer (premenopausal) 87.5% had elevated levels of LASA. Of the groups of 35 patients with breast cancer (postmenopausal), 32 had elevated levels of LASA 91.4%.

Table (2-6): Specificity and sensitivity of LASA in normal controls and in patients with breast cancer, other cancer and non-cancer.

Groups	No.	LASA≤19 mg/dL (normal)	Specificity % true negative	LASA>19 mg/dL (elevated)	Sensitivity % true positive
Normal	25	22	88	3	12

Breast cancer:					
Premenopausal	40	5	12.5	35	87.5
Postmenopausal	35	3	8.57	32	91.4
Benign	25	5	20	20	80
Other cancer:					
Endometerial carcinoma	15	0	0	15	100
Sarcoma	5	0	0	5	100
Non-cancer: (Asthma disease)	20	2	10	18	90

Note: Sensitivity and specificity of LASA measuring as follow: Sensitivity was calculated as % of patients with cancer who had higher LASA values than controls (normal). Specificity was calculated as % of patients with cancer who had normal LASA values than controls (normal).

For benign patients, 20 out of 25 patients included in group (80%) had elevated values of LASA. But in other cancer patients, test sensitivity of LASA had elevated values (100%).
On the other hand, in normal controls test specificity of LASA was 88% sothat only 12% of those tested were falsely positive. Figure (2.4) shows the distribution of the individuals values of LASA in sera of cancer patients, non-cancer patients (Asthma disease) and normal healthy.

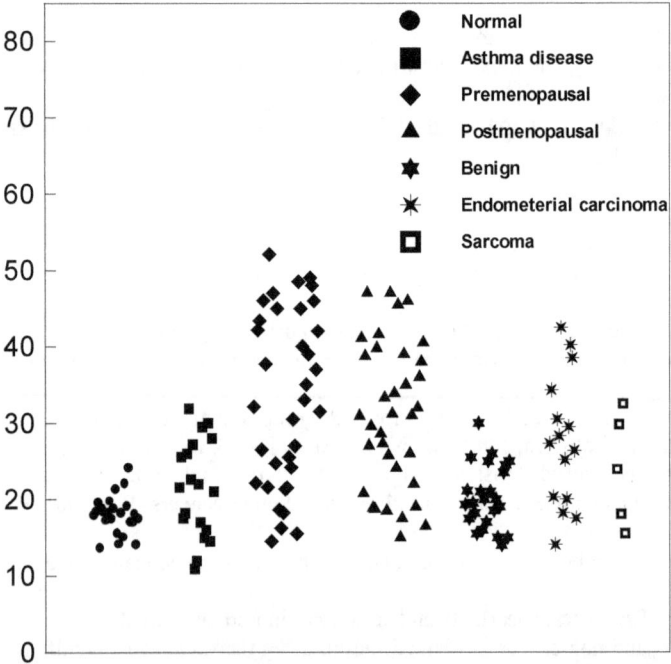

Figure (2.4): Distribution of the individual values of LASA in sera of breast cancer patients, other cancers, non-cancer (Asthma disease) and normal healthy.

Lipid-associated sialic acid (LASA) level is increased in a variety of tumors, and this level is directly related to cancer burden and disease recurrence[154].
In another studies LASA levels have been reported to be useful in monitoring patients with malignant melanoma [98]. However, LASA-P is a biomarker useful in a wide range of malignancies. It reflects alteration in the surface membrane of tumor cells. The LASA-P assay measures total gangliosides and glycoproteins[156]. Elevated LASA-P levels in breast cyst fluid have been associated with increased risk of breast cancer [157]. Also, LASA-P levels are higher in women with benign or malignant breast tumors than in

controls [158]. Some conditions other than cancer affect LASA-P. Among them, are myocardial infarction [159], infections, rheumatoid arthritis, and collagen degeneration. Polivkova et al [187], believe that the determination of LASA levels could be useful not only for cancer diagnosis but also prognosis.

In this investigation, it is established that the data for the TSA and LASA test confirms the previous observations, which have indicated that levels are significantly increased in the sera of breast cancer and other cancer patients, and both tests may prove to be of clinical value.

The increase of TSA are frequently modest in premenopausal patients and high in postmenopausal patients as compared to non-tumoral diseases and normal healthy controls and the demarcation between normal and abnormal levels is sufficiently sharp sothat with a 65 mg/dL cutoff (Table 2-5). The specificity and sensitivity data for the results was a favorable when compared to those of the most widely used immunodiagnostic test, CEA [160].

Sialic acid measurements appear to have a high sensitivity for a wide range of tumors.

However, the specificity of sialic acid measurements, especially the non-specific LASA measurements, is low since

Chapter Three

Abstract

The serum protein-bound hexose and seromucoid (galactose +

mannose) of the neutral glycoprotein fraction of human have been estimated. This test has been carried out on 165 human sera of which 95 of them were patients with three categories of malignancy, 25 from persons with benign disorders, 25 from normal healthy individuals and 20 patients with non-cancer (Asthma disease). Data analysis show a significant increase in protein-bound hexose serum levels in patients with cancer irrespective of tumor site when compared with the levels of normal individuals ($P < 0.01$) and also between patients with non-cancer diseases and normal healthy individuals ($P < 0.01$), but there is not statistically significant results between patients with benign tumor and normal healthy individuals ($P > 0.05$). However, the seromucoid values show a significantly higher increase in patients with cancer when compared with the levels observed in normal healthy individuals and non-cancer diseases ($P < 0.001$).

INTRODUCTION:

The measurement of biochemical markers is being increasingly used for early diagnosis and monitoring the progress of cancer. Increased levels of protein-bound carbohydrates have been shown to occur frequently in patients with neoplasms [161]. Glycoproteins play an important role in the cellular phenomena that undergo alterations during cancerours transformation.

Barlow and Dillard [162] demonstrated that serum L-fucose could be helpful as a means of assessing disease status in patients with breast and cervical cancer.

Sugars in glycoproteins, glycolipids, glycosaminoglycans, oligo-and polysaccharides have been identified in a variety of animal tissues [163]. Moreover, various carbohydrate fractions bound to the plasma proteins are elevated in patients suffering from cancer diseases and also certain non-cancer disease state [164].

Lawrence et al described a group of glycoproteins that are synthesized and released by human breast cancer maintained in organ culture and similar glycoproteins released by a human breast carcinoma cell line (BT-20). The electrophoretic mobility of these glycoproteins on cellulose acetate is consistent with increased glycoproteins staining material present in the α_2 to β-globulin region of serum glycoprotein electropherograms from patients with breast cancer[165].

However, the protein-bound carbohydrate and seromucoid of plasma have been demonstrated to be elevated in a wide variety of pathological conditions including spontaneous human carcinoma [166].

The present investigation was carried out to clarify further the possible usefulness of serum protein-bound sugars and seromucoid as for possible identification of the status of advanced breast and other cancers.

MATERIALS AND METHODS:

3.1 CHEMICALS:

All chemicals and reagents mentioned in section (2.1) were used in the experiments of this chapter.

3.2 INSTRUMENTS:

All instruments that described in section (2.2) were used in the experiments of this chapter.

3.3 PTIENTS AND BLOOD SAMPLES:

The same patients and specimens mentioned in section (2.3) were used in this chapter.

3.4 PREPARATION OF STOCK SOLUTIONS:

The stock solutions which that used in the experiments of this chapter are:

1. Perchloric acid solution (1.8 M): 16.6 ml of 72% perchloric acid was diluted to 100 ml with distilled water.

2. Phosphotungstic acid solution, 5% in 2N of HCl.

3. Sodium chloride solution, 0.85% (0.85 gm of NaCl was dissolved in 100 ml of distilled water).

4. Sodium hydroxide solution, 0.1 N (0.4 gm of NaOH was dissolved in 100 ml of distilled water).

5. H_2SO_4, 60% (v/v): 60 ml of H_2SO_4 was mixed with 40 ml of distilled water.

6. Orcinol reagent (2%): 2 gm of recrystallized orcinol was dissolved in 100 ml of 30% (v/v) H_2SO_4.

7. Stock standard sugars: 100 mg of galactose and 100 mg of mannose were dissolved in 100 ml of distilled water, then saturated with benzoic acid and stored in cold place.

8. Working standard: 1 ml of stock standard sugars was mixed with 9 ml of distilled water, this reagent was prepared freshly in the day of use.

3.5 DETERMINATION OF BIOCHEMICAL CONSTITUENTS IN SERA OF PATIENTS WITH BREAST CANCER, OTHER CANCERS AND NON-CANCER (ASTHMA-DISEASE):

3.5.1 Determination of Seromucoid:

The method of Weimer and Mashin [167], was used to determine serum seromucoid. This method includes the following steps:

(1) Half milliliter of serum was added to 4.5 ml of 0.85% NaCl, then mixed and 2.5 ml of 1.8 M perchloric acid was added, after mixing by inversion, the assay tubes were allowed to stand at room temperature for 10 minutes, then centrifuged for 15 minutes at 3500 r.p.m. to obtain clear supernatant.

(2) To five milliliter of the supernatant, 1 ml of phosphotungstic acid reagent was added, after mixing the tubes were allowed to stand for 10 minutes.

(3) The tubes were centrifuged, after removing of the supernatant, five milliliter of 95% ethanol was added, then strie, centrifuged and the supernatant was removed.

(4) The resulting precipitate was dissolved in 0.5 ml of (0.1 N) NaOH, this was considered as the unknown.

(5) Set up blank with 0.5 ml of distilled water, and standard using 0.5 ml of working standard.

(6) To each unknown, blank and standard, added 1.25 ml of orcinol reagent and 7.5 ml of (60% v/v) H_2SO_4.

(7) All tubes were placed in a water bath at (80±0.5°C) for 20 minutes, cooled and read against distilled water at 520 nm.

Calculations:

$$\text{mg seromucoid}/100\ \text{ml} = \frac{A_x - A_b}{A_s - A_b} \times 0.1 \times \frac{100}{0.333}$$

$$= \frac{A_x - A_b}{A_s - A_b} \times 30$$

Where:

A_x = The absorbance of unknown solution at 520 nm.

A_s = The absorbance of standard solution at 520 nm.

A_b = The absorbance of blank solution at 520 nm.

3.5.2 Estimation of Serum Protein-Bound Hexose:

The method includes the following steps:

(1) Hundred microliter of serum was added to 5 ml of 95% v/v ethanol and mixed carefully, then centrifuged for 15 minutes at 3500 r.p.m., after that the supernatant was decanted.

(2) The remaining precipitate was washed with 5 ml of 95% ethanol, then stired, after that centrifugation.

(3) The supernatant was decanted, the steps 4-7 was carried out as in section (3.5.1) [167].

Calculations:

$$\text{mg protein bound hexose}/100\ \text{ml} = \frac{A_x - A_b}{A_s - A_b} \times 0.1 \times \frac{100}{0.1}$$

$$= \frac{A_x - A_b}{A_s - A_b} \times 100$$

Where:

A_x = The absorbance of unknown solution at 520 nm.

A_s = The absorbance of standard solution at 520 nm.

A_b = The absorbance of blank solution at 520 nm.

RESULTS AND DISCUSSION

The level of protein-bound hexose (galactose and mannose) was estimated in the sera of normal volunteers and in patients with breast cancer, other cancers and non-cancer (Asthma disease), using the method of Weimer and Mashin [167].

The mean concentrations of protein-bound hexoses in sera of all patients and normal healthy are summarized in table (3-1). Figure (3.1) shows the distribution of the individual values of protein-bound hexoses in sera of all cancer patients, non-cancer and normal controls.

Table (3-1): Serum protein-bound hexose (galactose and mannose) in normal controls and in patients with breast cancer, other cancers and non-cancer (Asthma disease) ± standard error.

Groups	Number.	Protein-bound hexose mg/dl±S.E.
Normal	25	87.09±1.82
Breast cancer: Premenopausal Postmenopausal Benign	40 35 25	111.9±5.55 103.30±4.48 90.7±5.01
Other cancers: Endometerial carcinoma Sarcoma	15 5	127.75±3.25 131.5±4.87
Non-cancer: (Asthma disease)	20	101.66±4.1

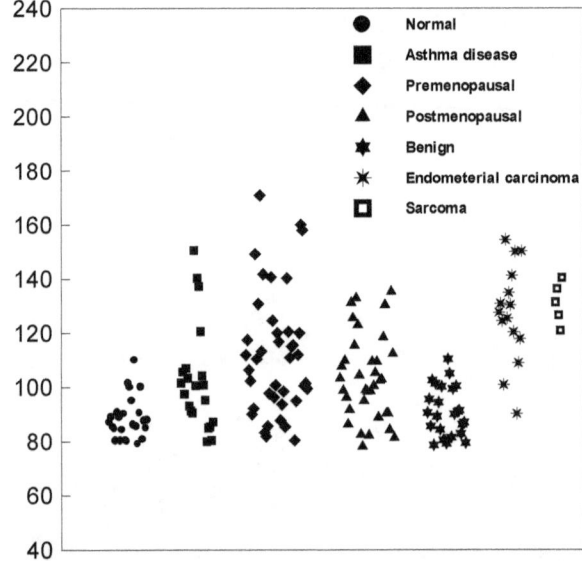

Figure (3.1): Distribution of the individual values of protein-bound hexoses in sera of patients with breast cancer, other cancers, non-cancer (Asthma disease) and normal healthy.

From the results presented in this table, reveal that the mean values of protein-bound hexose in sera of patients which suffer from breast and other cancers were elevated significantly in comparison to those of normal controls and patients with benign (P < 0.01).

The mean concentrations of protein-bound hexose reached to 111.9 ± 5.55 mg/dl in premenopausal patients, 103.3 ± 4.48 mg/dl in postmenopausal patients, whereas in other cancers it was 127.75 ± 3.25 mg/dl in endometerial carcinoma patients, 131.5 ± 4.87 mg/dl in sarcoma patients.

However, the mean level of protein-bound hexose in patients with non-cancer (Asthma disease) was significantly elevated when compared to those observed in normal healthy controls and with benign patients (P < 0.01), but the elevation was not significant when compared to the mean values of protein-bound hexose between patients with benign and normal healthy individuals (P > 0.05). The serum levels of protein-bound hexose in patients with other cancers were significantly higher (P < 0.01) than those in breast cancer patients and non-cancer, and also than those observed in normal healthy controls and benign patients (P < 0.001).

Table (3-2) shows the specificity and sensitivity of the protein-bound hexose in normal controls and in patients with breast cancer, non-cancer.

Table (3-2): Specificity and sensitivity of protein-bound hexose in normal controls and in patients with breast cancer and non-cancer (Asthma disease).

	No.	Protein-bound hexsoe≤89 mg/dL (normal)	*Specificity % true negative	Protein-bound hexsoe>89 mg/dL (elevated)	**Sensitivity % true positive
Normal	25	6	24	19	76
Breast cancer: Premenopausal Postmenopausal Benign	40 35 25	3 4 1	7.5 11.43 4	37 31 24	92.5 88.57 96
Non-cancer: (Asthma disease)	20	0	0.00	20	100

* Calculated as the number of cases having ≤ 89 mg/dl divided by the total number of cases by 100.

** Calculated as the number of cases having >89 mg/dl divided by the total number of cases by 100.

In this test using 89 mg/dl as the upper limit of normal, test sensitivity in those with breast cancer patients was varied from 88.57% to 92.5%, but this test had elevated in benign patients and non-cancer patients (Asthma disease) 96% and 100% respectively. The test was more sensitive for breast cancer patients and non-tumor disease (Asthma disease), as compared to the normal healthy controls.

Also, the results obtained revealed low specificity in different types of breast cancer.

On the other hand, the level of seromucoid was estimated in sera of normal healthy controls and in patients with breast cancer, other cancers and non-cancer (Asthma dusease).

The data obtained in this study are summarized in table (3-3). Figure (3.2) shows the distribution of the individual values of seromucoid in sera of all cancer patients, non-cancer and normal healthy controls.

Table (3-3): Serum seromucoid in normal controls and in patients with breast cancer, other cancers and non- cancer (Asthma disease) ± standard error.

Groups	Number.	Seromucoid mg/dl±S.E.
Normal	25	10.29±0.29
Breast cancer:		
Premenopausal	40	17.58±2.08
Postmenopausal	35	15.92±1.83
Benign	25	12.63±0.33
Other cancer:		
Endometerial carcinoma	15	18.16±1.3
Sarcoma	5	19.78±1.7
Non-cancer: (Asthma disease)	20	13.24±0.73

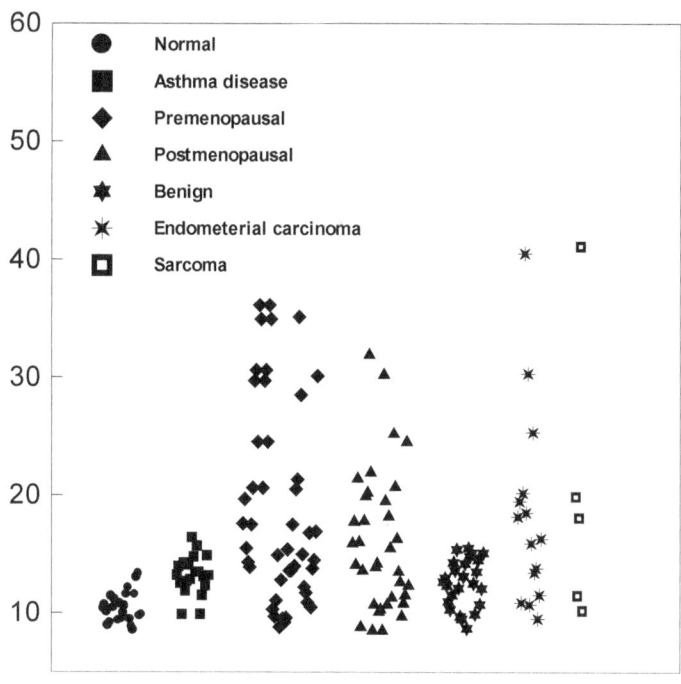

Figure (3.2): **Distribution of the individual values of seromucoid in sera of patients with breast cancer, other cancers, non-cancer (Asthma disease) and normal healthy.**

From these data it is apparent that the mean values of seromucoid level in hospitalized patients suffering from breast and other cancers were significantly elevated when compared to the benign, normal controls and patients with non-cancer (Asthma disease), (P < 0.001).

But there is no statistically significant differences between patients with benign and Asthma disease (P > 0.05).

However, the mean values of seromucoid level in other cancer patients show a significantly higher increase than in patients with breast cancer (P < 0.01) and also as compared with normal healthy controls, non-cancer and benign patients (P < 0.001).

The specificity and sensitivity of the seromucoid test are considered in table (3-4), using 11 mg/dl as the upper limit of normal, this test was sensitive for different types of breast cancer as compared to the normal healthy.

Table (3-4): **Specificity and sensitivity of seromucoid test in normal controls and in patients with breast cancer and non-cancer (Asthma disease).**

	No.	Seromucoid ≤11 mg/dL (normal)	*Specificity % true negative	Seromucoid >11mg/dL (elevated)	**Sensitivity % true positive
Normal	25	18	72	7	28

Breast cancer: **Premenopausal** Postmenopausal **Benign**	40 35 25	7 6 2	17.5 17.14 8	33 29 23	82.5 82.86 92
Non-cancer: **(Asthma disease)**	20	1	5	19	95

* Calculated as the number of cases having ≤ 11 mg/dl divided by the total number of cases by 100.

** Calculated as the number of cases having >11 mg/dl divided by the total number of cases by 100.

Test sensitivity in those with breast cancer patients (premenopausal and postmenopausal) was 82.5% and 82.86% respectively, but in benign patients and non-cancer patients had elevated, 92% and 95% respectively. Test specificity in different types of breast cancer ranged between 17.5% and 17.14%.

From these results, it possible to conclude that it was useful for differential diagnosis and disease monitoring, but not for the early diagnosis of these tumors.

Elevated levels of glycoproteins have been reported in the sera of patients with metastatic breast carcinoma [168]. Alterations in serum glycoproteins have been studied by determining the carbohydrate moieties, viz. fucose, hexose, hexosamine and sialic acid [169]. Variations in glycoproteins in human uterine cervical carcinoma has been reported [170]. These alterations are due to the exponential growth of malignant cells, which results in a rapid rate of membrane glycoprotein turnover and shedding of these excessive glycoproteins into the sera.

Patel, PS. et al [158] has observed significant increase in the levels of the protein-bound hexose and seromucoid in breast carcinoma patients compared with the normal controls, also the differences were significant when compared to the patients with benign breast diseases, it has also been suggested that the measurement of the two parameters be helpful in the diagnosis of breast carcinoma as well as in differentiating between lobular carcinoma and infilterating duct carcinoma patients.

Serum levels of tumor-associated glycoprotein-72 in patients with gynecological malignancies can be used as a clinical marker [171]. The potential role of sialic acid in the mechanism of tumor formation is indicated by the finding that sialic acid-rich glycoconjugates mask the surface of certain tumor cells by interfering with the immune response of the host [172], and that sialic acid content appears to be correlated with metastatic ability in a variety of tumor cells [173].

On the other hand, Yamamooto et al [174], reported that the increase in protein-bound hexose arises as a result of depolymerization of the ground substance connective tissue adjacent to cancer with release of these compound into circulation.

Bolmer et al [175] have suggested that the elevation of the plasma glycoprotein reflects merely the occurrence of tissue destruction, while Yaskhiko et al [176] have concluded that tissue proliferation or repair is a more probable etiological factor.

The results show that there is a significant increase in the level of glycoprotein such as protein-bound hexose and seromucoid in sera of various cancer patients and that the glycoprotein test may be used as a clinical marker

---

Chapter Four
Abstract

The present investigation was carried out to find some typical differences between the content of lipid-bound sialic acid (LBSA) and the activity of superoxide dismutase (SOD) in sera of normal donors (n = 10) and patients with breast cancer (premenopausal n = 20, postmenopausal n =20, and benign n =20). Data analysis show a lower SOD activity in the sera of breast cancer patients (premenopausal, 1.29 ± 0.14 and postmenopausal 1.05 ± 0.23) as compared to the benign patients 1.53 ± 0.17 and normal controls 1.65 ± 0.21. whereas, LBSA levels in sera of breast cancer patients show a significantly higher increase (premenopausal, 29.32 ± 3.55 and postmenopausal, 27.35 ± 5.29) than in benign patients 18.73 ± 2.29 and normal healthy 16.9 ± 2.13 respectively. The ratio of LBSA/SOD was higher in patients with breast cancer. On the other hand, a negative correlation ($P < 0.001$) exists between LBSA and SOD in the sera of breast cancer patients. From our findings SOD and LBSA appear to be putative markers of malignant disease with potential usefulness not only in breast cancer but also in other conditions associated with an increased risk of neoplastic development.

INTRODUCTION

An enzyme which catalyzes the dismutation of superoxide free radicals ($O_2^{-\bullet}$) according to the reaction:

$$O_2^{-\bullet} + O_2^{-\bullet} + 2H^+ \rightarrow O_2 + H_2O_2$$

It has been purified by a simple procedure from bovine erythrocytes [177]. This enzyme, called superoxide dismutase, contains 2 equivalent of copper per mole of enzyme. The copper may be reversibly removed, and it is required for activity. Superoxide dimutase has been shown to be identical with the previously described copper-containing erythrocuprein (human) and hemocuprein (bovine)[177]. The enzyme has since been detected in a large number of tissues and organisms, and it is thought that it is present to protect the cell from damage by the highly reactive superoxide free radical [178].

Superoxide is formed by the one-electron reduction of oxygen, and has been identified as a product in a number of biological reactions [178]. It is particularly likely to be formed in the red cell and has been shown to be produced when oxyhemoglobine is outoxidized to methemoglobine [179].

$$Hb - Fe^{2+} + O_2 \rightarrow Hb - Fe^{3+} + O_2^{-\bullet}$$

Other likely sources include reactions initiated by ionizing radiation.
Stable solutions of the superoxide radical were generated by the electrolytic reduction of O_2 in an aprotic solvent, dimethylformamide.

Slow infusion of such solutions into buffered aqueous media permitted the demonstration that $O_2^{-\bullet}$ can reduce ferricytochrome C and tetranitromethane, and that superoxide dismutase, by competing for the superoxide radicals, can markedly inhibit these reactions.

Superoxide dismutase was used to show that the oxidation of epinephrine to adrenochrome by milk xanthine oxidase is mediated by the superoxide radical[177].

Furthermore, there are numerous indications in human that prolonged increased cell proliferation is necessary for the development of tumors (180). On the other hand, the changes in the cell surface and the serum during malignant transformation have been established (181).

The content and composition of glycoproteins and glycolipids are affected, with an increase in sialic acid on the cell surface membrane and in the serum [143,182]. Decreased activity of the enzyme superoxide dismutase (SOD) has also been found in all malignant tumors investigated so for [183]. Such changes in the content of glycolipids (LBSA) and the activity of SOD are not found in patients with benign tumors [143,184]. The present investigation was carried out to compare the content of LBSA and activity of SOD in sera of patients with breast cancer as compared to the normal healthy individuals.

MATERIALS AND METHODS:

4.1 CHEMICALS:

All chemicals and reagents mentioned in the section (2.1) were used in the experiments of this chapter.

4.2 INSTRUMENTS:

All instruments that described in section (2.2) were used in the experiments of this chapter.

4.3 BUFFERS AND REAGENTS:

All buffer solutions were prepared by dissolving the appropriate amount of salts in distilled water and the required pH was adjusted. The following stock solutions are prepared in the experiments of this chapter:

- 0.067 M phosphate buffer, pH 7.8.

- 0.1 EDTA containing 1.5 mg of sodium cyanide per 100 ml.

- 0.12 mM riboflavin (4.5 mg per 100 ml), stored cold in a dark bottle.

- 1.5 mM NBT (12.3 mg per 100 ml), stored cold.

4.4 PATIENTS AND COLLECTION OF SPECIMENS:

Ten samples of blood were taken from healthy volunteers, these samples were used as a control groups (female volunteers aged between 25-35 years), who gave no history of previous diseases.

Two groups of breast cancer patients and one group of patients with benign breast tumors were included in this study.

Group 1 contained 20 premenopausal patients with breast cancer.

Group 2 consisted 20 postmenopausal patients with breast cancer.

Group 3 comprised 20 cases of benign breast tumors.

All patients were admitted for treatment to AL-Karama Teaching Hospital, patients suffered from any disease that may interfere with our study were excluded.

The blood samples were collected from these patients, left for 30 min. at room temperature; blood clots were separated at 3,000 r.p.m. for 10 min. by using centrifuge. Sera were aspirated and stored in sterilized tubes at $-20°C$ until time of analysis.

4.5 ESTIMATION OF SERUM SUPEROXIDE DISMUTASE ACTIVITY:

Superoxide dismuatse (SOD) activity was estimated spectrophotometery by the method of Winterbourn et al [185], with our modification for serum. The method is based on the ability of the enzyme to inhibit the reduction of nitroblue tetra-zolium (NBT) by superoxide generated during the reaction of photoreduction riboflavin and oxygen.

Procedure:

1. Addition of 0.2 ml of EDTA/NaCN solution to 0.1 ml of serum sample was carried out, then 0.1 ml of NBT solution were added.

2. The assay tubes were brought to a standard temperature (20-22°C), after that 0.05 ml of riboflavin solution were added to each tube. The final assay volume of 3 ml was made up with phosphate buffer 0.067 M, pH = 7.8.

3. Subsequent exposure to bright lighting was controlled by placing the assay tubes in a white-light box where they received uniform illumination for 20 minutes with 18W fluorescent tube attached to the lid, then the absorbance was red at 560 nm against distilled water.

4. To determine the control value, the absorbance for another set of tubes containing the same mixture was read at 560 nm against distilled water immediately after the addition of riboflavin. (riboflavin was added after the addition of buffer).

5. To determine SOD unit, ten tubes containing (10, 20, 40, 60, 80, 100, 200, 300, 400 and 500 μL) of normal serum samples, and another tube containing no serum were treated as described in the steps 1, 2, and 3.

Calculations:

Percentage inhibition was calculated from each absorbance in the presence and absence of the enzyme:

Inhibition % = $\left(A_E - A_{NE}\right) \times 100$

Where;

A_E: The absorbance at 560 nm of the tubes containing different amounts of the enzyme.

A_{NE}: The absorbance at 560 nm in the absence of the enzyme.

The percentages of inhibition were plotted against the corresponding amounts of serum (Figure 4.1).

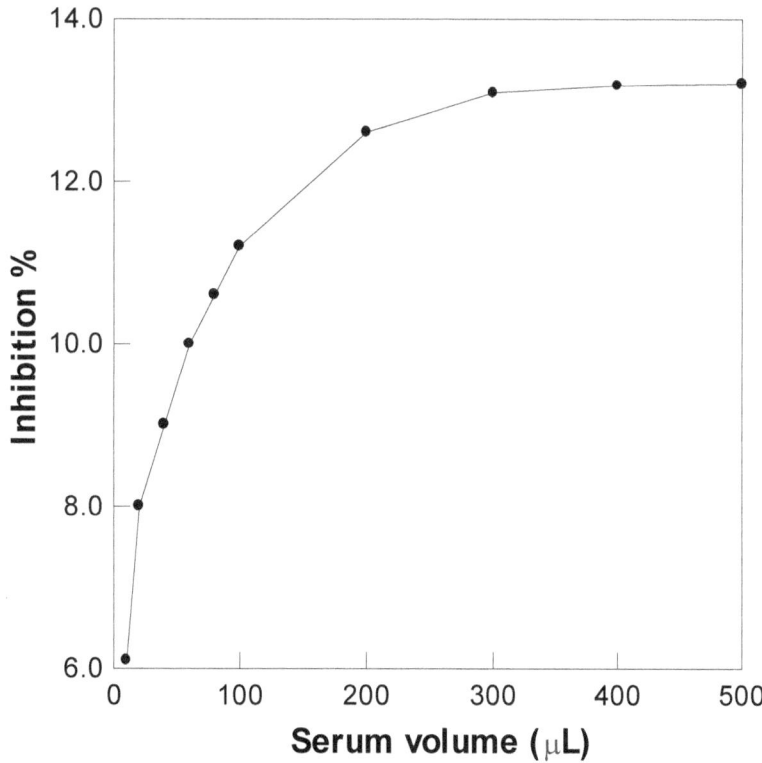

Figure (4.1): The Standard curve for the determination of SOD unit.

SOD unit was calculated from Figure (4.1) according to the following: the amount of serum (VμL) which gives half the maximum inhibition of NBT reduction (1 unit = 10.1 μL).

To calculate the SOD activity in sera of patients, the differences between absorbances before and after the light irradiation were, multiplied by the SOD unit.

347

4.6 DETERMINATION OF SERUM LIPID-BOUND SIALIC ACID:

Serum LBSA was determined by the method of Katopodis et al [143]. All details are explained in section (2.5.2).

4.7 STATISTICAL METHODS:

Statistical analysis was performed by student's t-test [146].

RESULTS AND DISCUSSION:

The individual values of SOD activity and individual levels of LBSA for the normal healthy controls and patients with breast cancer are summarized in Table (4-1).

Figure (4.2) shows the distribution of the individual values of SOD activity in different types of breast cancer patients.

SOD levels in sera of patients with breast cancer differ significantly (P < 0.001) from those of healthy and patients with benign tumor, but the SOD levels in sera of benign patients shows no significant differences from those of healthy individuals.

It is clear from table (4-1), patients with breast cancer had lower values of SOD activity compared with that of healthy individuals, (premenopausal, 1.29 ± 0.14, and postmenopausal, 1.05 ± 0.23). Very low SOD activity in comparison with healthy was observed in sera from patients with postmenopausal. However, a reliable decrease in the SOD activity was found in patients with breast cancer (premenopausal and postmenopausal) as compared to that of benign patients (P < 0.001), on the other hand, patients with breast cancer have a higher levels of LBSA (premenopausal, 29.32 ± 3.55 and postmenopausal, 27.35 ± 5.29), compared with healthy individuals and benign patients. No statistically significant difference was found between LBSA levels in patients with benign and healthy individuals, also there was no statistically significant difference among LBSA levels between the different subgroups of breast cancer (premenopausal and postmenopausal).

Furthermore, the ratio of LBSA/SOD was higher in patients with breast cancer compared with those with benign and normal healthy controls, the mean values of this index are shown in table (4-1)

Table (4-1): Comparison of mean values for the superoxide dismutase (SOD) activity and Lipid-Bound sialic acid (LBSA) levels in sera from healthy and patients with breast cancer. All details are explained in the text.

	No.	SOD activity (mean±SD)	LBSA mg/dL (mean±SD)	LBSA/SOD (mean±SD)
Breast cancer:				
Premenopausal	20	1.29±0.14	29.32±3.55	22.74±2.25
Postmenopausal	20	1.05±0.23	27.35±5.29	30.05±5.87
Benign	20	1.53±0.17	18.73±2.29	12.34±3.64
Healthy (normal)	10	1.65±0.21	16.9±2.13	10.15±2.38

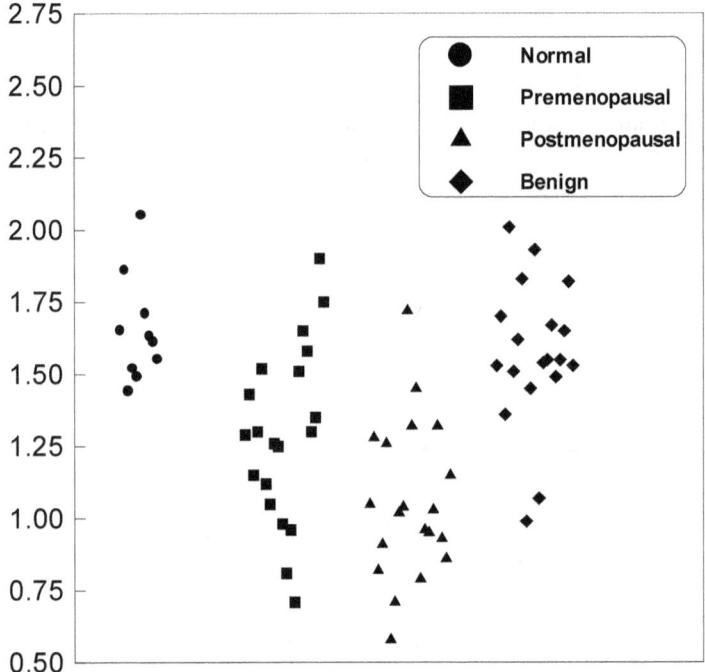

Figure (4.2): Distribution of the individual values of SOD activity in sera of patients with different types of breast cancer and normal healthy.

Knee et al [186] and Bolzan et al [187] reported that lower levels of SOD (especially the manganese dependent enzyme) are found in various malignant tumors compared with normal control cells.

High metastatic cell lines contain less SOD than low metastatic cell ones [186]. Prognosis for patients with low SOD levels in the leukaemic blasts is unfavourable [188], and a negative correlation has been established between SOD activity in the whole blood and the chromosomal sensitivity after x-ray irradiation of lymphocytes from patients with cancer of the mammary gland [187]. The results show that lower SOD activity in patients with tumors compared to the healthy controls (table 4-1), and the mean values for SOD activity in benign patients are near to those in healthy controls, these results are in agreement with the same results have been observed by Abella et al [189].

Table (4-2) shows the sensitivity and specificity of LBSA and SOD in normal controls and in patients with different types of breast cancer. Sensitivity and specificity of the determination of serum SOD activity in distinguishing patients with benign from premenopausal and postmenopausal patients are (85% and 95% respectively) for sensitivity, and it is possible that the lower specificity of this determination (15% and 5% respectively), while the sensitivity of LBSA for determination of benign from malignant is (95% for premenopausal and 100% for postmenopausal). Also, as with SOD activity measurement, the specificity of the determination (premenopausal 5%, and postmenopausal 0 %) is lower than its sensitivity.

Table (4-2): Specificity and sensitivity of LBSA and SOD activity in normal controls and in patients with breast cancer.

Groups	No.	Parameter	Negative results	Specificity % true negative	Positive results	Sensitivity % true positive
Healthy (normal)	10	*LBSA*	9	90	1	10
		SOD	8	80	2	20
Premenopausal	20	LBSA	1	5	19	95
		SOD	3	15	17	85
Postmenopausal	20	LBSA	0	0	20	100
		SOD	1	5	19	95
Benign	20	LBSA	12	60	8	40
		SOD	17	85	3	15

Note: Sensitivity and specificity of the SOD measuring as follows: sensitivity was calculated as % of patients with **breast cancers who had positive results (lower SOD activity than healthy). Specificity was calculated as % of patients with breast cancers who had negative results (normal or higher SOD activity than healthy).**

These results are due to the fact that patients with breast cancer have different diagnosis, also the differences in the serum activity of SOD in patients with malignant tumors and benign are probably due to the decreased enzyme level in the tumor cell [187]. Such deficiency appears, for example, when enhanced serum levels of gangliosides (e.g. LBSA) are released by the tumor cells. LBSA binds to the plasma membranes of the mononuclear cells and inhibits their functions, which may be an important mechanism for immunosuppression in malignant diseases [190].

However, from these results, it possible to conclude that the serum levels of SOD and LBSA reflect the changes in the content of membrane glycolipids and cellular activity of SOD.

Some authors admit that these changes in the tumor cell are in closed connection; the membrane of the malignant cell has an altered lipid content and structure organization, which leads to decreased antioxidant protection [191].

The loss of cell differentiation leads to an increase of the cell glycolipids, on the other hand, to decrease of the intracellular SOD [192]. Figure (4.3) shows a significant negative correlation between the two parameters in patients with breast cancer, and it is also indicates that patients with high levels of LBSA have low SOD content in the serum. Such negative correlation does not appear in benign. This is in accordance with the observations that changes in membrane glycolipids and cellular antioxidants occur in malignant, but not benign, tumors [193]. These findings suggest that SOD and LBSA are good tumor markers.

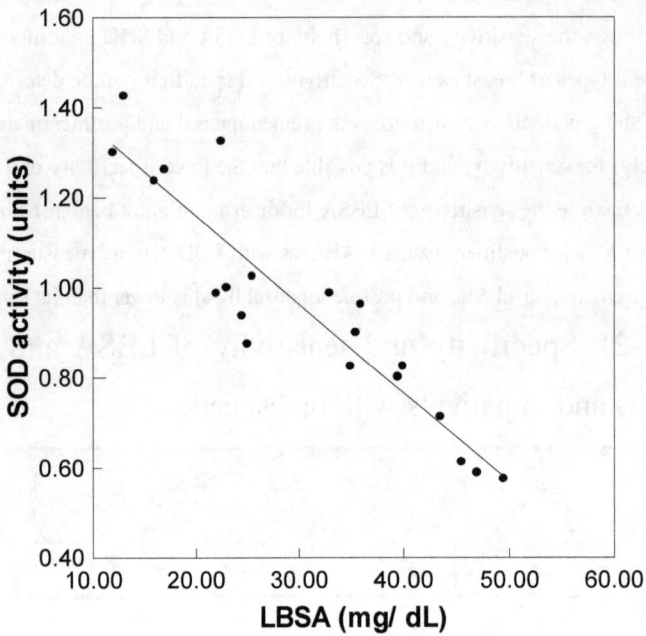

Figure (4.3): Correlation between superoxide dismutase (SOD) activity and the levels of lipid-bound sialic aci

Chapter Five

ABSTRACT

The purpose of this study is to investigate the binding characteristics (as expressed by hemagglutination) of human breast tumor homogenate lectin with glycoprotein present on the human erythrocyte surface. The results show that the non-specific binding was 16-20% of the total binding. Approximately (93 μg) lectin gave a 50% reduction in absorbance at 620 nm, and thus contained one hemagglutination unit (H.U). However, the effect of exogenous Ca^{++} ion concentration was examined, and the results indicate that the binding of lectin with glycoprotein was Ca^{++} ion dependent. The maximum hemagglutination activity of human breast lectin was occurred at pH 8.5, 25°C and Ca^{++} ion concentration of 10 mM. It is clear from these results that the hemagglutination (binding) seemed to be affected by ionic strength, and small amount of salts appeared to be necessary for optimum binding. On the other hand, the specificity of cancerous breast lectin was examined by inhibition tests. The results indicates that this lectin recognize the sialyl residue of the glycoprotein, and the binding is due to sugar specific interaction.

INTRODUCTION

Lectin, a general term applied to haemo-agglutinating substances presents in extracts of seeds of certain plants, which specifically agglutinate RBC of certain blood group, are also glycoproteins and glycolipids [125].

The ability of lectins to interact with soluble glycoproteins and glycolipids have been used to isolate and fractionate glycoproteins and glycolipids. This ability is due to binding of lectin to the oilgosaccharides moieties of glycoproteins and glycolipids [125]. Various lectins are known to be present in mammalian tissues and organs, the β-D-galactose and mannose-6-phosphate as a specific lectins from various tissues and organs have been studied extensively [194,195]. However, a few lectins with high specificity for sialic acid have been identified, such as limulin and carcinoscropine [196]. One major role of lectins, typified by the bacteria-legume symbiosis, appears to be to bind together cells of two different species. There is evidence that lectins acting in this way participate in both the prevention of plant infection, by binding to saccharides on bacteria or fungi, and in the promotion of bacterial infection of vertebrate cells. Bacterial lectins apparently mediate the adhesion of these microorganisms to oligosaccharides on animal cells, which could be a prelude to infection.

Thus, Escherichia coli contains a lectin that binds D-mannose and its α-glycosides, and that presumably mediates bacterial attachment to cells. Another evidence for binding of Vibrio cholera to intestinal cell surfaces by a reaction inhibited by L-fucose has been presented [197]. Although a specific cellular function cannot yet be unequivocably assigned to any lectin, a large body of evidence indicates that lectins have been adapted for a variety of cell surface and intercellular functions in which the specific carbohydrate-binding site of the lectin binds a complementary saccharisde-containing substance as a prelude to one of a number of biological actions. Some lectins, such as those in root hairs, apparently bind complementary saccharides in a fairly discriminating way, which leads to highly specific symbiosis. In this case the lectin is apparently playing a highly refined recognition function, although its binding site is not so exclusive as to reject receptors on test erythrocytes. In contrast, evidence for different localizations and functions of the same lectin in different animal cells suggests that the specificity of lectin function is not dictated solely by the precise nature of its binding sites, but also by opportunistic factors that determine which of many receptors of adequate complementarity are available [129].

Furthermore, lectin which binds sialic acid residue of glycoproteins has also been isolated from wheat germ [198]. Sialic acid occupies an outstanding position both sterically and with respect to biological function in various glycoproteins and glycolipids.

Also, a lectin which specificity recognizes terminal sialic acids residue is likely to be a useful in studying the biological functions of sialoglycoproteins [199].

In this chapter it has been attempted to develop a binding assay for lectin isolated from tumors of breast cancer patients and a characterization of the binding of specific lectin with glycoprotein of human red cell by different techniques were carried out.

MATERIALS AND METHODS

5.1 CHEMICALS:

All chemicals and reagents mentioned in the section (2.1) were used in the experiments of this chapter.

5.2 INSTRUMENTS:

All instruments that described in section (2.2) were used in the experiments of this chapter.

5.3 BUFFERS AND REAGENTS:

All buffer solutions were prepared by dissolving the appropriate amount of salts in distilled water and the required pH was adjusted. The following stock solutions are prepared in the experiments of this chapter:

- 0.02 M HCl.
- 0.02 M Tris (hydroxymethylaminomethane), (2.423 gm in 1000 ml distilled water).
- 0.075 M Na_2HPO_4 (8.98 gm in 1000 ml distilled water).
- 0.075 M KH_2PO_4 (13.25 gm in 1000 ml distilled water).
- 0.02 M $CaCl_2$ (2.12 gm in 1000 ml of buffer).
- 0.15 M NaCl (8.765 gm in 1000 ml of buffer).
- 0.002 M EDTA (0.75 gm in 1000 ml of buffer).

Working buffers:

1. Extraction buffer (phosphate saline buffer). Each liter of 0.075 M Na_2HPO_4/ 0.075 M KH_2PO_4, pH 7.2 containing 0.004 M β-mercapto ethanol, 0.002 M EDTA and 0.075 M NaCl.

2. Assay buffer (Tris-Saline buffer pH 8). Each liter of 0.02 M Tris-HCl pH 8 containing 0.15 M NaCl and 0.02 M $CaCl_2$.

5.4 Collection of specimens and preparation of tissue homogenate:

5.4.1 TISSUE COLLECTION AND PROCESSING:

Human tissues of breast cancer was removed by surgery from females patients, admitted for treatment at AL-Saddon Hospital, and diagnosed by specialist. Breast tissues were immediately immersed in ice-cold saline and then washed with phosphate-saline buffer pH 7.2 and kept at -20°C until time of homogenization process.

5.4.2 Preparation of Tissue Homogenate:

Three grams of frozen breast tissues were washed with 5 ml volume of 0.9% NaCl to remove surface mucus materials and contamination, then homogenized in 15 ml of 0.02 M phosphate buffered saline (0.075 M Na_2HPO_4/ 0.075 M KH_2PO_4 pH 7.2 containing 0.004 M of β-mercapto ethanol, 0.002 M EDTA and 0.075 M NaCl), using Tenbroeck ground-glass homogenizer to prepare the homogenate. The homogenate was centrifugated at 4000 r.p.m. for one hour. The supernatant was used as a source of lectin (crude or homogenate lectin) for binding (expressed as hemagglutination) studies. The supernatant was used through out study and stored at -20°C till using [200].

5.5 DETERMINATION OF BIOCHEMICAL CONSTITUENTS IN CANCEROUS BREAST HOMOGENATE:

5.5.1 Protein Determination:

Protein was determined by the method of Lowry et al [145], using bovine serum albumin as standard.

5.6 PRELIMINARY TEST FOR THE BINDING OF CANCEROUS BREAST LECTIN TO ERYTHROCYTE SUSPENSION:

5.6.1 Preparation of Standard erythrocytes Suspension for Hemagglutination:

Type A of human blood was obtained from physically normal volunteer. The erythrocytes were washed four times in 0.9% (w/v) NaCl and diluted with 0.9% NaCl to an absorbancy of about 2 at 620 nm [201].

5.6.2 The Hemagglutination Assay:

This assay is a modification of the procedure reported by Liener [202,203]. The binding of cancerous breast lectin to erythrocyte suspension was preliminary checked by hemagglutination according to the following modification.

1. Half ml of diluted cancerous breast lectin by ((Tris-Saline buffer, 0.02 M Tris-HCl buffer pH 8 containing 0.15 M NaCl and 0.02 M $CaCl_2$) was incubated with 0.5 ml of washed erythrocytes at room temperature for 30 min. Then the cells were pelleted by centrifugation for 3 min., the supernatant was decanted.

2. The cells were resuspended in the same buffer and aggregated cells allowed to settle for 5 min., then the absorbance at 620 nm of the upper layer (free lectin and cells) of the assay solution was measured.

Calculations:

Total binding (expressed as % hemagglutination) represents the amount of lectin, which binds the erythrocytes and causing hemagglutination.

$$\text{Hemagglutination activity } \% = \frac{A - A^*}{A} \times 100$$

(total binding %)

Where:

A = The absorbance of standard erythrocyte suspension at 620 nm.

A* = The absorbance of free (unbound) erythrocytes at 620 nm.

5.6.3 Determination of non-Specific Binding of Cancerous Breast Lectin Homogenate to Erythrocyte Surface Glycoconjugates:

The same steps mentioned in section (5.6.2) were followed to determine the percent of the non-specific binding, except, the whole blood were washed (3-4) times with normal saline (0.9% w/v NaCl), then two times with assay buffer (Tris-HCl buffer pH 8) and the suspension was prepared. Ten ml of this suspension placed in a test tube and 100 μL of neuraminidase was added (500 unit/ml) and the mixture was shaken for four hours at 25°C. Then centrifugation at 3000 r.p.m. for 3 minutes, was carried out. The supernatant was removed and the remaining blood cells were diluted with an appropriate amount of normal saline to an absorbancy of about 2 at 620 nm and then used for hemagglutination activity directly.

Calculations:

The percent of non-specific binding was calculated using the following equation:

$$\text{NSB } \% = \frac{A^\circ - A^*}{A^\circ} \times 100$$

Where:

NSB % = The percent of non-specific binding.

A° = The absorbance of neuraminidase-treated erythrocyte suspension at 620nm.

A* = The absorbance of free (unbound) erythrocytes at 620 nm.

5.6.4 Determination of The Specific Binding of Lectin to Erythrocyte Surface Glycoconjugates:

The percent of specific binding of lectin to glycoconjugates was calculated by subtracting the percent of non-specific binding from the percent of the total binding.

SB% = TB% - NSB%

Where:

SB% = The percent of specific binding of lectin to erythrocyte surface glycoconjugates.

TB% = The percent of total binding of lectin to erythrocyte surface glycoconjugates.

5.6.5 Determination of Hemagglutination Activity Unit:

The hemagglutination activity unit is the level of test solution (hence breast tumor homogenate concentration) which cause 50% of the standard cell suspension to sediment in (0.5-2 hours) as determined by Lis and Sharon method [203], or by plotting the hemagglutination activity data against the concentration of lectin, (Fig. 5.1).

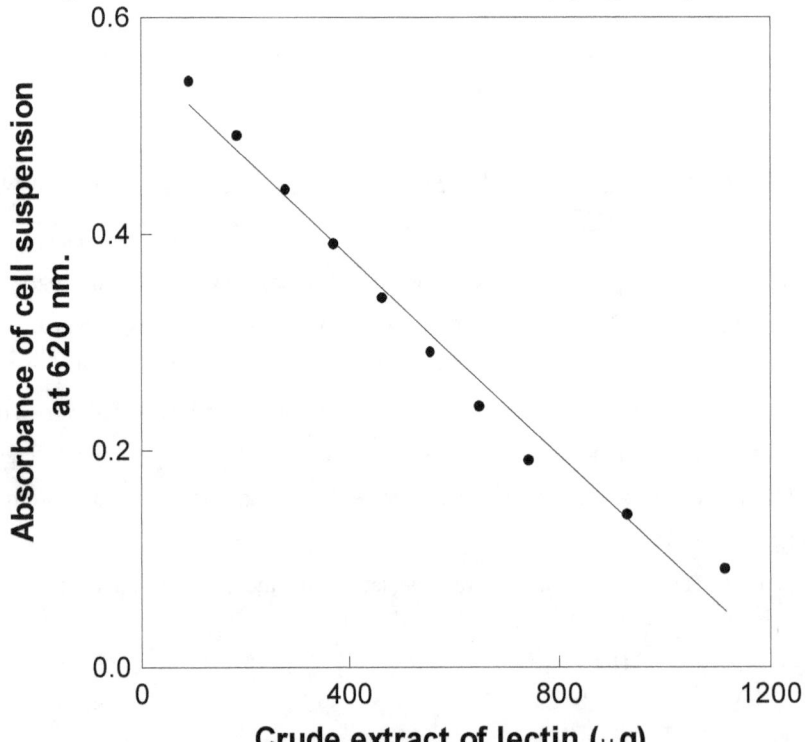

Figure (5.1): Hemagglutination assay of breast tumor homogenate lectin. All details are explained in text.

358

5.6.6 Factors Effecting on Lectin Binding to Erythrocyte Surface Glycoconjugates in Cancerous Breast Homogenate:

5.6.6.1 The Effect of Different Lectin Amounts on Its Binding To Erythrocyte Surface Glycoconjugates:

Half milliliter of erythrocyte suspension was incubated with different amounts of crude lectin (93, 186, 279, 372, 465, 558, 651, 744, 930 and 1116 µg) dissolved in assay buffer for 30 minutes at 22°C. The final reaction volume was 1 ml, then the cells were pelleted by centrifugation for 3 minutes. After that the step 2, of the experiment (5.6.2) was repeated.

Calculations:

1. The same equation mentioned in experiment (5.6.2) was used to calculate the percent of total binding.

2. The percent of specific binding (SB%) was calculated by using the equation mentioned in section (5.6.4).

3. The percent of specific binding (SB%) was plotted against their corresponding lectin amount, as shown in Figure (5.2).

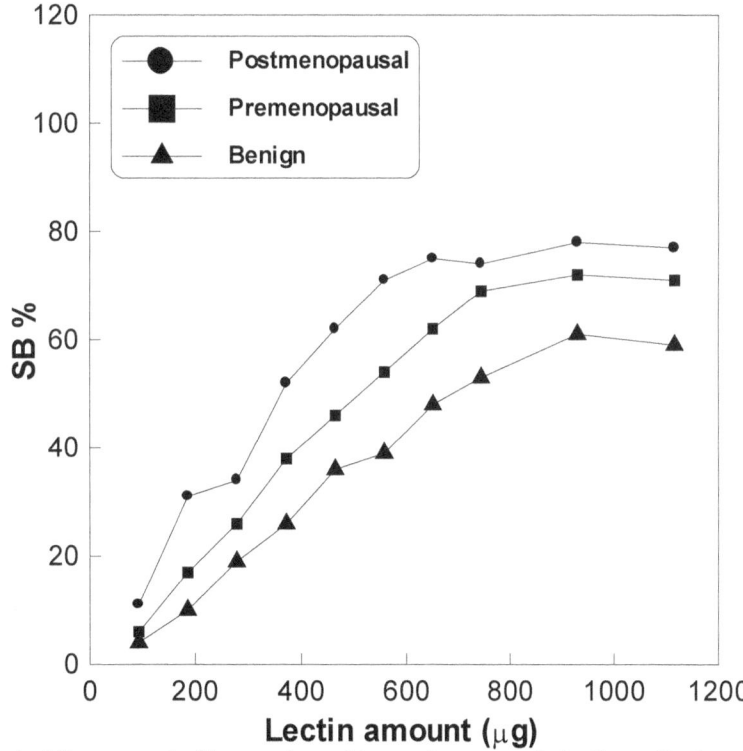

Figure (5.2): Effect of the amount of human breast tumor homogenate lectin on the hemagglutination activity. All details are explained in text.

5.6.6.2 Effect of pH on Hemagglutination Activity:

Fifty µL of human cancerous breast homogenate was incubated with 0.5 ml of erythrocyte suspension at 22°C for 30 minutes, using Tris-saline buffer of different pH from 7 to 9. The final reaction volume was 1

ml, then the cells were pelleted by centrifugation for 3 minutes. After that, the step 2 of the experiment (5.6.2) was repeated.
Calculations:
1. **The same equation mentioned in experiment (5.6.2) was used to calculate the percent of total binding.**

2. **The percent of specific binding (SB%) was calculated by using the equation mentioned in section (5.6.4).**

3. **The percentages of specific binding (SB%) were plotted against their corresponding pH values, as shown in Figure (5.3).**

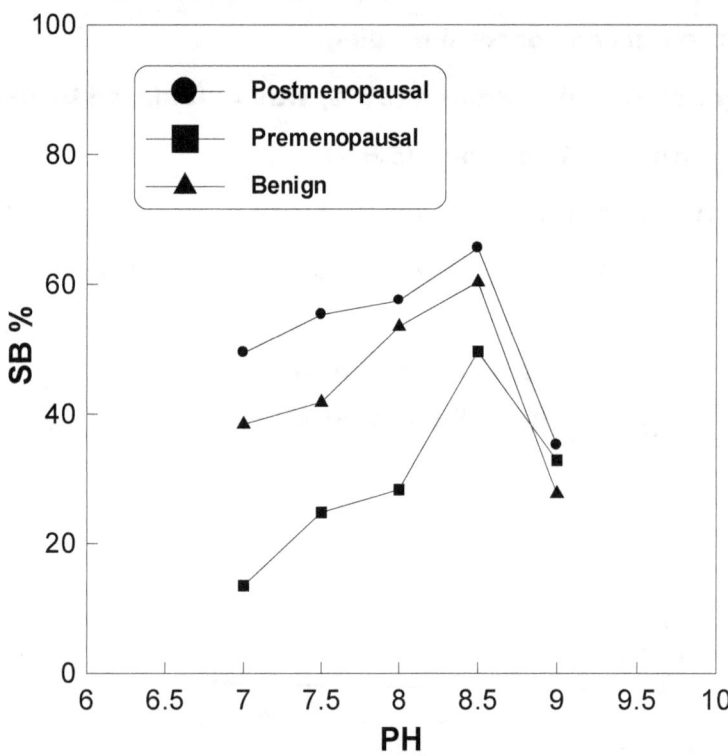

Figure (5.3): Effect of pH on the hemagglutination activity of human breast tumor homogenate lectin. All details are explained in text.

5.6.6.3 *Effect of Temperature on Hemagglutination Activity:*

Fifty µL of human cancerous breast homogenate was incubated with 0.5 ml of erythrocyte suspension for 30 minutes at different temperatures (5, 15, 20, 25, 30, and 35°C) using the assay buffer (pH 8.5). The final reaction volume was 1 ml, then the red cells were pelleted by centrifugation for 3 minutes. After that, the step 2 of the experiment (5.6.2) was repeated.
Calculations:
1. **The same equation mentioned in experiment (5.6.2) was used to calculate the percent of total binding.**

2. **The percent of specific binding (SB%) was calculated by using the equation mentioned in section (5.6.4).**

3. **The percentages of specific binding (SB%) were plotted against their corresponding temperatures, as shown in Figure (5.4).**

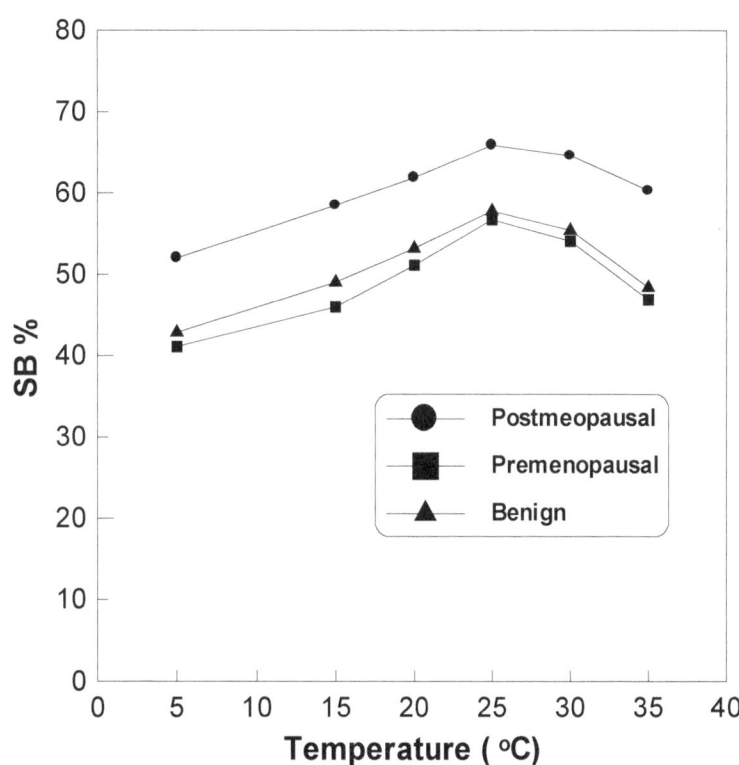

Figure (5.4): Effect of temperature on the hemagglutination activity of human breast tumor homogenate lectin. All details are explained in text.

5.6.6.4 Effect of Incubation Time on Hemagglutination Activity:
Fifty µL of human cancerous breast homogenate was incubated with 0.5 ml of erythrocyte suspension at 25°C for several time intervals (15, 30, 60, 90, 120, and 135 minutes) using the assay buffer (pH 8.5). The final reaction volume was 1 ml, then the red cells were pelleted by centrifugation for 3 minutes. After that, the step 2 of the experiment (5.6.2) was repeated.

<u>Calculations:</u>

1. **The same equation mentioned in experiment (5.6.2) was used to calculate the percent of total binding.**

2. **The percent of specific binding (SB%) was calculated by using the equation mentioned in section (5.6.4).**

3. **The percentages of specific binding (SB%) were plotted against their corresponding times, as shown in Figure (5.5).**

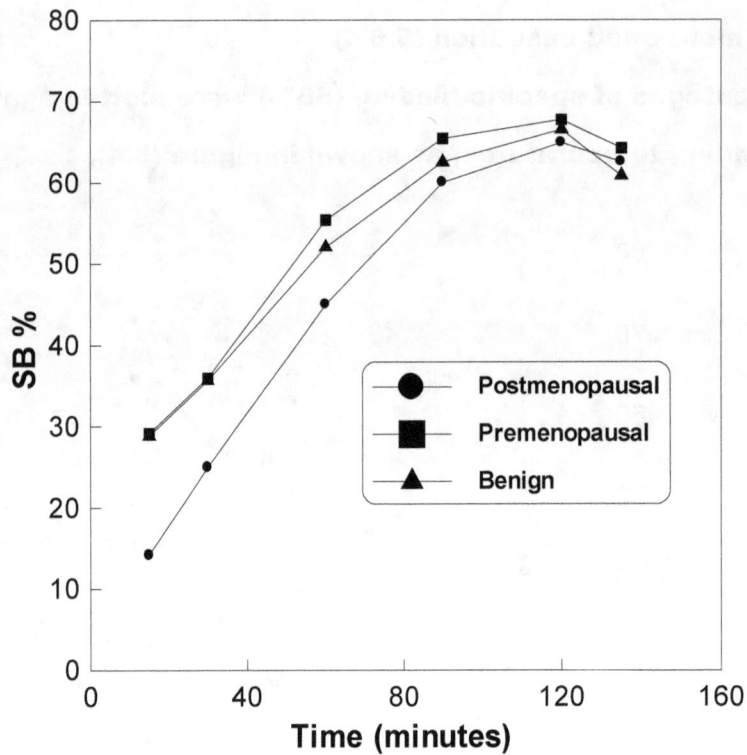

Figure (5.5): Dependence of hemagglutination activity of human breast tumor homogenate lectin on reaction time. All details are explained in text.

5.6.6.5 Effect of Exogenous Ca++ ions on Hemagglutination Activity:

Fifty μL of human cancerous breast homogenate was incubated with 0.5 ml of erythrocyte suspension at 25°C for 120 minutes using the assay buffer (pH 8.5), containing different Ca++ ions (2.5, 5, 10, 15, 20, 25,and 30 mM). The final reaction volume was 1 ml, then the red cells were pelleted by centrifugation for 3 minutes. After that, the step 2 of the experiment (5.6.2) was repeated.

Calculations:

1. The same equation mentioned in experiment (5.6.2) was used to calculate the percent of total binding.

2. The percent of specific binding (SB%) was calculated by using the equation mentioned in section (5.6.4).

3. The percentages of specific binding (SB%) were plotted against their corresponding Ca++ ions concentrations, as shown in Figure (5.6).

362

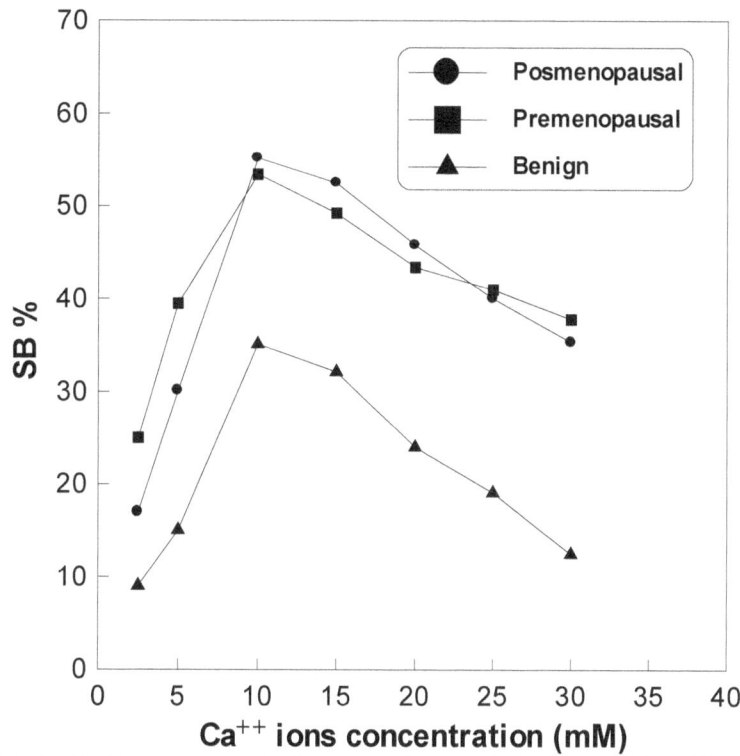

Figure (5.6): Effect of Ca⁺⁺ ions concentration on the hemagglutination activity of human breast tumor homogenate lectin. All details are explained in text.

5.6.6.6 Effect of Denaturating Agents on Hemagglutination Activity:

Fifty µL of human cancerous breast homogenate was incubated at 25°C for 120 minutes with different concentration of denaturating agents (Urea, PEG, NaOH, and HCl), dissolved in the assay buffer (pH = 8.5), then 0.5 ml of erythrocyte suspension was added. The final reaction volume was 1 ml, then the red cells were pelleted by centrifugation for 3 minutes. After that, the step 2 of the experiment (5.6.2) was repeated.

Calculations:

1. The same equation mentioned in experiment (5.6.2) was used to calculate the percent of total binding.

2. The percent of specific binding (SB%) was calculated by using the equation mentioned in section (5.6.4).

3. The data of percent specific binding (SB%) were summarized in Table (5.1):

Table (5-1): Effect of denaturating agents on the hemagglutination activity of breast homogenate lectin. All details are explained in the text.

Groups	Type of test	Reagent added (M)	Specific binding %
Premenopausal	Control	-	100
	urea	3	74.59
		4	66.77
		5	57.13
		6	56.54
Postmenopausal	=	3	75.46
		4	71.65

363

		5	58.48
Benign	=	6	47.58
		3	73.65
		4	58.44
		5	29.17
		6	20.19
Premenopausal	Polyethlene glycol	0.5%	76.82
		1%	75.43
		2%	73.82
		4%	67.1
Postmenopausal	=	0.5%	76.82
		1%	75.65
		2%	74.27
		4%	71.59
Benign	=	0.5%	75.35
		1%	71.67
		2%	70.85
		4%	68.01
Premenopausal	NaOH	0.15	74.25
	HCl	0.15	71.38
Postmenopausal	NaOH	0.15	74.62
	HCl	0.15	71.59
Benign	NaOH	0.15	73.03
	HCl	0.15	70.98

5.6.6.7 Inhibition Studies of Hemagglutination Activity:

A number of carbohydrate were used as inhibitors (sialic acid, glucuronic acid, fructose, mannose and xylose) for the binding of lectin to glycoconjugates of erythrocyte surface.

The first step in this type of assay was carried out by addition of high concentration of lectin, which gives more of hemagglutination. Before addition the sugar, which used as inhibitor, 50 µL of human cancerous breast homogenate was incubated with 0.5 ml of erythrocyte suspension at 25°C for 120 minutes. The final reaction volume was completed to 1 ml by adding Tris-saline buffer (pH 8.5), then the cells were pelleted by centrifugation for 3 minutes. After that, the step 2 of the experiment (5.6.2) was repeated.

The second step in this type of assay was carried out according to the following:

Fifty µL of human cancerous breast homogenate was incubated with 0.5 ml of erythrocyte suspension at 25°C for 120 minutes The final reaction volume was completed to 1 ml by adding Tris-saline buffer (pH 8.5) which contain the desired concentration of sugar used (inhibitor). Then the cells were pelleted by centrifugation for 3 minutes. After that, the step 2 of the experiment (5.6.2) was repeated.

<u>Calculations:</u>

1. The same mathematical formula mentioned in experiment (5.6.2) was used to calculate the percent of total binding before and after addition of the inhibitor.

2. The percent of specific binding (SB%) was calculated before and after addition of inhibitor, by using the equation mentioned in section (5.6.4).

364

3. **The percent of inhibition of hemagglutination represents the difference between the percent of specific binding (SB%) with lectin alone and that obtained with lectin plus the inhibitor. The data of % inhibition was summarized in table (5-2) and these data were plotted against their corresponding sugar concentration, as shown in Fig. (5.7).**

Table (5-2): Inhibition of hemagglutination activity of human breast homogenate lectin. All details are explained in the text.

Groups	Type of carbohy. test	Carbohy. conc. (mM)	Inhibition %
Premenopausal	Sialic acid	1	6.6
		1.5	10.4
		2	15.3
		2.5	19.7
		3	24.4
		3.5	29.1
Postmenopausal	=	1	15.45
		1.5	22.41
		2	30.6
		2.5	36.1
		3	44.85
		3.5	52.1
Benign	=	1	4.65
		1.5	6.61
		2	8.1
		2.5	10.55
		3	12.56
		3.5	14.65
Premenopausal	D-glucuronic acid	1	3.25
		10	6.1
		15	9.9
		20	12.2
Postmenopausal	=	1	4.1
		10	8.75
		15	11.1
		20	13.15
Benign	=	1	2.4
		10	5.9
		15	9.3
		20	11.9
Premenopausal	D-Fructose	30	59.15
	D-Mannose	30	55.8
	D-Xylose	30	51.05
Postmenopausal	D-Fructose	30	28.35
	D-Mannose	30	20.45
	D-Xylose	30	19.95
Benign	D-Fructose	30	-
	D-Mannose	30	-
	D-Xylose	30	11.6

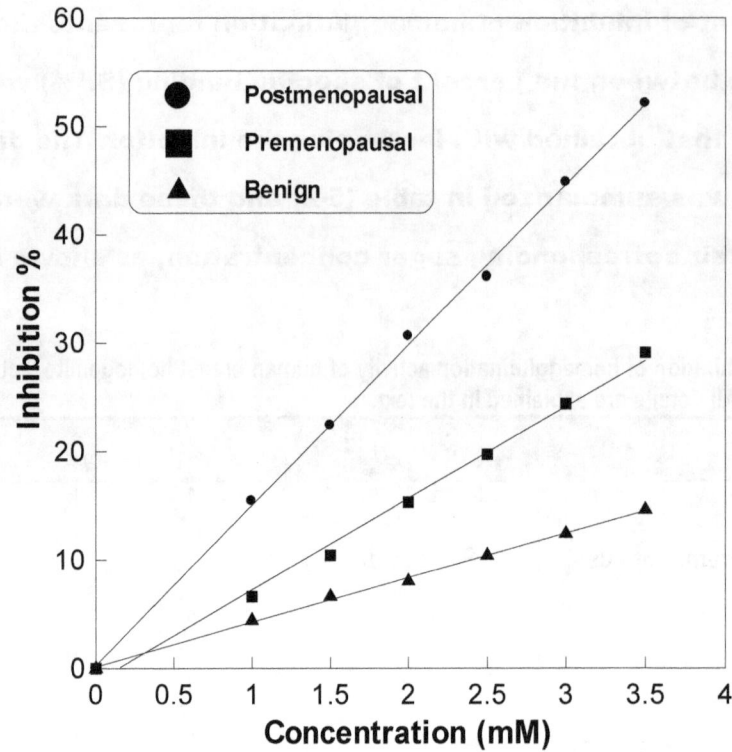

Figure (5.7): Effect of sialic acid on the hemagglutination activity of human breast tumor homogenate lectin. All details are explained in text.

5.6.6.8 Effect of Ionic Strength and Different Salts on Hemagglutination Activity:

5.6.6.8.1 Effect of monovalent salts on hemagglutination activity:

Fifty µL of human cancerous breast homogenate was incubated with 0.5 ml of erythrocyte suspension at 25°C for 120 minutes using the assay buffer (pH 8.5), which contains NaCl of various concentrations (0.05 M to 0.3 M). The total volume was 1 ml, then the cells were pelleted by centrifugation for 3 minutes. After that, the step 2 of the experiment (5.6.2) was repeated.

Calculations:

1. The same equation mentioned in experiment (5.6.2) was used to calculate the percent of total binding.

2. The percent of specific binding (SB%) was calculated by using the equation mentioned in section (5.6.4).

3. The percentages of specific binding (SB%) were plotted against their corresponding NaCl concentrations, as shown in (Figure 5.8 and table 5-3):

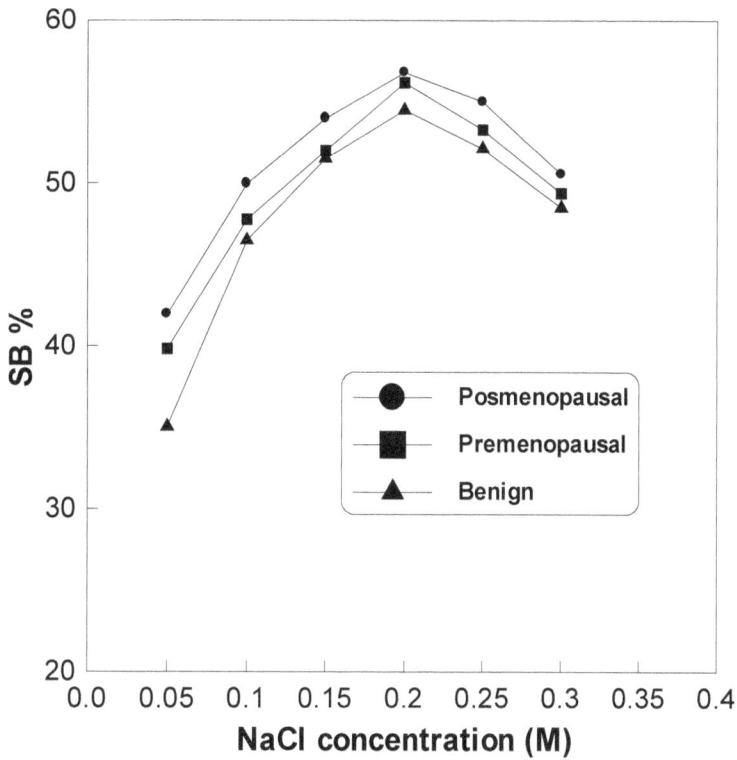

Figure (5.8): Effect of monovalent salt concentration on the hemagglutination activity of human breast tumor homogenate lectin. All details are explained in text.

5.6.6.8.2 Effect of divalent salts on hemagglutination activity:

Fifty µL of human cancerous breast homogenate was incubated with 0.5 ml of erythrocyte suspension at 25°C for 120 minutes in presence of 10 mM CaCl2 and different concentrations from (0.005 M to 0.02 M) of MgCl$_2$ (dissolved in the assay buffer pH 8.5). The final reaction volume was 1 ml, then the cells were pelleted by centrifugation for 3 minutes. After that, the step 2 of the experiment (5.6.2) was repeated.

Calculations:

1. The same equation mentioned in experiment (5.6.2) was used to calculate the percent of total binding.

2. The percent of specific binding (SB%) was calculated by using the equation mentioned in section (5.6.4).

3. The percentages of specific binding (SB%) were plotted against their corresponding MgCl$_2$ concentrations, as shown in (Figure 5.9 and table 5-3):

367

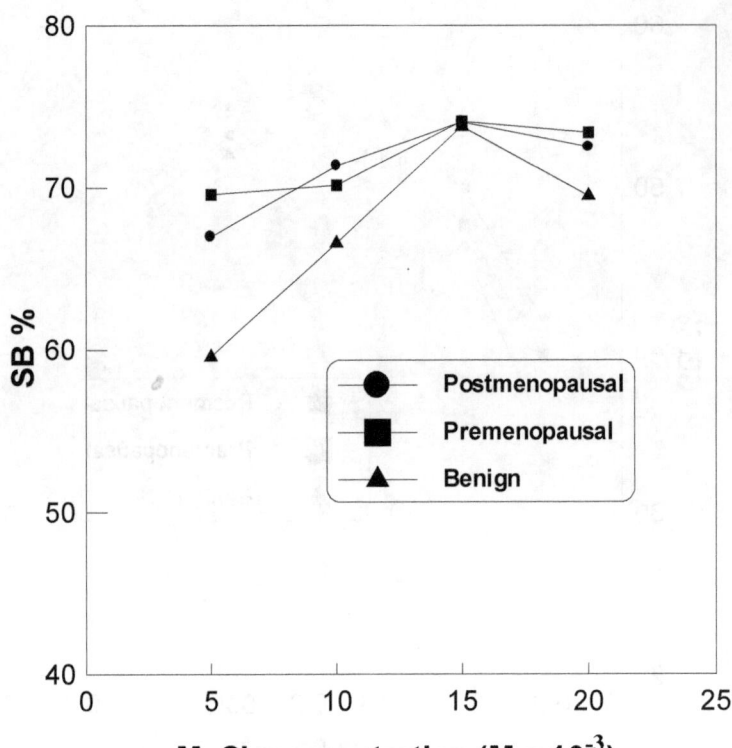

MgCl$_2$ concentration (M x 10^{-3})

Figure (5.9): Effect of divalent salt concentration on the hemagglutination activity of human breast tumor homogenate lectin. All details are explained in text.

Table (5-3): Effect of mono and divalent salts on the hemagglutination activity of breast homogenate lectin. All details are explained in the text.

Groups	Type of salt test	Salt conc. (M)	Specific binding %
Premenopausal	NaCl	0.05 0.1 0.15 0.2 0.25 0.3	36.82 47.78 52.05 56.17 53.3 49.39
Postmenopausal	=	0.05 0.1 0.15 0.2 0.25 0.3	41.93 49.95 54.01 56.81 55.03 50.56
Benign	=	0.05 0.1 0.15 0.2 0.25 0.3	35.05 46.5 51.52 54.96 52.13 46.19
Premenopausal	MgCl$_2$	5 x10^{-3} 10 x10^{-3} 15 x10^{-3} 20 x10^{-3}	69.58 70.16 74.09 73.4
Postmenopausal	=	5 x10^{-3} 10 x10^{-3} 15 x10^{-3} 20 x10^{-3}	66.98 71.35 74.06 72.52
Benign	=	5 x10^{-3} 10 x10^{-3} 15 x10^{-3} 20 x10^{-3}	59.6 66.58 73.76 69.49

RESULTS AND DISCUSSION:

The homogenate of human tissue of breast cancer was characterized by significant hemagglutination activity toward human red blood cells (group A). Homogenization of breast tumor tissues was carried out in cold medium to protect of protein from denaturation due to proteolytic enzyme activity [200]. However, addition of β-mercaptoethanol and EDTA to the extracting solution were necessary to prevent the inactivation of lectin and to achieve the maximal extraction of the lectin [204].

The total binding of lectin to glycoconjugates was estimated according to the hemagglutination assay [202,203]. Hemagglutination assay is a semiquantitaive procedure and has been widely used as a laboratory test because of its ease and versatility. It depends on aggregating and sedimentation of the erythrocytes after reaction with the bivalent or multivalent lectin [205].

However, the non-specific binding was determined by using neuraminidase to be incubated with the erythrocyte suspension before the assay, this enzyme is responsible for the release of terminal sialic acid residue from the erythrocyte surface glycoconjugates, and hence, the penultimate N-acetylgalactosamine will be exposed for the lectin binding.

Until now relatively few sialic acid specific lectins [206] have been identified, among which, only one is commercially a available. Most of them show a broad specificity. Basu et al [207] have reported a novel sialic acid binding lectin Achatinin$_H$ from the hemolymph of <u>Achatina Fulica</u> snail. The types [208,209] of sialic acid found on the erythrocytes, a striking correlation has been found between the ability of the agglutinin to agglutinate cells and the presence of O-acetylneuraminic acid residue on the mammalian cell surface [210]. Indeed the lectin agglutinated only those erythrocytes which contain 9-O-acetylneuraminic acid residues. Furthermore, human erythrocytes which contain only nueraminic acid [211], but no 9-O-acetylneuraminic acid were not agglutinable even by enzyme treatment [209]. Similarly horse erythrocyte known to contain high amount of 4-O-acetylneuraminic acid [210] were not affected.

From these results it was believed that this active agglutination may be due to the O or N-acetyl group which is present on the structure of sialoglycoproteins present on outer surface of the red cells, figure (5.1) shows a rapid steep decline in absorbance with an inflection point appeared in the curve was due to the lectin concentration used [212], found in 93 μg, which has a good agglutination activity.

<u>Optimum Conditions of Hemagglutination:</u>

Through figures (5.3, 5.4 and 5.5), the results show that the maximum hemagglutination activity of lectin required pH 8.5, the temperature at 25°C and the time for 120 minutes and it is clear that the lectin binding is dependent on pH and temperature. It should also be noted that the percent of specific lectin binding decreases with the change of pH (below 7 and upper 9.5), this suggests that the abundance of H$^+$ ions in the acidic medium may inhibit the binding sites on both glycoconjugate and lectin molecules, OH^- ions in more basic medium may influence in the same manner, and that the sialic acid which involved in binding is unstable to both acid and alkaline pH [213].

In general the results has led to the conclusion that the high binding process of lectin with glycoprotein is pH dependent and any shift in the pH of environment my affect the stability of the macromolecules involved in the binding. This effect includes the induction of protonation-deprotonation process occurring within the ionizable groups of the amino acids present in the binding groups of these macromolecules [214].

Effect of Ca⁺⁺ Ions Concentration on Hemagglutination Activity:

Figure (5.6) represents the effect of Ca⁺⁺ ions concentration on the binding activity of human breast homogenate lectin. The results in this figure show that the highest binding of lectin was found in the presence of 10 mM Ca^{+2} ions. However, the Ca^{+2} ions plays an important role in stabilization the complex formed between the lectin and the glycoprotein present on the red cell surface, also the stabilization is due to the conformational changes in the protein due to the binding of Ca^{+2} ions [215]. Dolichos [216], observed that the Ca^{+2} ion will stabilize the native structure of the protein itself.

Furthermore, there are many isolated lectins requires Ca^{+2} ions in their binding or their physiological roles [217], also, different Ca^{+2} ions dependent lectins have been purified from various sources and most of these possess multimeric structures and are capable of forming cross-linked complexes [218].

Effect of Denaturating Agents on the Hemagglutination Activity:

The data presented in table (5-1) shows the effect of different denaturating agents on the binding activity of breast lectin (Urea, PEG, NaOH and HCl). The results show that all of denaturating agents were effected on specific binding for lectin as compared to the control, but this effect was different between types of denaturating agents. An analysis of the data in table (5-1) shows that the percent of specific binding for lectin to glycoconjugates decreased with increasing urea concentrations, this effect can be attributed to the effect of urea on the hydrophobic forces between protein molecules.

Also, increasing concentrations of polyethylene glycol may results in precipitation of protein molecules which leads to decrease the interaction between lectin and glycoconjugates, and hence decrease the percent of specific binding.

Furthermore, the effect of 0.15 M of NaOH and HCl on the binding activity of breast lectin was investigated in this work, the results indicated that NaOH and HCl considerably reduce the percent of specific binding, their denaturating effect is due to great changes in pH of icubation medium.

Inhibition Studies of Hemagglutination Activity:

The inhibition percent of hemagglutination activity of human breast lectin by various carbohydrates (sialic acid, D-glucuronic acid, fructose, mannose and xylose) are summarized in table (5.2) and figure (5-7). The results in this table are classified according to their respective of group patients.

It is clear from this table, sialic acid and D-glucuronic acid were found to be the most potent inhibitors and gives high inhibition percent of hemagglutination at 3.5 mM and 20 mM respectively. However, the results obtained from this assay demonstrated that D-fructose, D-mannose and D-xylose tested at 30 mM in patients with breast cancer (premenopausal and postmenopausal) have the high activities to inhibit the binding of cancerous lectin, as compared to the benign patients. On the other hand, the inhibition of D-glucuronic acid suggests that it might be used as the eluting sugar in the purification of lectin.

Furthermore, 0.953 mM, 9.64 mM and 3.12 mM of sialic acid, sodium glucuronate and EDTA respectively were enough to produce 50% inhibition for lectin isolated from rat uterus [219,220].

Goebal et al [221], have suggested that the inhibition studies could be used as a technique for demonstrating and measuring the reactions occurred between lectin and its binding protein, for example, coupled disaccharides binds to protein via their p-aminophenyl glycosides and the terminal non-reducing hexose to play the predominate role in the specificity.

Effect of Mono and Divalent Salts on the Hemagglutination Activity:

The effect of NaCl and $MgCl_2$ as mono and divalent salts respectively on the binding of lectin to erythrocyte surface glycoconjugates were investigated. The results show that there is significant increases in specific binding percent in different types of breast cancer when using different $MgCl_2$ concentrations than those obtained in the presence of different NaCl concentrations, (table 5-3). However, an analysis of the Figures (5.8) and (5.9) shows that the highest binding of lectin was obtained in 0.2 M of NaCl and 0.015 M of $MgCl_2$. These data are in disagreement with the results reported by Nassir [222], who have found that there is no effect of such ions on the lectin binding in the presence of 15 mM Ca^{+2} ions in the range of concentriond (LBSA) in sera

Chapter six

Lectin which binds sialic acid residue of glycoproteins nature has been isolated and purified from human cancerous breast homogenate by a combination of adsorption to formalinized human red cells and gel filtration, elution with D-glucuroinc acid and subsequent fractionation on sephadex G_{150}. The recovery percent of lectin was determined and found in the range of 80%. The purification and identification of lectin was performed in the presence of Ca^{+2} ions, the elution volume (V_e) and then K_{av} value, for elution of breast homogenate which contain Ca^{+2} ions from sephadex G_{150}, was calculated and found to be 54 ml and 0.265 respectively.

The purified lectin appeared to be homogenous by electrophoresis in the presence of sodium dodecyl sulphate (SDS), was characterized by a single band whose molecular weight was (51,500) KDa.

However, the determination of physiochemical properties of human malignant breast lectin, such as stoke radius, purification folds and effect of Ca^{+2} ions were determined. On binding of glycoprotein to purified human cancerous breast lectin, heat stability and effect of EDTA were studied.

INTRODUCTION

There are several procedures available for the isolation and purification of lectin from human tissue [223,224], but most of these are conventional methods. However, other methods such as affinity chromatography are used for purification of lectin from mammalian tissues and organs [225]. Of known lectins which have been purified and characterized, few bind sialic acid [226]. Wheat germ agglutinin is the only plant lectin that binds to sialic acids, also it binds to oligosaccharides containing terminal N-acetylglucosamine and N-acetylgalactosamine [227].

Two lectins specific for sialic acids have been purified from animal sources, namely from the hemolymph of the horseshoe carbs of the class arachinda [228] and slug, Liman Flavus [229].

Recently two sialic acid-specific lectins have been purified from tritrichomonas mobilensis [230] and from the mushroom hericium exinaceum [231].

In general, this chapter was planned to develop a method for purification of lectin from breast cancer by using formalinized erythrocytes as affinity adsorbant accompanied by gel filtration.

MATERIALS AND METHODS

6.1 CHEMICALS:

All chemicals and reagents mentioned in the section (2.1) were used in the experiments of this chapter.

6.2 INSTRUMENTS:

All instruments that described in section (2.2) were used in the experiments of this chapter.

6.3 BUFFERS AND REAGENTS:

Buffers and reagents mentioned in section (5.3) were used in the experiments of this chapter, other solutions used were indicated in each experiment.

6.4 GEL PREPARATION AND COLUMN PACKING:

A. Preparation of the column: The dimension of the column were chosen according to the following equation [232]:

$$\text{diameter} = \sqrt[3]{m/10}\,(\text{cm}), \text{ where:}$$

m = amount of protein in mg.

$$\text{Length} = 30 \times \text{diameter}$$

B. Preparation of the gel: The gel (Sephadex G_{150}) was allowed to swell in excess of 0.02 M Tris-saline buffer pH 7.2 containing 0.01 M $CaCl_2$ and left to the stand for three days at room temperature without stirring, then the slurry was poured carefully into a vertical glass-column down the wall using a glass-rod. After the gel has settled, the column was equilibrated with Tris-saline buffer pH 7.2 for 24 hours with the dimension of (1.5 x 50 cm).

6.4.1 Void Volume (V_o) Determination:

The volume of the gel column was determined by using blue dextrin 2000 at concentration of 1mg/ml in Tris-saline buffer pH 7.2. One ml of blue dextrin solution was applied to the column surface carefully, then the elution was carried out with the same buffer using a flow rate of 8 ml/hour, fractions of three ml were then collected and their absorbance was measured at 600 nm to determine the void volume (V_o).

6.4.2 Column-Calibration:

The column was calibrated by the calibration Kits, purchased from Pharmacia Fine Chemicals using six standard proteins. Standard protein solutions were prepared according to manufactures instructions [233], then applied through two separate runs (2 ml portion). First run include the following proteins, ribonuclease A, ova albumin and aldolase, while the second run include the following proteins, ferritin, albumin and catalase. Elution was carried out by Tris-saline buffer pH 7.2 with flow rate of 8 ml/hour. The absorbance of the fractions collected were measured at 280 nm to determine the elution volume (V_e) of the standard proteins, as shown in Figure (6.1a).

Calculations:

1. The K_{av} values of the proteins eluted were determined by the following formula:

$$K_{av} = \frac{V_e - V_o}{V_t - V_o}$$

Where:

V_o = Void volume

V_e = Elution volume of each protein

V_t = Total gel-bed vloume: determined from the following equation:

$$V_t = \left(\frac{column \quad diametr}{2}\right)^2 \times \frac{22}{7} \times column \quad length$$

2. Tow calibrations curves were plotted:

a. K_{av} values against log. M.Wt. of the standard proteins, as shown in Fig. (6.1).

b. $\left(-\log.K_{av}\right)^{1/2}$ values against the stokes radius (R_s) of the standard proteins, as shown in Figure (6.2).

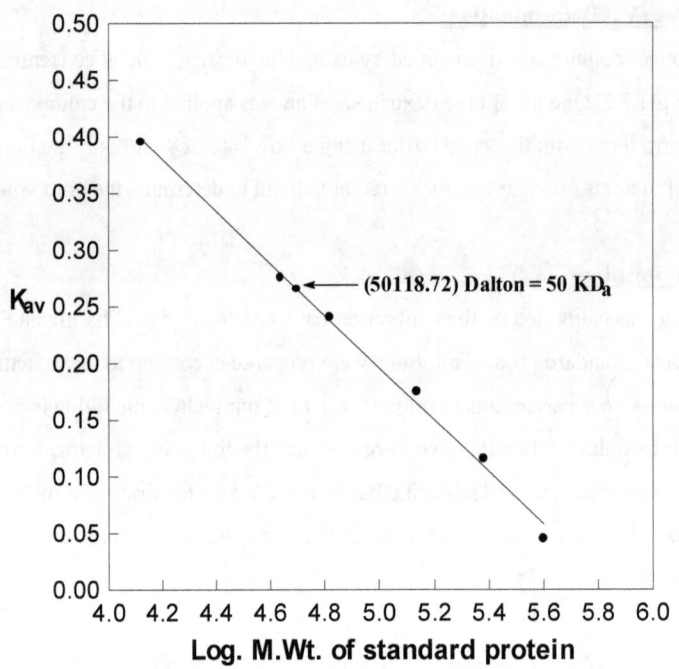

Figure (6.1): Calibration curve for estimation molecular weight of lectin by gel-filtration using standard proteins. All details are explained in the text.

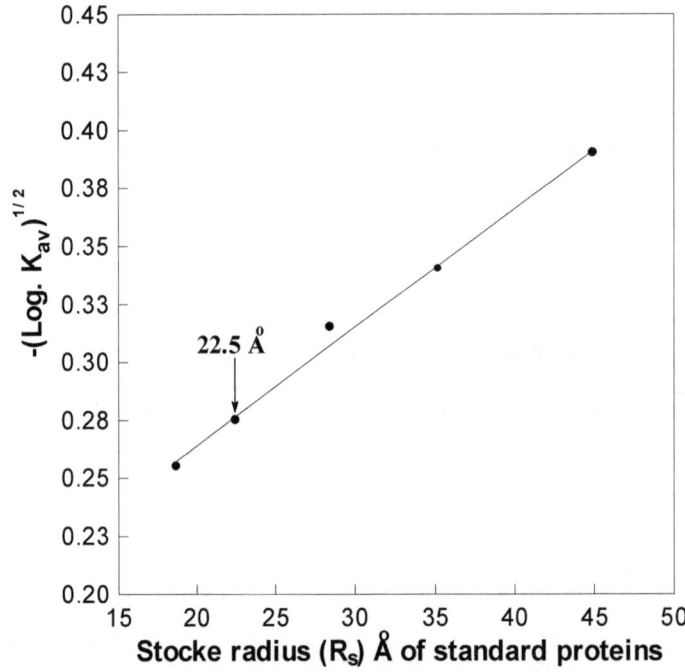

Figure (6.2): Calibration curve for estimation of Mol. Wt. and stocke radius (R_s) of cancerous breast lectin by gel-filtration technique using sephadex G-150 column (1.5x50cm). All details are explained in the text.

6.5 PURIFICATION AND IDENTIFICATION OF HUMAN CANCEROUS BREAST LECTIN USING FORMALINIZED ERYTHROCYTES AND GEL FILTRATION:

In each step of purification, the assay of binding was carried out and accompanied by protein

determination, the other calculations were also performed, as shown in table (6-1)

Table (6-1): Scheme of lectin purification form breast cancer. The hemagglutination activity of the lectin obtained by each step was determined as described under method.

Step	Volume of fraction (ml)	Protein (mg/ml)	Total amount of protein (mg)	Activity* Per mg protein	H.U** Total	purification	Recovery %
Crude cell free extract of breast cancer	10	18.6	186	11	2000	1	100
Elute from red cells	60	0.113	6.78	251	1700	23	85
Concentration elute	4	1.62	6.48	262	1700	24	85
Pooled peak from sephadex column	9	0.38	3.42	468	1600	43	80

Note: *   Hemagglutination activity was determined as in section (5.6.2).
** Hemagglutination activity unit was determined as Lis and Sharon method [208].

6.5.1 Preparation of Formalinized Erythrocytes as Affinity Matrix:

The method used for preparation of formalinized erythrocytes was essentially as described by Csizman [234] and Nowak et al [223].

Human erythrocytes (group A) were obtained from blood bank, washed four times in twenty volume of 75 mM NaCl, 75 mM Na_2HPO_4/75 mM K_2HPO_4 (phsphate-saline buffer, pH 7.2), per packed cell volume by centrifugation at 3000 r.p.m. for ten minutes. Twenty five ml of the packed red cells were resuspended to 200 ml in phosphate-saline buffer, pH 7.2 and placed in a 500 ml conical flask. Fifty ml of commercial formaline (40% formaldehyds) was introduced and the mixture was incubated at 37°C for 20 hours. The cells where then washed five times in five volumes phosphate-saline buffer, pH 7.2 per packed cell volume and kept at 4°C as 10% suspension in this buffer.

Step 1:

The stored formalinized cells prepared were used as an affinity reagent by washing the cells six times in ten volumes of 50 mM Tris-HCl, 100 mM NaCl (Tris-saline buffer). PH 7.2 containing 0.01 M $CaCl_2$. Ten ml of cancerous breast homogenate was added to thirty ml of cell suspension (10% suspension in buffer) for three hours at room temperature then washed three times with twenty volumes of Tris-saline buffer, pH 7.2 containing 0.01 M $CaCl_2$. Elution of the adsorbed lectin was accompanied by incubation the cells with 50 ml of 0.15 M D-glucuronic acid in Tris-saline buffer, containing 0.01 M $CaCl_2$ which has been brought to pH 7.2 with NaOH before the addition to the cells and placed in refrigerator for overnight. The elution mixture was then centrifuged for ten minutes at 3000 r.p.m., and the resultant supernatant was referred as fraction 2 whereas the breast homogenate being fraction 1.

Step 2:

Dialysis for concentration:

Dialysis tube of suitable length was prepared and used to concentrate fraction 2 to a volume of about 4 ml against solid sugar at 10°C for 24 hours. The resulted concentrated lectin was considered as fraction 3.

Step 3:

Gel Filtration:

Before applying the sample to a sephadex G_{150} column (1.5 x 50cm) the column had been equilibrated with Tris-saline buffer, pH 7.2, containing 0.01 M $CaCl_2$, after that the sample was transferred at the top surface of the column and then eluted with this buffer at an elution rate of 8 ml/hour. Fractions of three ml volume were collected then identified by the assay method as well as the absorbance at 280 nm and protein determination were carried out. The elution volume (V_e) of the lectin in each fraction was determined by the following formula:

V_e = Fraction volume (3 ml) x Fraction number containing the highest level of the lectin.

6.5.2 The Assay Methods:

In order to identify the fractions which contain lectin, the % of specific binding (S.B %) for each fraction was determined as follows:

a. Half ml of each fraction isolated by gel filtration was incubated with 0.5 ml of erythrocytes in a final volume of one ml at 25°C for 120 minutes, then determining absorbance at 620 nm of the upper layer (free lectin and cells) of the assay solution.

b. Parallel experiments was performed to determine the amount of non-specific binding for each fraction, as described in section (5.6.3).

Calculations:

The same equations mentioned in experiments (5.6.2, 5.6.3 and 5.6.4) were used to calculate the specific and non-specific binding of lectin.

6.5.3 Determination of Molecular Weight of Lectin Using K_{av} Values:

The same formula mentioned in section (6.4.2) were used to calculate the K_{av} value of lectin and then using Fig. (6.1) as calibration curve to determine molecular weight of lectin.

6.5.4 Determination of Lectin Stoke Radius:

$\left(-\log.K_{av}\right)^{1/2}$ values for standard proteins were determined and applied to stokes radius (R_s) in the standard curve (Fig. 6.2) to determine lectin stoke radius.

6.5.4 Determination of Purification Folds:

The purification fold for the lectin was determined using the following formula:

$$\text{Purification fold of lectin} = \frac{\text{Specific binding of purified lectin}}{\text{Specific binding of homogenate lectin}}$$

6.5.5 Determination of % Recovery of Lectin:

The % recovery was determined as follows:

$$\text{Recovery \%} = \frac{\text{Total unit of purified lectin}}{\text{Total unit of homogenate lectin}} \times 100$$

6.6 ANALYSIS OF THE PURIFIED FRACTION BY SDS POLYACRYLAMIDE GEL-ELECTROPHORESIS (SDS-PAGE):

Gel electrophoresis in the presence of sodium dodecyl sulphate (SDS) was performed according to the method of Laemmli [235]. Using 7.5% polyacrylamide separating gel. To determine the lectin molecular weight after comparison with molecular weights of standards proteins, the standard proteins and their molecular weights are detailed below:

Protein	M.Wt. (KD$_a$)
orylase b	94
	67
ɔumin	46
inhibitor	20
albumin	14

Procedure:

a. Polyacrylamide gel (concentration 7.5%) was prepared according to the application note 306 issued by LKB company [236].

b. Standard proteins solutions: Pharmacia electrophoresis calibration kit for the determination of native molecular weight of protein by polyacrylamide-gel electrophoresis was used. The content of each vial was redissolved in 0.3 ml of sample buffer (0.0625 M Tris-HCl, pH 6.8, containing 2% SDS, 10% glycerol, 5% 2-mercaptoethanol and 0.001% bromophenol blue as the dye).

c. Sample preparation: Purified lectin was concentrated by dialysis against sucrose, then diluted with sample buffer to protein concentration range (0.2-2 mg/ml).

The other steps were explained in the application note 306 of LKB company [236].

Calculations:

1. The relative mobility (R_m) of each protein was measured as follows:

$$R_m = \frac{\textbf{\textit{Distance of protein migration}}}{\textbf{\textit{Length after dying}}} \quad X \quad \frac{\textbf{Length before fixation}}{\textbf{\textit{Distance of dye migration}}}$$

379

2. Molecular weights of the marker standard proteins was plotted against relative mobilities (R_m), typically gives a straight line (Fig. 6.3) from which the lectin M.Wt. was estimated.

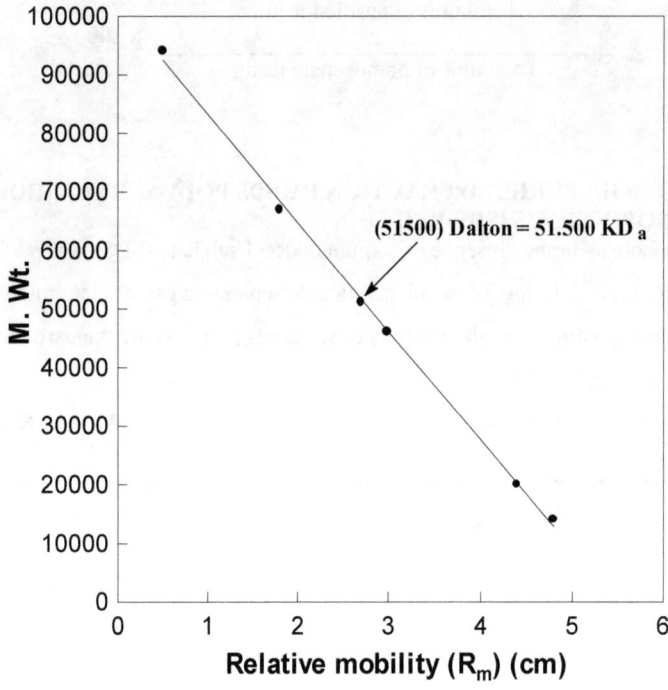

Figure (6.3): Molecular weight determination of purified breast lectin by sodium dodecyl sulphate-7.5% acrylamide gel analysis. All details are explained in the text.

6.7 THE EFFECT OF C^{a++} IONS ON BINDING OF GLYCOPROTEIN TO PURIFIED HUMAN CANCEROUS BREAST LECTIN:

a. Fifty microliters of purified cancerous breast lectin was incubated with 0.5 ml of red cell suspension at 25°C for 120 minutes. Different concentration of Ca^{++} ions (5, 10, 15, 20 and 25) $x10^{-3}$ M were dissolved in the assay buffer (pH = 8.5) and were added to the sample. The final reaction volume was one ml. Then the cells were pelleted by centrifugation for 3 min. The step two of the experiment (5.6.2) was repeated.

b. Parallel incubations were performed to determine the non-specific binding of the lectin, as described in section (5.6.3).

Calculations:

The same mathematical formula mentioned in experiment (5.6.4) was used to calculate the % of specific binding of purified lectin. The % specific binding was plotted against their corresponding Ca^{++} ions concentration, as shown in Fig. (6.4).

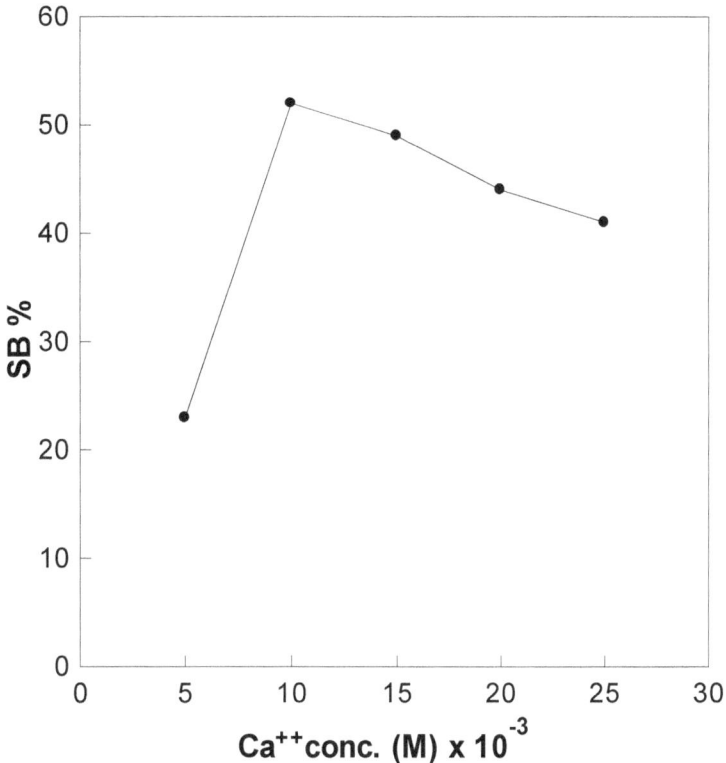

Figure (6.4): Effect of Ca⁺⁺ ions concentration on the hemagglutination activity of human purified cancerous breast lectin. All details are explained in the text.

6.8 THE EFFECT OF TEMPERATURE ON THE STABILITY OF THE PURIFIED CANCEROUS BREAST LECTIN:

The heat stability of lectin were carried out according to the following:

a. Fifty microliters of purified cancerous breast lectin was incubated with 0.5 ml of red cell suspension for 10 min. at different temperatures (15, 25, 37, 50, and 60°C), using the assay buffer (pH = 8.5). The final reaction volume was one ml, then cooled on ice-bath for 5 min. After that the cells were pelleted by centrifugation for 3 min. The step two of the experiment (5.6.2) was repeated.

b. Parallel incubations were performed to determine the non-specific binding of the lectin, as described in section (5.6.3).

Calculations:

The same mathematical formula mentioned in experiment (5.6.4) was used to calculate the % of specific binding of purified lectin. The % specific binding was plotted against their corresponding temperatures, as shown in Fig. (6.5).

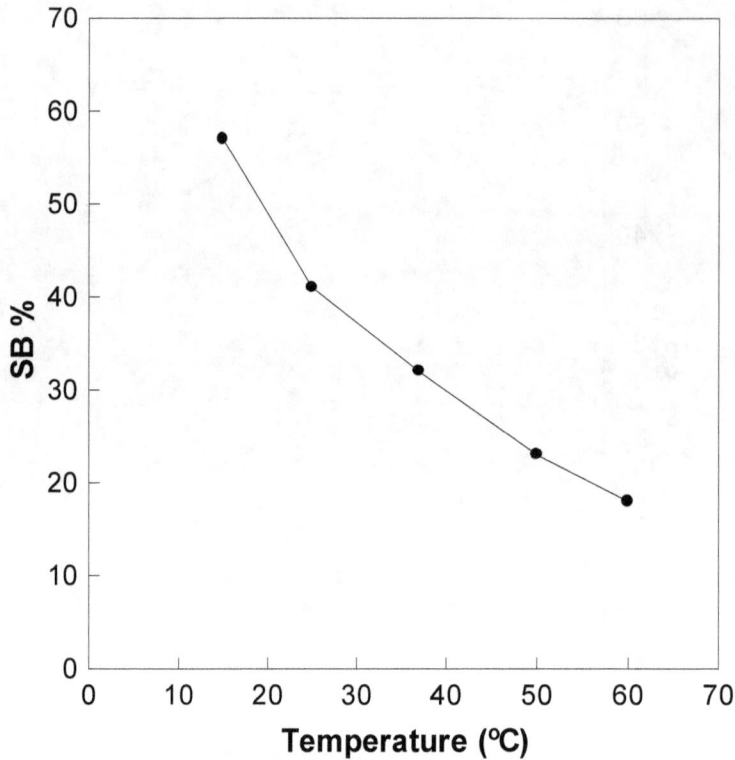

Figure (6.5): The effect of temperature on the stability of purified cancerous breast lectin. All details are explained in the text.

6.9 EFFECT OF EDTA ON HEMAGGLUTINATION ACTIVITY OF PURIFIED LECTIN:

a. Fifty microliters of purified human cancerous breast lectin which gave a high percent hemagglutination was incubated with 0.5 ml of red cell suspension at 25°C for 120 min. The final reaction volume was completed to one ml, by adding Tris-saline buffer pH 8.5 (Assay buffer) which contains the desired concentration of EDTA (0.1, 0.15, 0.2, 0.25, 0.3 and 0.35 mM). Then the red cells were pelleted by centrifugation for 3 min. The step two of the experiment (5.6.2) was repeated.

b. The same experiment was carried out in the absence of the EDTA.

<u>Calculations:</u>

The same mathematical formula mentioned in experiment (5.6.2) was used to calculate the % hemagglutination before and after addition of EDTA. The % inhibition of hemagglutination represents the difference between the % hemagglutination with lectin alone and that obtained with lectin plus EDTA. The data of % inhibition was shown in table (6.2).

RESULTS AND DISCUSSION

Purification and identification of lectin from human breast cancer tissue depends on the existence formalinized erythrocytes as affinity adsorbant accompanied by gel filtration , elution with D-glucuronic acid and subsequent fractionation on sephadex G_{150}.

The purification of lectin was performed in the presence of Ca^{+2} ions. Typical yields and purity data from the different steps in the purification of lectin from human breast cancer tissue are summarized in table (6-1). On the other hand, it has estimated the elution volume (V_e) and K_{av} value for elution of breast homogenate from sephadex G_{150} and found to be 54 ml and 0.265 respectively, whereas this K_{av} value of the lectin was obtained from the calibration curves in Figures (6.1 and 6.2), to determine molecular weight and stoke radius for lectin. These parameters were estimated and found to be 50118.72 Dalton and 22.5 Å respectively. Furthermore, other investigators have purified lectin form different human tissues using the same method and their results revealed that the M.wt. and stoke radius (R_s) of the purified lectin were (60000 dalton, 34 Å) and (40180 Dalton, 28.8 Å) respectively [222,237].

Analysis of Purified Fractions by SDS-PAGE:

Molecular weight of subunit for the purified human breast lectin was determined by sodium dodecyl sulphate-7.5% acrylamide gel electrophoresis [235], figure (6.3) represents the calibration curve for SDS-PAGE (7.5%), using low molecular weight groups of known subuints proteins as markers (14-94 KDa). From this curve, application of the relative mobility (R_m) values for each standard protein, molecular weight of purified lectin was estimated and found to be 51500 Dalton. However, and from the results obtained, there is only one lectin in breast homogenate specific for sialic acid.

On the other hand, the results were confirmed by Nassir [222] and Huda [237] where they found one lectin in human kidney and thyroid gland respectively, using the same method. Also, their results revealed that the molecular weight of purified lectin was 64000 and 44157.05 Dalton respectively.

Moreover, other investigators [238], were able to purify human kidney lectin with molecular weight of 63000 Dalton and whom using gel chromatogarphy and then adsorption with immobilized glycoconjugates and ion-exchange gel.

Stability of Cancerous Breast Lectin:

Figure (6.5) illustrates the effect of temperature on the stability of purified cancerous breast lectin, it is clear form this figure that there was only slight decrease in stability on increasing the temperature up to 60°C, whereas, above this temperature the stability of purified lectin may decrease due to the denaturation process of lectin.

Effect of Ca^{+2} Ions and EDTA on Hemagglutination Activity of Purified Lectin:

Figure (6.4) represents the effect of Ca+2 ions concentration on the hemagglutination activity of human purified cancerous breast lectin.

The results indicated that the breast lectin requires Ca^{+2} ions for binding activity. It was found that Ca^{+2} ions had a marked effect on the hemagglutination activity of the lectin [218], and this effect was proved by loosing the hemagglutination activity, when the chelating agent EDTA was added which may remove the available Ca^{+2} ions, this effect is illustrated in Table (6.2).

Table (6-2): Inhibition of hemagglutination activity of purified human cancerous breast lectin by EDTA.

EDTA conc. (mM)	Inhibition %	Conc. Of EDTA that gives 50 % inhibition (mM)
0.10	22.18	
0.15	27.91	
0.20	44.37	0.3
0.25	46.08	
0.30	53.43	
0.35	62.87	

frOer for diagnosis of the d

The present investigation is carried out to describe the kinetics and thermodynamic properties associated with the binding of lectin to glycoprotein, and then determination of their parameters.

The results show that the association constant (K_a) and the specific binding (B) were increased as temperature increased. However, the time course data for the binding follows, the pseudo-first order kinetics accompanied with increase in K_{obs}. Whereas, the Hill plot data revealed that there was a weak positive cooperativity between lectin binding sites.

Also, the Van't Hoff plot demonstrated a linear relationship between $\ln K_a$ and $\dfrac{1}{T}$.

On the other hand, the parameters for the equilibrium reaction described by ΔG°, ΔH° and ΔS° were determined. Furthermore, the Arrhenius plot indicates that there was a linear relationship between $\ln k_{+1}$ and $\dfrac{1}{T}$ from which the transition state thermodynamic parameters for the formation of the lectin-glycoprotein complex represented by; E_a, ΔG^*, ΔH^* and ΔS^* were determined.

Chapter Seven

Lectins are carbohydrate-binding proteins that are grouped together. They agglutinate cells or other materials that display more than one saccharide of sufficient complementarity. They are found in many categories of living things[239]. The major role of lectins are the ability to bind saccharides specifically, and interact with RBC of certain blood group, also glycoproteins and glycolipids [123]. However, binding of lectins at the cell surface may cause other changes in cell function than mitogenic stimulation or the release of stimulating factors [240].

In this investigation it has attempted to explain the mechanism of binding lectin to glycoprotein to form lectin-glycoprotein complex, and then to determine the kinetics and thermodynamic parameters, and describe the molecular basis of lectin interaction, through the effect of time-course, temperature and the other factors.

MATERIALS AND METHODS

7.1 CHEMICALS:

All chemicals and reagents mentioned in the section (2.1) were used in the experiments of this chapter.

7.2 INSTRUMENTS:

All instruments that described in section (2.2) were used in the experiments of this chapter.

7.3 KINETIC STUDIES:

7.3.1 The Time-Course of cancerous breast lectin binding to glycoprotein present on red cell surface:

Reagents:

The standard erythrocyte suspension and the assay buffer (Tris-HCl pH 8.5) containing 10 mM $CaCl_2$ and 0.2 M of NaCl were prepared as described in sections (5.6.1 and 5.3).

Procedure:

1. At zero time, 0.5 ml of erythrocyte suspension was added to 50 μL of purified lectin (930 μg protein), the final volume of the assay mixture was made up to 1 ml with Tris-HCl buffer pH 8.5. The reaction mixture was incubated at 25°C for several time intervals (15, 30, 60, 90, 120 and 150 minutes).

2. **After each time interval the assay tubes were treated according to steps mentioned in section (5.6.2).**

3. Parallel experiments were carried out according to section (5.6.3) to determine the amount of non-specific binding.

4. To determine the time-course of lectin binding to glycoprotein present on red cell surface at different temperatures, the above steps were performed at (4, 11, 18 and 25°C).

Calculations:

1. The molar concentration of lectin involved in total binding to erythrocyte surface glycoconjugates, was calculated according to the following formula:

$$\text{The concentration of lectin (M) involved in total binding} = \frac{A - A^*}{A} \times \text{The total concentration of lectin (M) used in the assay}$$

Where:

A: The absorbance of standard erythrocyte suspension at 620 nm.

A*: The absorbance of unbound (free) erythrocytes at 620 nm.

2. The concentration of lectin in molar involved in non-specific binding to erythrocyte surface glycoconjugates, was calculated according to the following formula:

$$\text{The concentration of lectin (M) involved in non-specific binding} = \frac{A' - A^*}{A'} \times \text{Total lectin concentration (M) used in the assay}$$

Where:

A': The absorbance of neuraminidase-treated erythrocytes suspension at 620 nm.

A*: The absorbance of unbound (free) erythrocytes at 620 nm.

3. The concentration of specifically bound lectin in molar was calculated by subtracting the concentration of lectin involved in non-specific binding from the concentration of lectin involved in total binding:

Concentration of specifically bound lectin (M)	=	Concentration of lectin (M) involved in total biding	-	Concentration of lectin (M) involved in non-specific binding

4. The concentrations of specifically bound lectin (lectin-glycoconjugate) complex in molar were plotted against their corresponding incubation times (Fig. 7.1).

7.3.2 Scatchard Analysis:

Reagents:

The tris-HCl buffer pH 8.5 and the standard erythrocyte suspension were prepared as previously described in sections (5.3 and 5.6.1).

Procedure:

1. Half ml of erythrocyte suspension was added to increasing amounts of lectin $(0.37-3.7 \times 10^{-7}$ M), the final volumes were made up to 1 ml with tris-HCl buffer (pH 8.5, contains 0.2 M NaCl and 10 mM $CaCl_2$).
2. The assay tubes were incubated for 120 minutes at 25°C, then they treated as mentioned in steps of section (5.6.2).
3. The previous steps were repeated at different temperatures (4, 11 and 18°C).

Calculations:

1. The concentration of specifically bound lectin (molar) was calculated for each tube according to the calculations of section (7.3.1).
2. The concentration of free (unbound or unreacted) lectin was calculated by subtracting the concentration of lectin (M) involved in total binding from the total concentration of lectin (M) used in each experiments:

The concentration of free lectin (M)	=	Total concentration of lectin (M)	-	The concentration of lectin (M) gives total binding

3. The concentration of lectin binding sites (B_{max}) and the affinity constant (K_a) were determined according to Scatchard equation [241].

$$\frac{B}{F} = \frac{1}{K_d} \times \left(B_{max} - B\right)$$

$$K_a = \frac{1}{K_d}$$

Where:

B: The concentration of specifically bound lectin.

F: The concentration of free lectin.

K_a: The affinity constant.

B_{max}: the maximal binding capacity.

K_d: The dissociation constant.

4. The plot of B/F values against the values of B, gives liner relationship. The total concentration of lectin binding sites (B_{max}) was calculated from the intercept on the x-axis, while the value of affinity constant was calculated from the slope of the straight line.

5. The K_a and B_{max} values were also determined from the Eadie-Hofstee plot of data getting from Scatchard plots, using the following equation:

$$B = -K_d \frac{B}{F} + B_{max}$$

The values of K_a and B_{max} were calculated from the slope of the straight line and the intercept on Y-axis respectively.

7.3.3 Determination of Hill-Coefficient (n) of Lectin binding to Glycoconjugates:

Calculations:

1. All data were obtained from the experiment mentioned in section (7.3.2).

2. The value of Hill-coefficient (n) was calculated according to Hill equation;

$$\log\left(\frac{B}{B_{max} - B}\right) = n \log F - \log K_d$$

3. The values of log (B/B_{max}-B) were plotted against the values of logF, the Hill coefficient (n) was calculated from the slope of the straight line.

7.4 THE THERMODYNAMIC STUDIES:

Procedure:

The same steps mentioned in section (7.3.1) and section (7.3.2) were performed.

Calculations:

1. The thermodynamic parameters of standard state were obtained from Van't Hoff plot, the values of the natural logarithm of affinity constant (K_a) obtained at different temperatures were plotted against the reciprocal values of the absolute temperature in Kelvin ($1/T$), according to the following equation:

$$\ln K_a = \frac{\Delta S^\circ}{R} - \frac{\Delta H^\circ}{RT}$$

Where:

ΔH° : The enthalpy change of the standard state.

ΔS° : The entropy change of the standard state.

R: The gas constant ($8.31441 \ J.K^{-1}.mol^{-1}$).

ΔH° value was obtained from the slope of the linear relationship of the plot.

The change in Gibbs free energy of the standard state (ΔG°) was obtained from the following equation:

$$\Delta G^\circ = -RT \ln K_a$$

While the entropy change of the standard state ΔS° was obtained from:

$$\Delta S^\circ = \frac{\Delta H^\circ - \Delta G^\circ}{T}$$

2. The thermodynamic parameters of the transition state were obtained from Arrhenius plot of $\ln k_{+1}$ values against $1/T$ values, that gives a linear relationship according to the following equation:

$$\ln K_{+1} = \ln A - \left(\frac{E_a}{RT}\right)$$

Where:

A: Arrhenius constant.

The value of apparent energy of activation (E_a) of the binding reaction can be determined from the slope of the straight line. The enthalpy of the transition state ΔH^* was obtained from:

$$\Delta H^* = E_a - RT$$

The free energy change of the transition state ΔG^* is calculated from the following equation:

$$\Delta G^* = -RT \ln K_{+1} + RT \ln\left(\frac{KT}{h}\right)$$

Where:

K: is Boltzmann constant ($1.38 \times 10^{-23} \ Jdeg^{-1}$)

h: is Plank constant ($0.622 \times 10^{-33} \ Js^{-1}$).

The change in entropy of the transition state ΔS^* is calculated from the following formula:

$$\Delta S^* = \frac{\Delta H^* - \Delta G^*}{T}$$

$$\Delta S^* = \frac{\Delta H^* - \Delta G^*}{}$$

7.3 Kinetics of lectin binding to erythrocyte surface glycoconjugate:

7.3.1 The Time-Course of Cancerous Breast Lectin Binding to Erythrocyte Surface Glycoconjugates:

Figure (7.1) shows the time course of the formation of lectin binding to erythrocyte surface to glycoconjugate complex at four different temperatures (4, 11, 18 and 25°C), in breast homogenate sample.

The concentration of lectin-glycoconjugate complex that formed after time (t) was calculated from the following equation:

The concentration of lectin-glycoconjugate complex formed after time (t) in molar	=	The concentration of lectin involved in total binding (M)	-	The concentration of lectin involved in non-specific binding (M)

The results of time-course pattern at different temperatures indicated that the lectin binding to erythrocyte surface glycoconjugate is a temperature and time dependent process, since a maximum binding can be obtained at 25°C after incubation for 120 minutes, there is no analogous studies are available to compare our results.

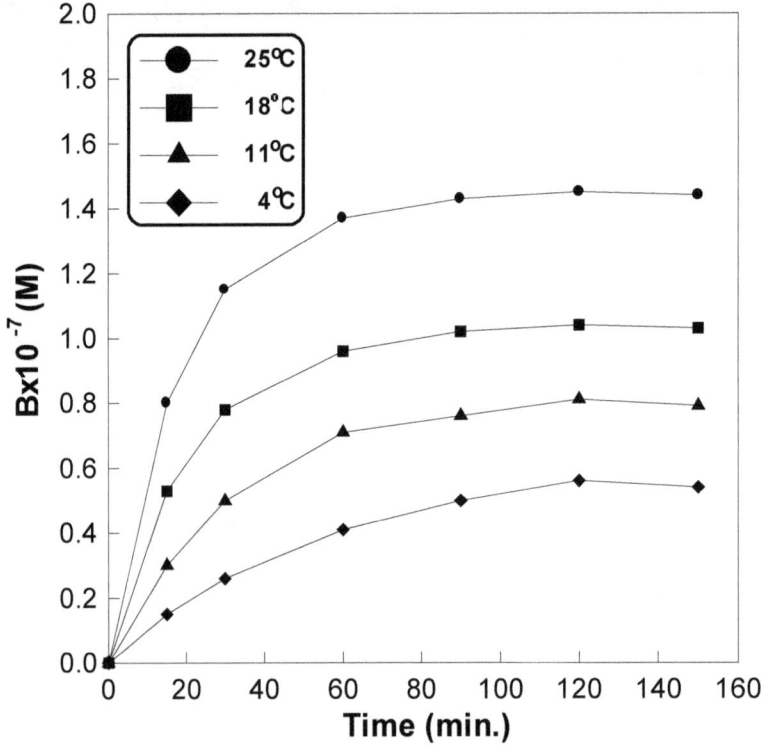

Figure (7.1): Time-course of lectin binding to erythrocyte surface to glycoconjugates at different four temperatures. All details are explained in the text.

Determination of Kinetic Parameters of Lectin-Glycoconjugate Complex Formation:

The time-course of lectin-glycoconjugate complex formation was carried out to describe the kinetic parameters of the binding (expressed as specific binding). The simplest proposed model representing this interaction was:

$$\text{Lectin + glycoconjugate} \underset{K_{-1}}{\overset{K_{+1}}{\rightleftharpoons}} \text{[Lectin-glucoconjugate]} \tag{1}$$

Where:

K_{+1}: is the association rate of lectin to glycoconjugate.

K_{-1}: is the dissociation rate of lectin-glycoconjugate complex.

At equilibrium:

K_a= [Lectin-glycoconjugate]/[Lectin][glycoconjugate] $\qquad\qquad$ (2)

K_d= [Lectin][glycoconjugate]/[Lectin-glycoconjugate] $\qquad\qquad$ (3)

Thus;

$$K_a = \frac{1}{K_d} = \frac{K_{+1}}{K_{-1}} \tag{4}$$

Where:

K_a: is the equilibrium constant (affinity constant).

K_d: is the equilibrium constant of dissociation of the complex.

The values of K_a and total concentration of lectin binding sites (B_{max}) were calculated from Scatchard and Eadie-Hofstee plots (Figures 7.2 and 7.3) respectively at four different temperatures (4, 11, 18 and 25°).

It is clear from table (7-1), the results show that the affinity constant (K_a) is depended on temperature (K_a increased from 0.847×10^7 M^{-1} at 4°C to 0.926×10^7 M^{-1} at 25°C). Whereas the value of dissociation constant (K_d) was calculated by using equation (4), and show that the lowest K_d value of lectin-glycoconjugate complex occurs at 25°C after incubation for 120 minutes.

Table (7-1): The kinetic parameters of lectin binding to erythrocyte surface glycoconjugates. All details are explained in the text.

	$K_d \times 10^{-7}$ (M)	$K_a \times 10^7$ (M^{-1})	$B_{max} \times 10^{-7}$ (M)
4	1.18	0.847	0.97
11	1.14	0.877	1.05
18	1.11	0.901	1.32
25	1.08	0.926	1.47

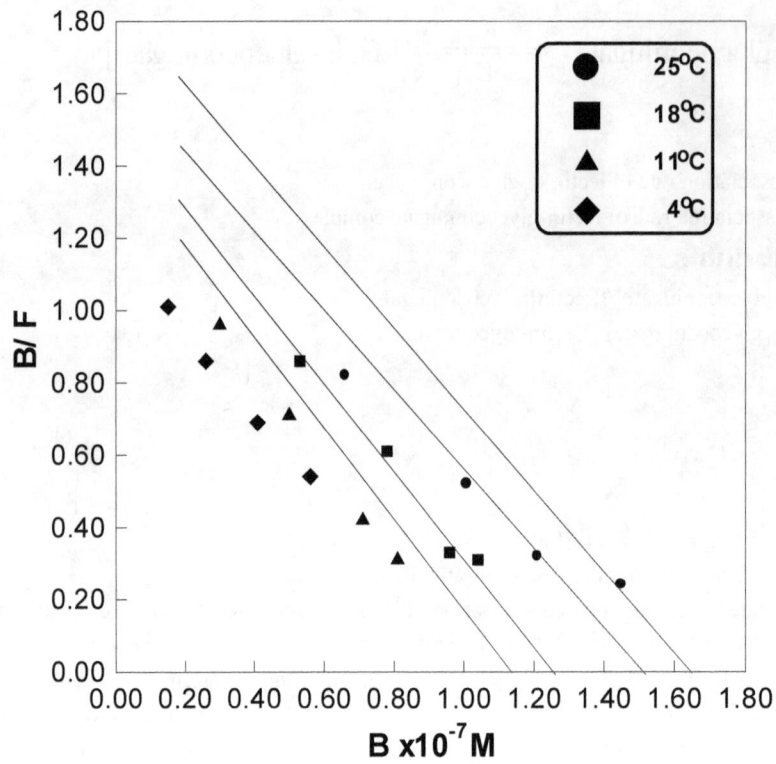

Figure (7.2): Scatchard plot of lectin binding to erythrocyte surface glycoconjugate at different four temperatures. All details are explained in the text.

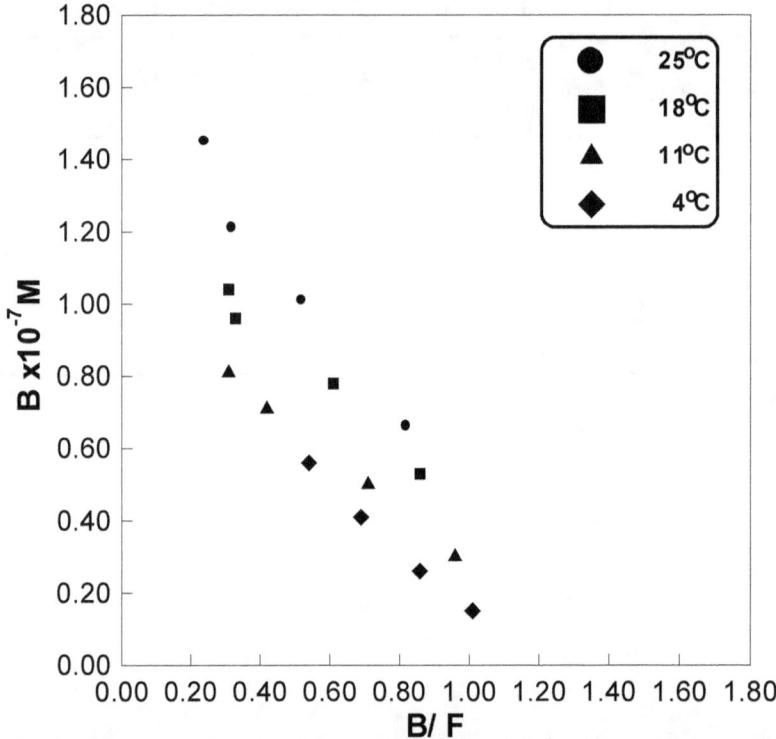

Figure (7.3): Eadie-Hofstee plot of data from Scatchard plot of lectin binding to erythrocyte surface glycoconjugate at different four temperatures. All details are explained in the text.

However, the time-course data shown in Figure (7.1) could be used to determine the reaction order of lectin binding to erythrocyte surface glycoconjugates using the following equation:

394

$$Ln[lectin\text{-}G]_e \left[\frac{(lectin)_t - (lectin\text{-}G)_t \; (lectin\text{-}G)_e / (G)_t}{(lectin)_t \; [(lectin\text{-}G)_e - (lectin\text{-}G)_t]} \right] =$$

$$K_{+1}t \left[\frac{(lectin)_t \, (G)_t}{(lectin\text{-}G)_e} - (lectin\text{-}G)_e \right] \quad (5)$$

Where:

K_{+1}: is the kinetic association constant in M^{-1} min.$^{-1}$.

(Lectin-G)$_e$: is the concentration of the complex formed at equilibrium.

(Lectin-G)$_t$: is the concentration of the complex formed after time (t).

The equation (5) represents the second order kinetics, but in this work the percent of specific binding was in some cases, small and most of the lectin remains free and only small fraction binds even at equilibrium, i.e, (lectin)$_t$ >> (lectin-G)$_e$ thus, (lectin)$_t$ >>

and $\dfrac{(lectin)_t \, (G)_t}{(lectin\text{-}G)_e} >> (lectin\text{-}G)_e$ $\dfrac{(lectin\text{-}G)_t \, (lectin\text{-}G)_e}{(G)_t}$

So that the following equation could be used in order to fit the pseudo-first order kinetics:

$$Ln \frac{(lectin\text{-}G)_e}{(lectin\text{-}G)_e - (lectin\text{-}G)_t} = K \frac{(lectin)_t \, (G)_t}{(lectin\text{-}G)_e} \quad (6)$$

On the other hand, Figure (7.4) shows the plot of $\ln \dfrac{(lectin - G)_e}{(lectin - G)_e - (lectin - G)_t}$ against

time (t) gives a straight line with a slope equal to the observed value of first rate constant (K_{obs}) in min^{-1}. The rate constant (K_{+1}) in M^{-1} min^{-1} was calculated at four different temperatures by using the following formula:

$$K_{obs} = K_{+1} \frac{[lectin]_t [G]_t}{[lectin - G]_e}$$

$$\therefore K_{obs} = K_{+1} [lectin]_t \quad (7)$$

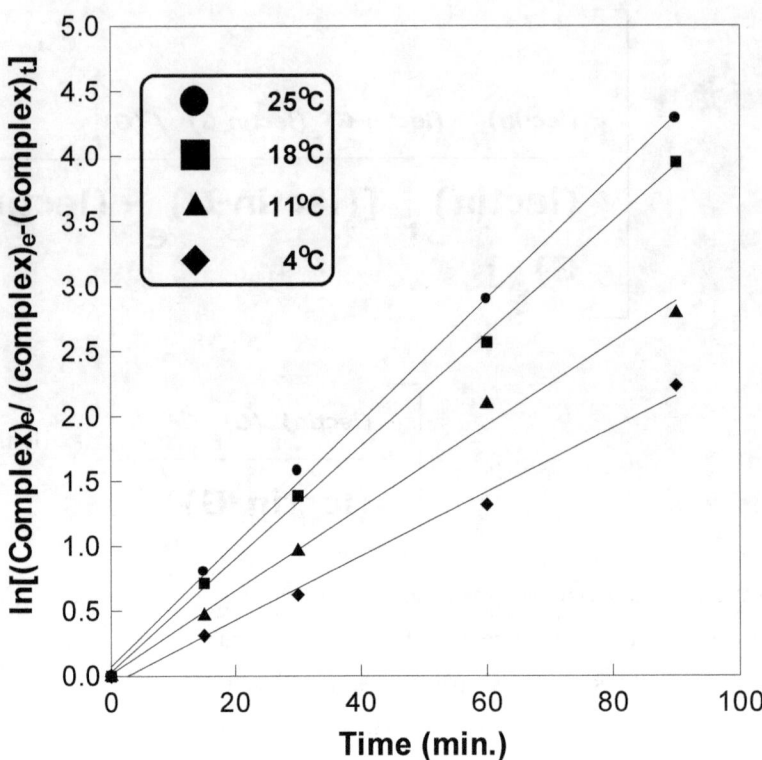

Time (min.)

e (7.4): Kinetics of lectin binding to erythrocyte surface to glycoconjugate at different four temperatures. All details are explained in the text.

Also, the value of K_{-1} at four temperatures were calculated by using equation (4). Whereas, the half life time of association $(t\frac{1}{2})_{ass.}$, which represents the time needed for the formation of half amount of the complex at equilibrium, was determined from the concentration of the complex at equilibrium and the time-course curve. Also, the half life time of dissociation $(t\frac{1}{2})_{diss.}$, was calculated from the following relation:

$$(t\frac{1}{2})_{diss.} = \ln\frac{2}{k_{-1}} = \frac{0.693}{k_{-1}} \qquad (8)$$

The values of $K_{obs.}$, $K_{+1.}$, $K_{-1.}$, $(t\frac{1}{2})_{ass.}$, and $(t\frac{1}{2})_{diss.}$ at four different temperatures are summarized in table (7-2). Data analysis of this table show that the highest rate for the association reaction occurs at 25°C, while the lowest rate occurs at 4°C, where the reaction temperature was increased from 4°C to 25°C, the value of K_{+1} increased from $(0.676\text{x}10^5\ M^{-1}.min^{-1})$ to $(0.973\ \text{x}10^5\ M^{-1}.min^{-1})$ which means the dependence of reaction rate on temperature. Also, the rate of dissociation of lectin-glycoconjugate complex (K_{-1}) is temperature dependent.

Table (7-2): The effect of temperature on the kinetic parameters of lectin binding to erythrocyte surface glycoconjugates. All details are explained in the text.

T° C	$K_{obs.}$ (min^{-1})	K_{+1} (M^{-1}. min^{-1})x10^5	K_{-1} (min^{-1})x10^{-3}	$(t\frac{1}{2})_{ass.}$ (min)	$(t\frac{1}{2})_{diss.}$ (min)

4	0.025	0.676	7.98	30	86.84
11	0.031	0.838	9.56	26	72.49
18	0.033	0.892	9.90	15	70
25	0.036	0.973	10.51	12	65.94

7.3.2 Scatchard Analysis:

Figure (7.2) shows Scatchard plot of lectin binding to erythrocyte surface to glycoconjugate in the presence of 10 mM Ca^{++} ions at different temperatures (4, 11, 18 and 25°C) after incubation for 120 minutes. This figure could be used to determine the kinetic parameters of lectin binding such as, the equilibrium constant of dissociation of the complex (K_d) and total concentration of lectin binding sites (B_{max}) of human cancerous breast lectin by using the following equation:

$$\frac{B}{F} = \frac{1}{K_d} \times \left(B_{max} - B\right)$$

The values of these parameters at different temperatures in presence of Ca^{++} ions are summarized in table (7-1). Through analysis the results in this table show that the total concentration of lectin binding sites (B_{max}) is temperature dependent, when the temperature was increased from 4°C to 25°C, B_{max} was increased from (0.97×10^{-7} M to 1.47×10^{-7} M), this fact could be explained according to the number of molecules possessing the activation-energy for interaction, increase with increase the temperature. On the other hand, the affinity constant (K_a) is also depended on temperature, this indicates that the reaction is slightly endothermic and explained by the fact that affinities of endothermic reactions enhanced by increasing temperatures. However, the values of B_{max} and K_d for human cancerous breast lectin at different temperatures obtained from Scatchard analysis were similar to those obtained from the Eadie-Hofstee plot (Fig. 7.3) [242].

7.3.3 Determination of Hill-Coefficient (n) of Lectin Binding to Glycoconjugates:

Figure (7-5) represents the Hill plot of lectin binding to erythrocyte surface glycoconjugate in the presence of 10 mM Ca^{++} ions at four different temperatures (4, 11, 18 and 25°C), The value of Hill-coefficient (n) equals the slope of the resulting straight line. The values were (1.54, 1.63, 1.71, 1.8) respectively. However, on application of the Hill equation [243,244] and using the results obtained from Scatchard analysis, it would be possible to evaluate the cooperativity of lectin binding sites through the determination of Hill coefficient (n). Furthermore, the results obtained in this work indicates that the cooperativity of lectin binding sites was low and affected by temperatures.

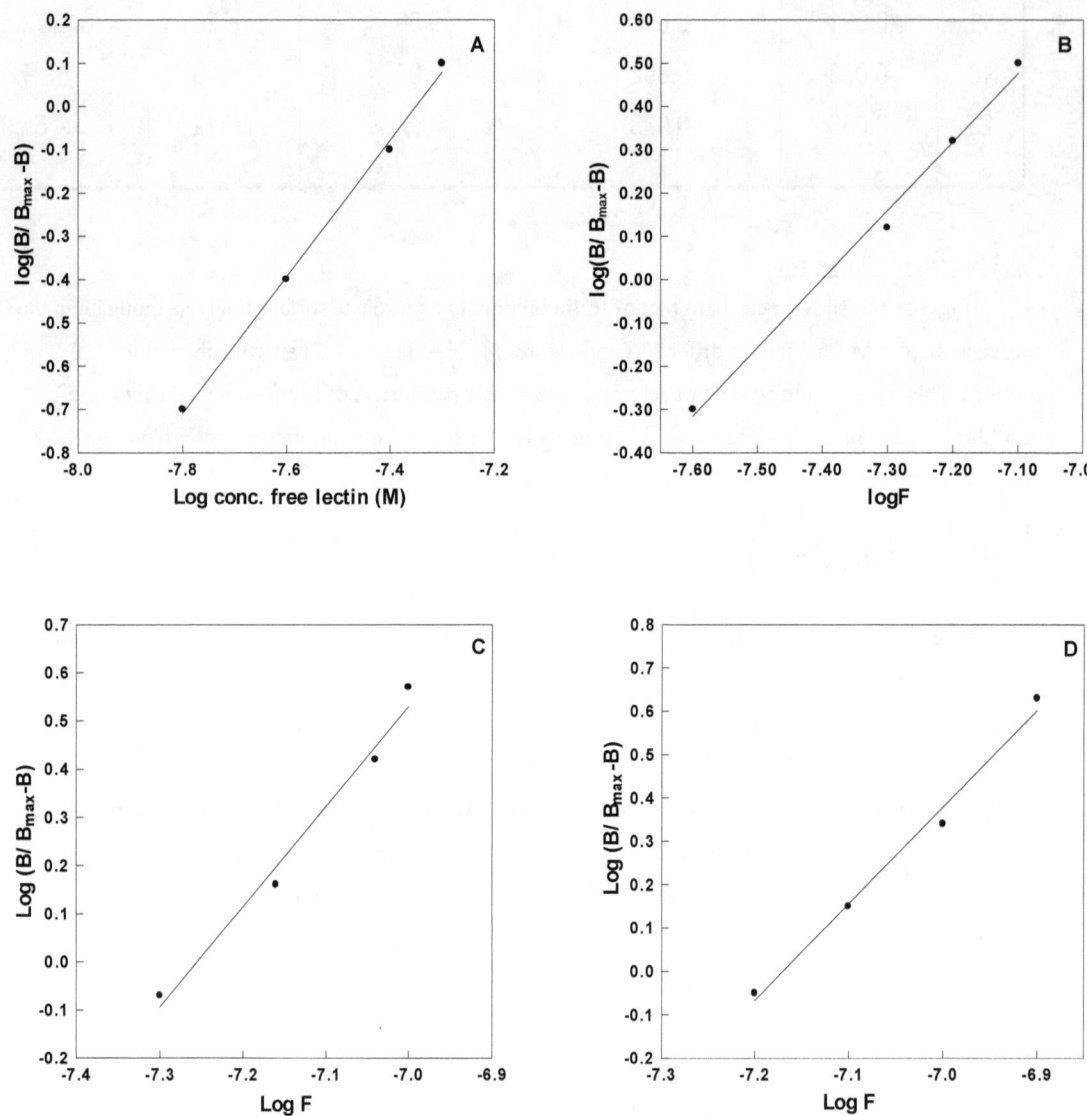

(7-5): Hill plots of lectin binding to erythrocyte surface glycoconjugates at different temperatures, -A-4°C, -B-11°C, -C-18°C, -D-25°C. All details are explained in the text.

7.4 THE THERMODYNAMICS OF THE LECTIN BINDING TO ERYTHROCYTE SURFACE GLYCOCONJUGATES:

A.   Thermodynamic parameters of standard state:

Figure (7.6) shows Van't Hoff plot for the binding of lectin to erythrocyte surface glycoconjugates at different temperatures (4, 11, 18 and 25°). This figure revealed that the equilibrium binding constant (affinity constant) for lectin binding to glycoconjugates is temperature dependent. The results obtained from Van't Hoff plot indicated that ΔH° in general had a positive value of 9.15 KJ/mol, and that the reactions were nearly endothermic. However, the small positive value of ΔH° may indicate a favorable interaction between the lectin and glycoconjugate subgroups. These include the non-covalent interaction which are fundamentally electrostatic in nature such as charge-charge, charge-dipole, dipole-dipole, charge-induced

398

dipole, dipole-induced dipole interactions and hydrogen bonds. The sum of these types of interactions can yield some stabilization to the folded structure of the complex.

The other values of thermodynamic parameters of standard state at four different temperatures (4, 11, 18 and 25°C), such as ΔG° values and ΔS° are summarized in table (7-3). From the analysis, the results in this table shows that the ΔG° values increases with decreasing temperatures, since the lectin binding to erythrocyte surface glycoconjugates needs higher energy at low temperatures. Whereas, the negative values of ΔG° indicates the stability of lectin glycoconjugate complex, subsequently the high affinity of the reactant.

Also, the high negative values of ΔG° indicates that the binding of lectin to glycoconjugates is a spontaneous reaction. Furthermore, these values are controlled by a high positive ΔS° values, table (7-3). The results show that the values of ΔS° decrease with increasing temperatures, this can be attributed to the more stable and more arranged status of lectin-glycoconjugate complex. On the other hand, the high positive value of ΔS° may be indicated that the binding spontaneity was enropically driven.

Entropy was the driven force for the occurrence of the binding, this shows that the hydrophobic interactions played an important role in the stability of complex formation [245].

Table (7-3): Thermodynamic parameters at standard state of lectin binding to erythrocyte surface glycoconjugates. All details are explained in the text.

T°C	$\Delta H°$ (KJ/mol.)	$\Delta G°$ (KJ/mol.)	$\Delta S°$ (J/mol.K)
4	9.15	-36.74	165.67
11	9.15	-37.75	165.14
18	9.15	-38.74	164.57
25	9.15	-39.74	164.06

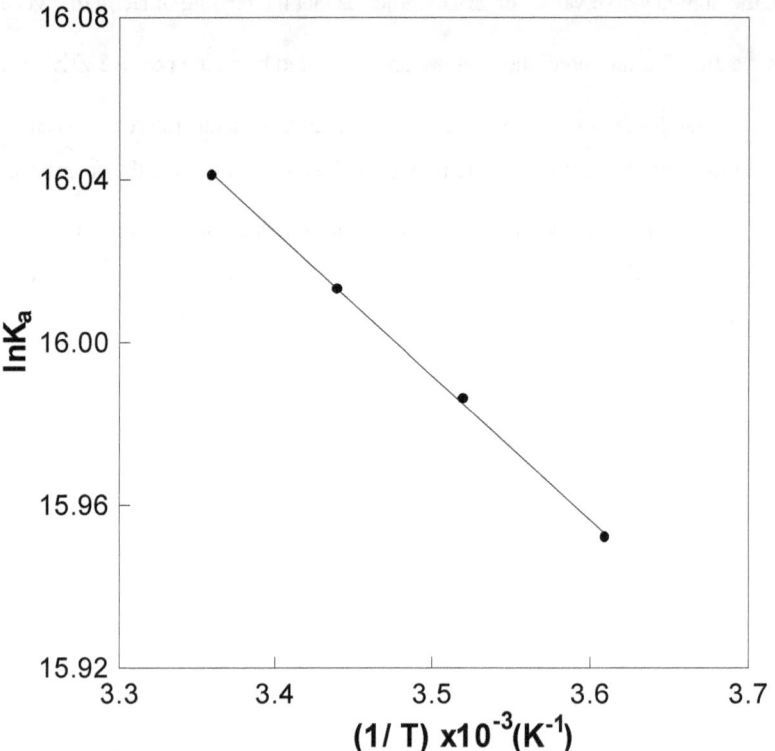

e (7-6): Van't Hoff plot for the binding of lectin to erythrocyte surface glycoconjugates. All details are explained in the text.

B. <u>**Thermodynamic parameters of transition state:**</u>

Through the transition state, the interaction of two substances leads to the formation of an activated complex (transition state), then the formation of the final product, i.e.: (the association of lectin with erythrocyte surface glycoconjugates) can be represented as follows:

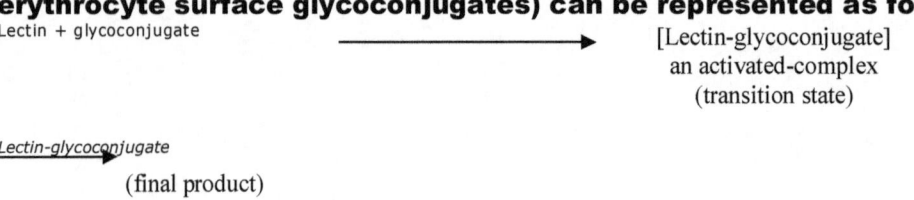

Lectin + glycoconjugate ⟶ [Lectin-glycoconjugate]
an activated-complex
(transition state)

Lectin-glycoconjugate
(final product)

According to Arrhenius equation and kinetic constant, it could be calculated the thermodynamic parameters of the transition state (ΔH^*, ΔG^* and ΔS^*) at four different temperatures (4, 11, 18 and 25°C). Figure (7.7) shows Arrhenius plot for the binding of lectin to erythrocyte surface glycoconjugates, the slope of the straight line represents the activation energy (E_a) of the binding reaction.

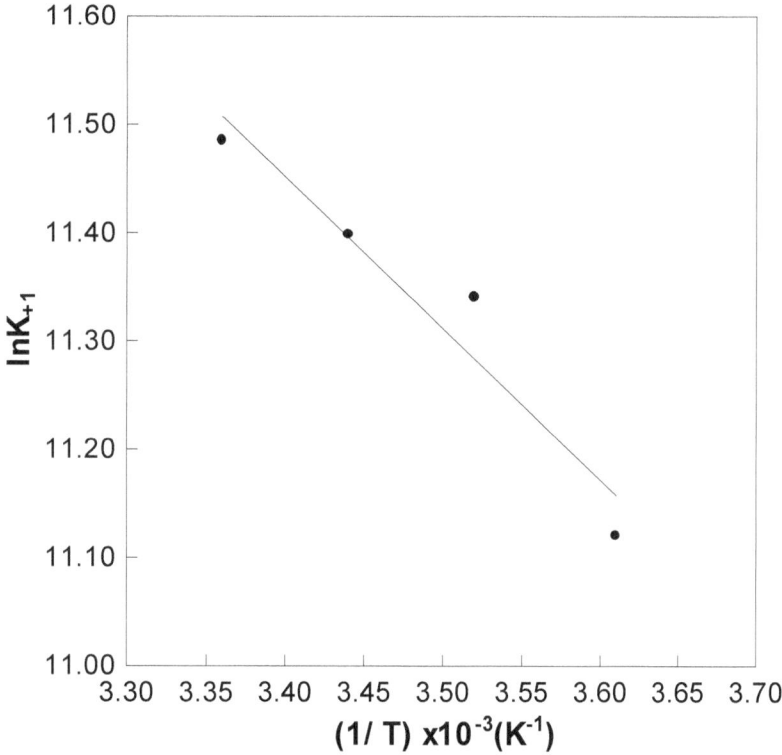

Figure (7.7): Arrhenius plot for the binding of lectin to erythrocyte syrface glycoconjugates. All details are explained in the text.

The values of thermodynamic parameters of the transition state (E_a, ΔH^*, ΔG^* and ΔS^*) are summarized in table (7-4). It is clear from this table, the high value of activation energy (13.05 KJ/mol) represents the required energy to overcome the transition state energy barrier and then giving the final product (lectin-glycoconjugate complex).

However, the value of activation energy is accordance with the high positive values of ΔG^* indicates that the formation of an activated complex is a non-spontaneous process.

Also, table (7-4) shows the values of ΔH^* at four different temperatures (4, 11, 18 and 25°C), the results revealed that the ΔH^* values decreased with increasing temperature. The slight changes in the values of ΔH^* at different temperatures could be attributed to the dependence on ΔH^* an activation energy (E_a) through the equation:

$\Delta H^* = E_a - RT$

Since the numerical value of RT is too small in comparison with the value of activation energy for the binding of lectin to glycoconjugates.

401

An analysis of the data in table (7-4), the results show that the ΔS^* values increases with decreasing temperature, was (-113.03 J/mol.K) at 4°C, -113.31 J/mol.K at 11°C, -113.44 J/mol.K at 18°C and –113.95 J/mol .K at 25°C). On the other hand, the high negative values of ΔS^* revealed that the activated complex had a more arranged structure than the reactants.

Finally, it could be concluded that the values of the thermodynamic parameters obtained from the study of lectin binding to erythrocyte surface glycoconjugates, give a distinct idea about the nature of forces that regulate the fromation of the complex.

Table (7-4): Thermodynamic parameters at transition state of lectin binding to erythrocyte surface glycoconjugates. All details are explained in the text.

T°C	E_a (KJ/mol.)	ΔH^* (KJ/mol.)	ΔG^* (KJ/mol.)	ΔS^* (KJ/mol.K)
4	13.05	10.75	42.06	-113.03
11	13.05	10.69	42.87	-113.31
18	13.05	10.63	43.64	-113.44
25	13.05	10.57	44.53	-113.95

In order to compare the values of transition state with those of standard state, it is suggested to have the thermodynamic model to describe the formation of the complex.

This model is illustrated in Figure (7.8). The thermodynamic model proposes that the formation of lectin-glycoconjugate complex undergoes three thermodynamic states.

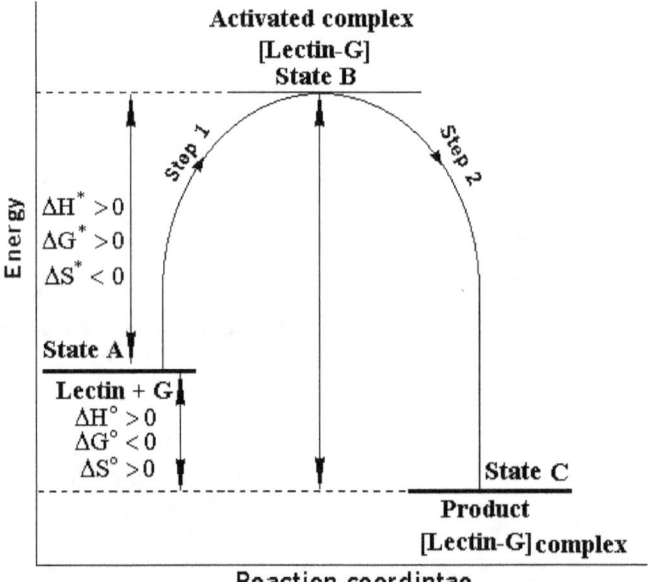

Figure (7.8): General energy diagram and thermodynamic model applied to the complex formation between lectin and erythrocyte surface glycoconjugates.

The thermodynamic state A, represents the initial energy level of lectin and glycoconjugate. The thermodynamic state B, represents the association of the two species to form the activated complex (lectin-glycoconjugate). The thermodynamic state C, represents the complete binding of lectin with glycoconjugate and formation of the complex.

(lectin-glycoconjugate complex). However, this model involves two steps, in step 1 of the reaction, the binding of lectin to glycoconjugates is associated with positive ΔG^* value and thus requires external energy. Also, in step 1, the lectin binding, shows negative value for entropy change (ΔS^*), this negativity indicates the alteration in the structure of lectin-glycoconjugate transition complex to a more arranged structure. At step 2, the contribution of more interactions, gives a fully interacting complex (lectin-glycoconjugate). The formation of a protein-ligand complex is proposed to occur in two steps, the first is, the stabilization of the complex by hydrophobic interactions and the second is the stabilization by short range interactions, such as electrostatic interactions, protonation, hydrogen bonding and Van der Waals interactions [240].

Hydrophobic interactions contribute to the complex stability via high positive entropy ($\Delta S^\circ > 0$), whereas, the electrostatic interactions, protonation, hydrogen bonding and Van der Waals interactions contribute to the complex stability via negative entropy change ($\Delta S^\circ < 0$) [246,247].

It is clear from the thermodynamic data, the binding of lectin with glycoconjugate is entropy driven and it is in agreement with the concept that the hydrophobic interactions play an important role in such reactions.

1. Total sialic acid (TSA) and lipid associated sialic acid (LASA) measurements appear to have a high sensitivity for a wide range of tumors such as breast cancer, endometerial carcinoma and sarcoma.

2. The level of protein-bound hexose was increased in sera of patients with breast and other cancers, but it may not specific enough to be a good diagnostic test for these cancers.

3. The estimation of seromucoid was more specific than protein-bound hexose, and it could be useful with other diagnostic tests for diagnosis of breast cancer.

4. The superoxide dismutase (SOD) and lipid-bound sialic acid (LBSA) appear to be putative markers of malignant diseases with potential usefulness not only in breast cancer, but also in other conditions (Endometerial carcinoma and sarcoma) associated with an increased risk of tumors development.

5. The results also indicates that lectin may recognizes the sialyl residue of the glycoprotein and the binding is due to sugar specific interaction.

According to the results obtained in this thesis, the following works are suggested for the future:

1. Study of comparison between serum levels of carcinoembryonic antigen, sialic acid and phosphohexose isomerase in patients with breast cancer.

2. Estimation the levels of free sialic acid in the urine of patients with breast cancer.

3. Determination and biochemical studies on the enzyme neuraminidase in the sera of breast cancer.

4. Study the effect of radiation on serum glycoproteins and glycosidases in patients with breast carcinoma.

5. Study other physiochemical properties of human malignant breast lectin such as amino acid composition, isoelectric point and sedimentation velocity behaviour of the lectin.

References

1. Disaia, P.J., and Creasman, W.T. (1989) In Breast Diseases and Colorectal Cancer in Clinical Gynecologic Oncology. Third ed., pp. 1-5, The C.V. Mosby Company, Toronto.

2. Parkin, D.M, Idara F., and Muir, C.S. (1988) Int. Cancer **41**, 184-97.

3. Coleman, M.P., Esteve, J., Damiecki, P., Arston, A., and Renard, H. (1993) Trends in cancer incidence and mortality, Lynos: International Agency for Research on Cancer (IARC. Scientific Publication No. 121).

4. Steven, A., Schroeder, Marcus, A., Krupp, Lawrence, M., Tierney, and Stephen, J. Mcphee (1989) In Current Medical Diagnosis & Treatment, pp 429-435.

5. Wooster, K., Neunausen, S.L., Mangion, J. et al (1994) Science **265**, 2088-2090.

6. Nowak, R. (1994) Science **265**, 1796-1799.

7. McPherson, K., Steel, C.M., and Dixon, J.M. (1994) BMJ **309**, 1003-1007.

8. Wealheral, Ledinglaic, and Warrel (1984) Oxford Text Book of Medicine, Oxford University press.

9. Anderson, D.E., and Badzioch, M.D. (1985) Cancer **56**, 383.

10. Gilbertsen, V.A. (1975) Semin. Oncol. **1**, 87-90.

11. Canellos, G.P., Hellman, S., and Veronesi, U. (1982) N. Engl. J. Med. **306**, 1430.

12. Calson, H.E. (1980) N. Engl. J. Med. **303**, 795-799.

13. Trialists Collaborative Group (1992) In: Early Breast Cancer Lancet 339, 1-15, 71-85.

14. Singleton, W.V., and Mecarty, K.S. (1987) Gyncol. Oncol. **26,** 271-275.

15. Gelber, R.D., and Goldhirsch, A. (1986) J. Clin. Oncol. **4,** 1696.

16. Donegan, W.L., and Spratt, J.S. (1979) In Cancer of the breast, W.B. Saunders, Philadelphia.

17. Horris, J.R, Leven, M.B., and Helman, S. (1987) Semin Oncol. **5,** 403.

18. Butta, A., Maclennan, K., Flanders, K.C., and et al. (1992) Cancer Res. **52**, 4261.

19. Slevin, M.L., Stubbs, L., Plant, H.J., and et al. (1990) Br. Med. J. **300,** 1458.

20. Wilson, J.D., Aiman, J., and MacDonald, P.C. (1980) Adv. Inter. Med. **25,**1.

21. Bailey & Love's, In Short Practice of Surgery, 21st edition (1991) pp. 806-810.

22. Helman, S. (1983) In Controversies in Breast Cancer, Conference Sponsored by the MD Aderson Hospital and Tumor Institute, Chicago, Year Book, Medical Publishers, Inc.

23. Chetty, U. (1979) Br. J. Surg. **67**, 789.

24. Humphrey, U. et al (1983) Contemp Surg. **23,** 97.

25. Cole, P., Elwood, J.M., and Kaplan, S.D. (1978) AM. J. Epidemiol. **108**, 112.

26. Oluwole, S.F., and Freeman, H.P. (1979) AM. J. Surg **37**, 786.

27. Rao, B.R.(1981) Cancer **47**, 2016.

28. Devitt, J.E. (1972) Surg. Gynecol Obstet. **134**, 803.

29. Yamamota, K. (1985) Ann. Rev. Genet. **19**, 209-252.

30. Reddik, K, and Holland, J.F. (1976) Proc. Natl. Acad. Sci. USA, **73**, 2308.

31. Wood, C.H., Varela, V., and Palmquist, M. et al (1977) J. Surg. Oncol. **5**, 251.

32. Secreto, G., Recchione, C., and Caralleri, A., et al (1983) Br. J. Cancer, **47**, 269.

33. Sharon, P. (1984) Scientific American **6**, 86-95.

34. Handas, S., and Nakamura, K. (1984) J. Biol. Chem. **95**, 1323-9.

35. Magdelenal, I.I. (1992) J. Immunol Meth. **150**, 133-143.

36. Suresh, M.R.(1944): Cancer markers In: The immunoassay handbook (Wild, D. ed.) pp. 441-460 NY. Stockton Press.

37. Suresh, M.R., Noujam, A.A., and Longeneeker, B.M. (1991): Recent developments in monoclonal antibodies In: biotechnology Current Progress, Vol. **1**, (Cheremisinoff, P.N., and Ferrante, L.M. eds), pp. 83-101, USA Technomic Publ.

38. Sano, T, Smith, L.S., and Cantor, C.R. (1992) Conjugates Science **258**, 120-122.

39. Suresh, M.R. (1996) Anticancer Research **16**, 2273-2278.

40. Suresh, M.R. (1991): Immunoassay for cancer-associated carbohydrate antigens In, Seminars in cancer Biol., Vol. **2**, pp. 367-377, London W.D. Saunders Publ.

41. American Society of Clinical Oncology (1996) J. Clin. Oncol. **14,** 2843-2877.

42. Greene, G.L., Sobel, N.B., and King, W.J. et al (1984) J. Steroid Biochem. **20**, 51-56.

43. Green, S., Gronemeyer, H., and Chambon, P. (1987) In Structure and function of steroid hormone receptors in Sluyser M. ed.: Growth Factors and Oncogenes in Breast Cancer, England, Mlis pp. 728, Horwood.

44. Walter, P., Green, S., and Green, G. et al (1985) Proc. Nail. Acad. Sci. U.S.A. **82**, 7889-7893.

45. Orti, E., Bodwell, J.I., and Munck, A. (1992) Endocr. Rev. **13**, 105-128.

46. Ruh, T.S., Ruh, M.F., and Singh R.K. (1988) In Nuclear acceptor Sites: Interaction with estrogen- versus antiestrogen-receptor complexes, in Moudgil V.K. ed.: Steroid Receptors in Health and Disease, pp. 233-250, New York, N.Y. Plenum.

47. Bauer, K.D., Bagwell, C.B., and Giaretti, W. et al. (1993) Cytometry **4**, 486-491.

48. Tandon, A.K., Clark, G.M., and Chamness, G.C. et al. (1990) N. Engl. J. Med. **322**, 297-302.

49. Reid, P.E., Culling, C.F.A., and Dunn, W.L. (1978) J. Histochem. Cytochem. **26**, 187-192.

50. Culling, C.F.A., and Reid, P.E. (1979) J. Histochem. Cytochem. **27**, 1177-1179.

51. Bouchier, I.A.D., and Clamp, J.R. (1971) Clin. Chim. Acta., **35**, 219-224.

52. Horowitz M.I.(1977), "Gastrointestinal glycoproteins' In: "The glycoconjugates" eds., Horowitz M.I. and Pigman W., 1st ed. Vol. **1**, pp. 189-213 Academic Press.

53. Tuppy, H., and Gottschalk, A. (1972) "The structure of sialic acids and their quantitation" In: "Glycoproteins" ed. Gottschelk A., 2nd ed., pp. 403-449 Amsterdam.

54. Schauer, R. (1982) Adv. Carbohydr. Chem. Biochem., Vol. **40**, in press.

55. Schauer, R. (1978) Methods Enzymol. **50C**, 64-89.

56. Buscher, H.P., Casals-Stenzel, J., and Schauner, R. (1974) Eur. J. Biochem. **50**, 71-82.

57. Reuter, G., Vielgenthart, J.F.G., Wember, M., Schauer, R. and Howard, R.J. (1980) Biochem. Biophys. Res. Commun. **94**, 567-572.

58. Sheshadri, N. (1994) Annals. of Clinical and Laboratory Science, Vol. 24 No. 4, 376-384.

59. Warren, L. (1959) J. Biol. Chem. **234**, 1971-5.

60. Shukla, A.K., and Schauer, R. (1981) Physiol. Chem. **362**, 236-7.

61. Shamberger, R.J. (1986) Anticancer Res. **6**, 717-20.

62. Hara, S., Takemori, Y., Yamaguchi, M., Nakamura, M., and Ohkura, Y. (1987) Anal. Biochem. **164**, 138-45.

63. Pigman W., (1977): Blood group glycoproteins" In: "The glycoconjugates" eds. Horowitz M.I. and Pigman W., 1st ed., Vol. **1**, pp. 181-188 Academic press.

64. Schwick, H.G., Heide, K., and Haupt, H. (1977) "Plasma" In: "The glycoconjugates" eds., Horowitz M.I. and Pigman W., 1st ed., Vol. **1**, pp. 261-321 Academic press.

65. Patel, V. (1978) "Degradation of glycoprotein" In: "the glycoconjugates" eds., Horowitz M.I. and Pigman W., 1st ed., Vol. **2**, pp. 185-229 Academic press.

66. Baumann, H. and Doyle, D. (1984) "Determination of carbohydrate structures in glycoproteins and glycolipids", Molecular and chemical characterization of membrane receptors, pp. 125-160.

67. Horowitz M.I., (1978) "Immunological aspect and lectins" In: "The glycoconjugates" eds., Horowitz M.I. and Pigman W., 1st ed., Vol. **2**, pp. 387-425 Academic press.

68. Yogeeswaran, G., and Salk, P. (1981) Science **212**, 1514-6.

69. Bahl, O.P., and Shah, R.H. (1977) "Glycoenzymes and glycohormons" In: "The glycoconjugate" eds., Horowitz M.I. and Pigman W., 1st ed., Vol. **2**, pp. 385-422 Academic press.

70. Schauer, R. (1985) Trends Biochem. Sci. **10**: 357-60.

71. Reid, P.E., Culling, C.F.A., Dunn, W.L., and Clay M.G. (1978) J. Histochem. Cytochem. **26**, 1033-1041.

72. AL-Suhail, A., Reid, P.E., Yeung, M., Corret, S., Frolich, J., and Brooks, D.E. (1981) "Serum level of sialic acid: Effect of age, smoking and heart disease". C.S.C.C. 25th annual convention No. 5.

73. McNeil, C., Berrett, C.R., Lucia, Y.S.U., Trentelman, E.F., and Helmick, W.M. (1965) Am. J. Cli. Path. **43**, 130-133.

74. Carter, A., and Martin, N.H. (1962) J. Clin. Path. **15**: 69-72.

75. Erbil, K.M., Jones, J.D., and Klee, G.G. (1986) Cancer **57**, 1889.

76. Khanderia, V., and Keller, J. (1983) J. Surg. Oncol **23**, 163-166.

77. Stefnelli, N., Klotz, H., Engel, A., and Bauer, P. (1985) J. Cancer Res. Clin. Oncol. **109**, 55-59.

78. Shamberger, R.J. (1984) J. Clin. Chem. Clin. Biochem. **22**, 647-651.

79. Silver, H.K.B., Rangel, D.M, and Morton, D.L. (1978) Cancer **41**, 1497-1499.

80. Brozmanova, E., and Skrovina, B. (1972) Neoplasma **19**, 115-123.

81. Silver, H.K.B., karim, K.A., Archibald, E.A., and Salinas, F.A. (1979) Cancer Res. **39**, 5036-5042.

82. Moss, A.J., Bissada, N.K., Bayed, C.M., and Hunter, W.C. (1979) Urology **13**, 182-184.

83. Silver, H.K.B., Karim, K.A., Salinas, F.A., and Swenterton, K.D. (1981) Surg. Gynecol. Obstect. **153**, 203-213.

84. Harvey, H.A., Lipton, A., White, D., and Davidson, E. (1981) Cancer, **47**, 324-327.

85. Bradely, W.P., Blasco, A.P., Weiss, J.F., Alxander, J.C., Silverman, N.A., and Chretien, P.B. (1977) Cancer, **40**, 2264-2272.

86. Culling, C.F.A., Reid, P.E., Burton, J.D., and Dunn, W.L. (1975) J. Clin. Pathol. **28**, 650-656.

87. Culling, C.F.A., Reid, P.E., Worth, A.J., and Dunn, W.L. (1977) J. Clin. Pathol. , **30**, 1056.

88. Corfield, A.P., Michalski, J.C., and Schauer, R. (1981) "In perspective in inherited metabolic diseases", (Tettamanti G., Durand P. and Di Donato S., eds.), Vol. **4**, pp. 3-70, Edi Ermes Publ., Milano.

89. Varki, A., and Kornfeld, S. (1980) J. Exp. Med. **152**, 532-544.

90. Culling, C.F.A., Reid, P.E., Clay, M.G., and Dunn, W.L. (1973) J. Histochem. Cytochem. **22**, 826-831.

91. Casals-Stenzel, J., Buscher, H.P., and Schauer, R. (1975) Anal. Biochem. **65**, 501-507.

92. Erbil, K.M., Jones, J.D., and klee, G.G. (1985) Cancer **55**, 404-409.

93. Silver, H.K.B., Rangel, D.M., and Morton, D.L. (1978) Cancer **41**, 1497-1499.

94. Fukushima, K., Hirota, M., and Terasaki, P.I., et al. (1984) Cancer Res. **44**, 5279-5285.

95. Kessel, D., and Allen, J. (1975) Cancer Res. **35**: 670-672.

96. Berg, E.L., Robinson, M.K., and Mansson, O., et al. (1991) J. Biol. Chem. **266**, 14869-14872.

97. Lundblad, A. (1980) Scand. J. Clin. Lab. Invest. **40**, 3-11.

98. Reintgein, D.S., Cruse, C.W., Wells, K.E., Saba, H.I., and Fabri, P.J. (1992) Ann. Plast. Surg. **28**, 55-9.

99. Patel, P.S., Adhvaryn, S.G, and Baxi, R.B. (1991) Int. J. Biol. Markers **6**, 177-82.

100. Xing, R.D., Chen, R.M., Wang, Z.S., and Zhang, Y.Z. (1991) J. Oral Maxillofac. Surg. **49**, 843-7.

101. Horowitz, M.I., and Pigman, W. (1977) "The glycoconjugate" eds., Horowitz, M.I., and Pigman, W, 1st ed., Vol. **1**, pp. 1-10, Academic Press.

102. Martin, D.W. Jr. (1985) "Glycoproteins, proteoglycons, and glycosaminoglycans" In.: "Harper's review of biochemistry" eds., Martin, D.W.Jr., Mayer, P.A., Rodwell V.W., and Granner, D.K. 20th ed., pp. 464-479, Lange Medical Publications.

103. Filipe, M.I., and Fengar, C. (1979) Histochem. J. **11**, 277-287.

104. Shehan, D.G., and Jervis, H.R. 1(1976) Am. J. Anta. **146**, 103-132.

105. La Mont, J.T., Smith, B.F., and Moore, J.R.L. (1984) Hepatology **4**, 515-565.

106. Zinn, A.B., Plantner, J.J., and Carlson, D.M. (1977) "Nature of linkages between protein core and oligosaccharides": In "The glycoconjugates" eds., Horowitz, M.I., and Pigman, W. 1st ed., Vol. **1**, pp. 69-85, Academic Press.

107. Lenten, L.V., and Ashwell, G. (1970) J. Biol. Chem. **246**, 1889-1894.

108. Glick, M.C., and Flowers, H. (1978) "Surface membrane" In: "The glycoconjugates" eds., Horowitz, M.I., and Pigman, W. 1st ed., Vol. **2**, pp. 337-384, Academic Press.

109. Hoskins, L.C. (1978) "Degradation of mucus glycoproteins in the gastrointestinal tract" In: "The glycoconjugates" eds., Horowitz, M.I., and Pigman, W. 1st ed., Vol. **2**, pp. 235-253, Academic Press.

110. Holden, K.G., and Griggs, L. (1987) "Respiratory tract" In: "The glycoconjugates" eds., Horowitz, M.I., and Pigman, W. 1st ed., Vol. **1**, pp. 215-237, Academic Press.

111. Ferguson, R.N., Edelhoch, H., and Saroff, H.A. (1975) Biochem. **14**, 282-289.

112. Wasserman, R.L., and Capra, J.D. (1977) "Immunoglobulins" In: "The glycoconjugates" eds., Horowitz, M.I., and Pigman, W. 1st ed., Vol. **1**, pp. 323-348, Academic Press.

113. Glandemans, C.P.J. (1975) Adv. Carbohydr. Chem. Biochem. **31**, 313-346.

114. Pazur, J.H., and Forsberg, L.S. (1978) Carbohydr. Res. **60**, 167-178.

115. Stern, P.L., Willison, K.R., Lennox, E., Calfre, G., Milstein, C., Secher, D., Ziegler, A., and Springer, T. (1978) Cell **14**, 775-785.

116. Granner, D.K. (1985) "Hormone action" In: "Harper's review of biochemistry" eds., Martin, D.W., Mayes, P.A., Rodwell, V.W., and Granner, D.K. 20th ed., pp. 505-515, Lange Medical Publications.

117. Macbeth, R.A.L., and Bekesi, J.G. (1962) Cancer Res. **22**, 1170-1175.

118. Seibert, F.B., Seibert, M.V., Atno, A.J., and Campbell, H.W. (1974) J. Clin. Invest. **26**, 90-102.

119. Nigelson, G.L., and Postle, G. (1976) N. Engl. J. Med. **295**, 253-258.

120. Cunietti, E., Vaiani, G., Gandini, M, Monti, M, Locatelli, E., Gandini, R., and Reggiani, A. (1985) Cancer Detec. Prev. **8**, 222-232.

121. Schachter, H. (1978) "Glycoprotein biosynthesis" In: "The glycoconjugates" eds., Horowitz, M.I., and Pigman, W. 1st ed., Vol. **2**, pp. 87-181, Academic Press.

122. Both, S.N., King, J.P.G., Leonard, J.C., and Dykes, P.W. (1973) Gut **14**, 794-799.

123. Barondes, S.H., (1981) Annual Review of Biochemistry **50**, 207-231.

124. Toyoshima, S., Osawa, T., and Tonomura, A. (1970) Biochem. Biophys. Acta, **221**, 514-521.

125. Kornfeld, S., and Kornfeld, R. (1978) "Use of lectins in the study of mammalian glycoproteins" In: "The glycoconjugates" eds., Horowitz, M.I., and Pigman, W. 1st ed., Vol. **2**, pp. 437-449, Academic Press.

126. Powell, J. (1980) Biochem. J. **187**, 123.

127. Sammel, H. (1984) Science **223**, 4639.

128. Springer, G., and Desai, P. (1971) Biochemistry **10**, 3749.

129. Lis, H., and Sharon, N. (1981) Ann. Rev. Biochem. **50**, 207-31.

130. Miller, J.B., Hsu, R., Heinrikson, R., and Yachnin, S. (1975) Proc. Natl. Acad. Sci. USA, **72**, 1388-91.

131. Ceri, H., Kobiler, D., and Barondes, S.H. (1981) J. Biol. Chem. **256**, 390-94.

132. Pereira, M.E.A., and Kabat, E.A. (1979) Crit. Rev. Immunol. **1**, 1-73.

133. Barondes, S. (1981) Annu. Rev. Bioche. **50**, 207.

134. Boldt, D.H., et al. (1975) J. Immunol. **114**, 1532-1536.

135. Scott, R.E., and Rosenthal, A.S. (1977) J. Immunol. **119**, 143-148.

136. Alhadeff, J.A. (1989) Crit. Rev. Oncol. Hematol. **9**, 37-47.

137. Taner, O., Nursen, E., and Limrin, A. (1990) Clin. Chem. **36**, 393-397.

138. Shamberger, R.J. (1984) J. Clin. Chem. Clin. Biochem. **22**, 647-651.

139. Vegh, Zs., Kremmer, T., Boldizsar, M., Gesztes, K.A., and Szajani, B. (1991) Clin. Chim. Acta **203**, 259-268.

140. Raynes, J.G. (1983) Biomed. Pharmacother. **37**, 136-138.

141. Lipton, A., Harvey, H.A., and De Long, S. et al. (1979) Cancer **43**, 1766-1771.

142. Dnistria, A.M., Schwartz, M.K., and Katopodis, N. (1982) Cancer **50**, 9.

143. Katopodis, N., Hirshaut, Y., and Geller, N.L. (1982) Cancer Res. **42**, 5270-5.

144. Rothenberg, R.E., La Ruja, R.D., Mueller, O.T., and Pryce, E.H. (1994) Breast Dis. **7**, 3, 197-202.

145. Lowry, O.H., Rosebrough, N.J., Farr, A.L., and Randall, R.J. (1951) J. Biol. Chem. **193**, 265-75.

146. Wilkinson, L. SYSTAT. (1990) The system for Statistics Evanston, II: SYSTAT Inc.

147. Crook, M. (1993) Clin. Biochem. **26**, 31-38.

148. Cohen, S.L., Lincoln, S.T., and Rosen, S.T. (1986) Cancer Invest. **4**, 305-327.

149. Patel, P.S., Baxi, B.R., Desal, S.S., and Balar, D.B. (1990) Ind. J. Pathol. Microbiol. **33**, 124-128.

150. Fukushima, K. (1991) J. Exp. Med. **163**, 17-30.

151. Itai, S., Arii, S., Tobe, R., Kitahara, A., and Kim, Y.C. et al. (1988) Cancer **61**, 775-87.

152. Hakomori, S. (1985) Cancer Res. **45**, 2405-14.

153. Warren, L., Fuhrer, J.P., and Buck, C.A. (1972) Proc. Natl. Acad. Sci. **69**, 1838-42.

154. Schutter, E.M.J., Vissr, J.J., and Van kamp, G.J. et al. (1992) Tumor Biol. **13**, 121-32.

155. Kakari, S., Stirngou, E., Toumbis, M., Ferderigos, As., and Poulaki, I. et al. (1991) Anticancer Res. **11**, 2107-10.

156. Dnistrian, A.M., and Schwartz, M.K. (1981) Clin. Chem. **27(10)**, 1737-1739.

157. Mannello, F., Bocchiotti, G., Troccoli, R., and Gazzanelli, G. (1993) Breast Cancer Res. Treat. **24**, 167-170.

158. Patel, P.S., Baxi, B.R., Adhvaryu, S.G., and Balar, D.B. (1990) Cancer Lett. **51**, 203-208.

159. Haq, M., Haq, S., Tutt, P., and Crook, M. (1993) Ann. Clin. Biochem. **30**, 383-386.

160. Hansen, H.J., Snyder, J.J., Miller, E., Vandevoorde, J.P., Miller, O.N., Hines, L.R., and Burns, J.J. (1974) Hum. Pathol. **5**, 139-147.

161. Harshman, S., Reynolds, V.H., Neumaster, T., Patikas, T., and Worrall, T. (1974) Cancer **34**, 291.

162. Barlow, J.J., and Dillard, P.H. (1972) Obstet. Gynecol. **39**, 727.

163. Aronson. N.N., and De Duve. C. (1978) J. Biol. Chem. **243**, 4564.

164. Robert, A. (1962) Cancer Res. **22**, 1170.

165. Lawrence, M., Gerald, B., and Zoltan, A. (1977) Clin. Chem. **23**, 2055.

166. Yogeswarar, G. (1983) Advance Cancer Res. **38**, 289.

167. Weimer, and Mashin, J. (1965) Clin. Chem. Principle and Techniques.

168. Waalkes, P.T., Mrochek, J.E., Dinsmore, S.R., and Tormey, D.C. (1978) J. Natl. Cancer Inst. **6**, 703.

169. Bradley, W.P., Blasco, A.P., and Weiss, J.F. (1977) Cancer **40**, 2264-2272.

170. Bhuvarahamurthy, V., Balasubramanian, N., Subramanian, S., and Govindasamy, S. (1992) Bichem. Int. **28**, 105.

171. Scambia, G., Panici, B., Perrone, L., Sonsini, C., Giannelli, S., and Gallo, A. et al. (1990) Br. J. Cancer **62**, 147.

172. Sherblow, A.P., Buck, R.L., and Carraway, K.L. (1980) J. Biol. Chem. **255**, 783.

173. Yogeeswaran, G., and Salk, P.L. (1981) Science **212**, 1514.

174. Yamamooto, K. (1984) Eur. J. Biochem. **143**, 133.

175. Bolmer, S., and Davidson, E. (1981) Biochemistry **20**, 1047.

176. Yaskhiko T., Wataru I. and Mitsunon Y., Am. J. Nephrol 1988; **2**: 21.

177. McCord, J.M., and Fridovich, I. (1969) J. Biol. Chem. **244**: 6049-6055.

178. Fridovich, I. (1972) Acc. Chem. Res. **5**, 321-326.

179. Wever, R., Oudega, B., and Van, Gelder, B.F. (1972) Biochem. Biophys. Acta, **302**, 475-478.

180. Batalie, R., Klein, B., Durie, B., and Sany, J. (1989) Clin. Exp. Rheumatol. **7**, 319-28.

181. Sigureirsson, B., Lindelof, B., Edhag, O., and Allander, E. (1992) N. Engl. J. Med. **326**, 363-7.

182. Polivakova, J., Vosmikova, K., and Horak, L. (1992) Neoplasma **39(4)**, 233-6.

183. Borrello, S., De Leo, M.E., Wohirab, H., and Galeoti, T. (1992) FEBS Lett. **310**, 249-54.

184. Van Balgooy, J.N.A, and Roberts, E. (1979) Comp. biochem. Physiol. **62B**, 263-8.

185. Winterbourn, C.C., Hawkins, R.E., Brian, M., and Carrell, R.W. (1975) J. Lab. Clin. Med. **85(2)**, 337-341.

186. Knee, J.K., Mitidieri, E., and Affonso, O.R.b (1991) Cancer Lett **57**, 199-202.

187. Bolzan, A.D., Bianchi, M.S., and Bianchi, N.O. (1993) Cancer Res. **6**, 142-6.

188. Yoshimitsu, K., Kobayashi, Y., and Usui, T. (1984) Acta Paediatr. Scand., **73**, 92-6.

189. Abella, A., Clerc, D., Chalas, J., Baret, A., Leluc, R., and Lindenbaum, A. (1987) Ann. Biol. Clin. **45**, 152-5.

190. Tsurn, S., Nomoto, K., Aiso, S., Ogata, T., and Zinnaka, Y. (1983) Int. Arch. Allergy appl. Immunol. **71(1)**, 88-92.

191. Galeotti, T., Masotti, L., Borrello, S., and Casali, E. (1991) Xenobiotica **21(8)**, 1041-51.

192. Oberley, L., and Oberley, T.D. (1988) Mol. Cell Biochem. **84**, 147-53.

193. Wong, Y.F., Wong, W.S.H., and Fung, Y.H., et al. (1993) Med. Sci. Res. **21**, 397-8.

194. Mizuno, K., and Kozutsumi, T. (1981) J. Biol. Chem. **256**, 4247.

195. Briles, E., and Gregory, W. (1979) J. Cell Biol. **81**, 528.

196. Bishayee, S., and Dorai, D. (1980) Biochem. Biophys. Acta **623**, 89.

197. Bohlool, B.B., Schmidt, E.L. (1976) J. Bacteriol. **125**, 1188-94.

198. Burger, M., and Goldberg, A. (1967) Proc. Nat. Acad. Sci. USA **57**, 359.

199. Kaplan, R., Li, S., and Kehoe, J. (1977) Biochemistry **16**, 4297.

200. Hans J., Attila, B., Sigrun, G., and Michael, K. (1989) Biochem. Biophys. Res. Commun. **163**, 506.

201. Ronald, L., James, F., and Wayne, W. (1982) J. Biol. Chem. **257**, 7574.

202. Liener, I. (1955) Arch. Biochem. Biophys. **54**, 223.

203. Lis, H., and Sharon, N. (1972) Methods Enzymol. **28**, 360.

204. Gabius, H., Bandlow, G., Schirramacher, V., and Vehmeyer, K. (1987) Int. J. Cancer **39**, 634.

205. Kaplan, A. (1998) Clinical Chemistry Theory, Analysis and Correlation, 2nd edition pp. 180.

206. Ahmed, H., Chatterjee, B.P., Klem, S., and Schauer, R. (1986) Biol. Chem. Hoppeseyler **367**, 501-506.

207. Basu, S., Sarkar, M., and Mondal, C. (1986) Mol. Cell biochem. **71**, 149-157.

208. Varki, A., and Kornfeld, S. (1980) J. Exp. Med. **152**, 532-544.

209. Schauer, R., Shukla, A.K., Schroder, G., and Muller, E. (1984) Pure and Appl. Chem. **57**, 907-921.

210. Shukla, A.K., and Schauer, R. (1982) Hoppe-Seyler's and Physiol. Chem. **363**, 255-262.

211. Ravindranath, M.H., Higa, H.H., Copper, E.L., and Paulson, J.C. (1985) J. Biol. Chem. **260**, 8850-8856.

212. Elvin, A., and Manfred, M. (1967) Experimental Immuno Chemistry, Second edition, Illinois, USA.

213. Gottschalk, A. (1960) In: The chemistry and biology of sialic acids and related substances, Cambridge University Press, London.

214. Bakhtear, M. (1992) Ph.D. thesis, College of Science, Baghdad University.

215. Wild, J., Robinson, D., and Winchester, B. (1983) Biochem. J. **21**, 167.

216. Dolichos, R. (1983) Biochemistry **22**, 2741.

217. Finstand, C., Good, R., and Litman, C. (1974) Ann. N. Y. Acad. Sci. **234**, 170.

218. Dipti, G., Fred, C., and Brewer, J. (1994) Biochemistry **33**, 5526-5530.

219. Pemberton, R. (1970) Vax Song. **18**, 74.

220. Goldstein, I., and Hays, C. (1978) Adv. Carbohydr. Chem. Biochem. **35**, 127.

221. Goebel, W., et al. (1934) J. Exper. Med. **60**, 599.

222. Nassir M. (1995) Ph.D. thesis, College of Science, Baghdad University.

223. Nowak, T., and Barondes, S.(1975) Biochimica et Biophys. Acta **393**, 15.

224. Finstad, C., Litman, G., Finstand, J., and Good, R. (1972) J. Immunol. **108**, 1704.

225. Oppenheim, J., Nachbar, M., Salton, M., and Aull, F. (1974) Res. Commun. **58**, 1127.

226. Peters, B.P. et al. (1979) Biochemistry **18**, 5505-5511.

227. Roche, A.C. et al. (1975) FEBS Lett. **57**, 245-249.

228. Mohan et al. (1982) Biochem. J. **203**, 253-261.

229. Miller, R.L. (1982) J. Invertebr. Pathol. **39**, 210-214.

230. Babal, P. (1994) Biochem. J. **229(2)**, 341.

231. Kawagishi, H. (1994) FEBS Lett. **340**, 56.

232. Scopes, R. (1982) Protein purification principles and practice, Springer Verlag, pp. 162, New York Heidelber Berlin.

233. Pharmacies fine chemical: Gel filtration calibration kit instruction mannual for protein molecular weight determination.

234. Csizman, L. (1960) Proc. Soc. Exp. Biol. N. Y. **103**, 157.

235. Laemmli, U.K. (1970) Nature **227**, 680.

236. Hans, F. (1977) Application note 306 LKB-Produkter AB., pp. 1-15, Bromma, Sweden.

237. Huda, H. (1998) M.Sc. thesis, College of Science, Baghdad University.

238. Isamu, M., Heruko, K., Naoko, I., and Yukiko, S. (1986) Carb. Res. **151**, 261.

239. Goldstein, I.J., Hughes, R.C., Monsigny, M., Osawa, T., and Sharon, N. (1980) Nature **285**, 66.

240. Czech, M.P., Lynn, W.S. (1973) Biochem. Biophys. Acta **297**, 368-377.

241. Scatchard, G. (1949) Ann. N. Y. Acad. Sci. **51**, 660.

242. Emil, L. (1985) In: Principles of biochemistry, seven edition, pp. 289, International Student Edition.

243. Rae-Venter, B., and Dao, T. (1982) Biochem. Biophys. Res. Commun. **107**, 624.

244. Hussain, M. (1990) M.Sc. thesis, College of Science, Baghdad University.

245. Waelbroeck, M., Van Obeerghen, E., and Demeyts, P. (1979) J. Biol. Chem. **254**, 7736.

246. Blumenthar, D.K., and Stul, J.T. (1982) Biochemistry **21**, 2386.

247. Laport, D.C., Wierman, E.M., and Storm, D.I. (1980) Biochemistry **19**, 3814.